Sustainable food processing and engineering challenges

Sustainable food processing and engineering challenges

Edited by

Charis M. Galanakis
Research & Innovation Department, Galanakis Laboratories, Chania, Greece
King Saud University, College of Science, Riyadh, Saudi Arabia
Food Waste Recovery Group, ISEKI Food Association, Vienna, Austria

ELSEVIER

ACADEMIC PRESS
An imprint of Elsevier

Academic Press is an imprint of Elsevier
125 London Wall, London EC2Y 5AS, United Kingdom
525 B Street, Suite 1650, San Diego, CA 92101, United States
50 Hampshire Street, 5th Floor, Cambridge, MA 02139, United States
The Boulevard, Langford Lane, Kidlington, Oxford OX5 1GB, United Kingdom

Library of Congress Cataloging-in-Publication Data
A catalog record for this book is available from the Library of Congress

British Library Cataloguing-in-Publication Data
A catalogue record for this book is available from the British Library

ISBN: 978-0-12-822714-5

For information on all Academic Press publications
visit our website at https://www.elsevier.com/books-and-journals

Publisher: Emma Hayes
Acquisitions Editor: Nina Rosa de Araujo Bandeira
Editorial Project Manager: Lena Sparks
Production Project Manager: Swapna Srinivasan
Designer: Miles Hitchen

Typeset by Thomson Digital

Contents

Contributors

Sara Aghajanzadeh, Department of Food Process Engineering, Gorgan University of Agricultural Sciences and Natural Resources, Basij Square, Gorgan, Iran

Samreen Ahsan, Department of Food Science and Technology, Khwaja Freed University of Engineering and Technology, Rahim Yar Khan, Pakistan

Ayesha Ali, Department of Food Science and Technology, Khwaja Freed University of Engineering and Technology, Rahim Yar Khan, Pakistan

Norah Alshammari, School of Medicine, University of Nottingham, University Park Campus, United Kingdom

Serafim Bakalis, Department of Food Science, University of Copenhagen, Frederiksberg, Copenhagen, Denmark

A.A. Bashir, ICAR—Central Institute of Post-Harvest Engineering and Technology (CIPHET), Ludhiana, India

Marta Biegańska, Department of Industrial Product Quality and Packaging Development, Institute of Quality Science, Poznań University of Economics and Business, Poznań, Poland

Barbara Bigliardi, Department of Engineering and Architecture, University of Parma, Parma, Italy

João P.F. Carvalho, CICECO–Aveiro Institute of Materials, Department of Chemistry, University of Aveiro, Aveiro, Portugal

Tanaporn Chatkitanan, Department of Packaging and Materials Technology, Faculty of Agro-Industry, Kasetsart University, Bangkok, Thailand

Muhammad Farhan Jahangir Chughtai, Department of Food Science and Technology, Khwaja Freed University of Engineering and Technology, Rahim Yar Khan, Pakistan

Ryszard Cierpiszewski, Department of Industrial Product Quality and Packaging Development, Institute of Quality Science, Poznań University of Economics and Business, Poznań, Poland

Nathália M. Coutinho, Faculty of Veterinary Medicine, Federal Fluminense University (UFF), Niterói, Brazil

Adriano Gomes da Cruz, Department of Food, Federal Institute of Education, Science and Technology of Rio de Janeiro (IFRJ), Rio de Janeiro, Brazil

Noémie David-Rogeat, Division of Plant and Crop Sciences, University of Nottingham, NIAB EMR, Crop Science and Production System, United Kingdom

Maria Carmela K.H. Duarte, Faculty of Veterinary Medicine, Federal Fluminense University (UFF), Niterói, Brazil

Erick A. Esmerino, Faculty of Veterinary Medicine, Federal Fluminense University (UFF), Niterói, Brazil

Tim J. Foster, School of Biosciences, University of Nottingham, Sutton Bonington Campus, United Kingdom

Carmen S.R. Freire, CICECO–Aveiro Institute of Materials, Department of Chemistry, University of Aveiro, Aveiro, Portugal

Mônica Q. Freitas, Faculty of Veterinary Medicine, Federal Fluminense University (UFF), Niterói, Brazil

Ourania Gouseti, Department of Food Science, University of Copenhagen, Frederiksberg, Copenhagen, Denmark

Nathdanai Harnkarnsujarit, Department of Packaging and Materials Technology, Faculty of Agro-Industry, Kasetsart University, Bangkok, Thailand

Navnath Indore, ICAR—Central Institute of Post-Harvest Engineering and Technology (CIPHET), Ludhiana, India

Bogdan Janusz, Department of Food Hygiene and Public Health Protection, Institute of Veterinary Medicine, WULS-SGGW, Nowoursynowska, Warsaw, Poland

Adnan Khaliq, Department of Food Science and Technology, Khwaja Freed University of Engineering and Technology, Rahim Yar Khan, Pakistan

Yogesh Kumar, ICAR—Central Institute of Post-Harvest Engineering and Technology (CIPHET), Ludhiana, India

Kaja Kristensen, School of Biosciences, University of Nottingham, Sutton Bonington Campus, United Kingdom

Anusz Krzysztof, Department of Food Hygiene and Public Health Protection, Institute of Veterinary Medicine, WULS-SGGW, Nowoursynowska, Warsaw, Poland

Yeyen Laorenza, Department of Packaging and Materials Technology, Faculty of Agro-Industry, Kasetsart University, Bangkok, Thailand

Atif Liaqat, Department of Food Science and Technology, Khwaja Freed University of Engineering and Technology, Rahim Yar Khan, Pakistan

Qingsu Liu, Faculty of Engineering, University of Nottingham, University Park Campus, United Kingdom

Luca Marciani, School of Medicine, University of Nottingham, University Park Campus, United Kingdom

Zarzynska Joanna Magdalena, Department of Food Hygiene and Public Health Protection, Institute of Veterinary Medicine, WULS-SGGW, Nowoursynowska, Warsaw, Poland

Tariq Mehmood, Department of Food Science and Technology, Khwaja Freed University of Engineering and Technology, Rahim Yar Khan, Pakistan

Molly Muleya, School of Biosciences, University of Nottingham, Sutton Bonington Campus, United Kingdom

Syahrizal Muttakin, School of Chemical Engineering, University of Birmingham, Edgbaston Campus, United Kingdom; Indonesian Agency for Agricultural Research and Development, Indonesia

Muhammad Nadeem, Department of Environmental Sciences, COMSATS University Islamabad, Vehari, Pakistan

Tatiana C. Pimentel, Federal Institute of Paraná, Paranavaí, Brazil

Reverson F. Quero, Department of Analytical Chemistry, Institute of Chemistry, University of Campinas (UNICAMP), Campinas, Brazil

Junaid Ur Rehman, Department of Food Science and Technology, Khwaja Freed University of Engineering and Technology, Rahim Yar Khan, Pakistan

Ramon S. Rocha, Faculty of Veterinary Medicine, Federal Fluminense University (UFF), Niterói; Department of Food, Federal Institute of Education, Science and Technology of Rio de Janeiro (IFRJ), Rio de Janeiro, Brazil

Kanza Saeed, Department of Food Science and Technology, Khwaja Freed University of Engineering and Technology, Rahim Yar Khan, Pakistan

Nimra Sameed, Department of Food Science and Technology, Khwaja Freed University of Engineering and Technology, Rahim Yar Khan, Pakistan

Marcia Cristina Silva, Department of Food, Federal Institute of Education, Science and Technology of Rio de Janeiro (IFRJ), Rio de Janeiro, Brazil

Marcello R. Silveira, Faculty of Veterinary Medicine, Federal Fluminense University (UFF), Niterói, Brazil

R.K. Singh, ICAR—Central Institute of Post-Harvest Engineering and Technology (CIPHET), Ludhiana, India

Atcharawan Srisa, Department of Packaging and Materials Technology, Faculty of Agro-Industry, Kasetsart University, Bangkok, Thailand

Mariusz Tichoniuk, Department of Industrial Product Quality and Packaging Development, Institute of Quality Science, Poznań University of Economics and Business, Poznań, Poland

Carla Vilela, CICECO–Aveiro Institute of Materials, Department of Chemistry, University of Aveiro, Aveiro, Portugal

R.K. Vishwakarma, ICAR—Central Institute of Post-Harvest Engineering and Technology (CIPHET), Ludhiana, India

Phanwipa Wongphan, Department of Packaging and Materials Technology, Faculty of Agro-Industry, Kasetsart University, Bangkok, Thailand

Aman Mohammad Ziaiifar, Department of Food Process Engineering, Gorgan University of Agricultural Sciences and Natural Resources, Basij Square, Gorgan, Iran

Preface

Food processing ensures that the resources required to produce raw food materials and ingredients for food manufacturing are used most efficiently. Nowadays, changes in the legislative frameworks, environmental concerns, and increasing attention toward sustainability have stimulated the food industry to reconsider their practices and food processing challenges in spite of reduced energy and resource consumption. Responding to the goals of sustainability requires the maximum utilization of all raw materials produced and integration of activities throughout all the production-to-consumption stages. To maximize the conversion of raw materials into consumer products, food engineering, and food processing challenges should be met. Indeed, with the recent advantages in food processing (e.g., non-thermal technologies and industry), new developments, data, and the state of the art come up in the field. Subsequently, there is a need for a further reference highlighting the current challenges of food processing and engineering in spite of sustainability.

Food Waste Recovery Group provides insights into the food and environmental science and technology sectors and publishes relevant books. Our portfolio of books includes reference materials dealing with sustainable food systems, nutraceuticals, and non-thermal processing, innovations in traditional foods, innovations strategies in food and environmental science, food waste recovery technologies, saving food efforts, biobased products, and biobased industries, the valorization of food processing by-products (e.g., from olives, meat, coffee, cereals, and grapes). The group has also prepared books for non-alcoholic drinks, personalized nutrition, food quality and shelf-life, innovative food analysis, and specific food components, such as carotenoids, dietary fiber, glucosinolates, polyphenols, lipids, and proteins.

Following the above considerations, the current book covers sustainable food processing and urgent food engineering challenges. The ultimate goal is to support the scientific community, professionals and enterprises that aspire to develop sustainable industrial food processes.

The book consists of 10 chapters. In **Chapter 1**, a systematic literature review on the implementation of the Industry 4.0 paradigm in the food industry is provided. The Fourth Industrial Revolution, or Industry 4.0, (Big data management, Internet of Things, cloud computing, automation process control) includes a variety of new enabling technologies (i.e., Big data, Internet of Things, cloud computing, automation, and so on) that allow the digitization of a manufacturing environment. Consequently, the way companies operate is drastically changing, regardless of the industry they belong to. As for the food industry in particular, its increasingly customer-oriented characterization implies the need for faster response times in order to address dynamic and collaborative supply chains.

The next chapters of the book deal with non-thermal processing technologies that are known for their minimal, safe and sustainable treatment of foods and juices. For example, pasteurization is the primary process to guarantee the safety, quality, and shelf life of juices. While thermal pasteurization is the most common method to inactivate the microorganisms and enzymes, it causes some undesirable effects that reduce the juice quality. In **Chapter 2**, non-thermal pasteurization methods, including pulsed electric field, high-pressure, high-pressure carbon dioxide, sonication, ultraviolet and ozone treatments, are described. Their equipment, process conditions and mechanism(s) of the inactivation of microorganisms and enzymes are also studied, whereas their effect on the juice nutritional values, appearance, sensorial properties, advantages and limitations are investigated.

Chapter 3 covers the principles, applications, impact, and prospects of the high-pressure processing (HPP) technique. The latest is an emergent food processing and preservation technique that extends shelf life without altering the nutritional and health attributes of food commodities. In HPP, a high hydrostatic pressure (between 100 and 1000 Mpa) is applied to food material consistently and simultaneously from all directions that will eventually destroy the detrimental microbes and indigenous enzymes due to high pressure by providing the pasteurization effect without thermal treatment. This technique can be applied to both liquids and solid foods with comparatively higher moisture content.

Chapter 4 provides an overview of recent research on the effects of cold plasma on the quality of food products, considering a microbiological, physical-chemical, sensory, and toxicological perspective. The advantages, disadvantages, the regulatory context, and the challenges of its scalability from the laboratory to the industry are also presented. **Chapter 5** provides an overview of the current knowledge on the use of pulsed electric field (PEF) technique in food processing. PEF technique is economical, environmentally safe, and can be up-scaled for sustainable industrial applications. In recent times, this technique has been evaluated for inactivation of microorganisms, improvement in the mass transfer, and modifications in the properties of bioactive compounds including enzymes, etc. The factors which affect the efficiency of PEF in food processing are also summarized. Latest applications of PEF technique for microbial inactivation, extraction of bioactive compounds, drying, and modification of food properties are also discussed.

Chapter 6 deals with the application of ultraviolet light assisted titanium dioxide photocatalysis for food safety applications. Titanium dioxide, which was previously successfully used in cosmetics or paintings, based on its characteristic active properties, generates reliable oxidizing power under UV light treatment. This photocatalytic activity is tested in many areas of the food sector—photocatalytic surfaces (self-cleaning and self-disinfecting), modified active packaging, air, and water purification. The goals are effectiveness, low costs, and competitiveness to traditional methods, as well as environmental remediation. At the same time, it is essential to remember about correct risk assessments and the needs of appropriate law regulations.

Chapter 7 describes the development of bioplastics for food packaging including fresh produce, bakery, meat, and seafood products. Bioplastics have been processed into multilayer materials via coating, lamination, and co-extrusion that improve barrier and functional properties. Moreover, reinforcement of bioplastic materials with organic and inorganic fillers, or other functional modification such as plasma treatment that induces cross-linking between polymers, effectively enhances barrier properties of packaging. Improved barrier properties and addition of active functions, especially antimicrobial performance, are significant developments in investigations of bioplastic for food packaging. Efficacy to increase the shelf-life of packaged food via functional bioplastic packaging can reduce food loss and food waste, thereby facilitating sustainable food production.

Chapter 8 presents intelligent packaging that has gained in importance in recent years. The main purpose of using intelligent packaging for each food product is to increase the functionality of the product, to facilitate product quality control in the supply chain, and to reduce food waste. The chapter discusses in detail the most important indicators, such as time temperature indicators, food freshness indicators, as well as integrity (gas) indicators. Sensors and data carriers such as bar codes and RFID tags are also presented, whereas the latest achievements in the field of intelligent packaging are analyzed both in research and commercial application. Finally, considerable attention is given to discussing the contribution of intelligent packaging to the sustainable food processing and distribution.

The inevitable augment of food demand, due to the growing world population, is contributing to the expansion of the global market of packaging materials. Thus, the active packaging concept emerges as an instrument to boost the safety, quality, and shelf life of the packaged foods. In this context **Chapter 9** revises the latest advances in active agents for active food packaging, giving emphasis on active functions, such as antimicrobial and antioxidant activities, carbon dioxide emitting, and oxygen and ethylene scavenging.

The final **Chapter 10** deals with engineering challenges at the final consumption level, and particularly during digestion. It presents the physiology of the digestive system and its components, before exploring different methods used to engineer it. In vivo (meaning "in the living"), in vitro (meaning "in the glass"), and in silico (referring to computer models) methods are all used in food engineering laboratories, more or less commonly, and the advantages and drawbacks of each of them are presented.

Conclusively, the current book assists food engineers, food scientists, and technologists working in the whole food science field as well as researchers, academics, specialists, chemical engineers, and professionals who are seeking to improve efficiency of production systems. Likewise, University libraries and Institutes could use it as a textbook in any course dealing with food science and technology, food processing and food engineering, especially in postgraduated programs.

At this point I want to thank all contributors and authors for their fruitful collaboration and their dedication to the editorial guidelines and timeline. I would also like to acknowledge the acquisition editor Nina Bandeira and the book managers Nina Bandeira and Lena Sparks, and all colleagues from Elsevier's production team, for their assistance during the preparation of this book. Last but not least, a message for you, the reader. These collaborative projects contain hundreds of thousands of words, and thus may contain minor errors or gaps. Constructive comments and even criticism are always welcome. In that case, please contact me to discuss any issues.

Charis M. Galanakis
Research & Innovation Department
Galanakis Laboratories
Chania, Greece
King Saud University
College of Science
Riyadh, Saudi Arabia
Food Waste Recovery Group,
ISEKI Food Association,
Vienna, Austria
charismgalanakis@gmail.com

Chapter 1

Industry 4.0 Applied to Food

Barbara Bigliardi
Department of Engineering and Architecture, University of Parma, Parma, Italy

1 Introduction and background

The developments that occurred in production processes and their automation in the last decade, have led to the Fourth Industrial Revolution, the so-called "Industry 4.0" (Galati and Bigliardi, 2019). Industry 4.0 is a broad domain which refers to a trend in industrial automation integrating new production technologies with a threefold aim, that is, to improve working conditions, to increase productivity, and to improve the quality of plants production (Lu, 2017). Industry 4.0 takes its name from the German government's industrial plan presented in 2011 and implemented at the end of 2013, concerning in investments in infrastructure, schools, energy systems, research institutions, and companies to modernize the German production system. The term 4.0 derives from the fact that earlier there have been three other industrial revolutions called 1.0, 2.0, and 3.0 (Lukač, 2015). Industry 1.0 (1784) is the revolution in manufacturing concerning the use of energy. It has been characterized by the development of the steam engine that has allowed the translation from purely artisanal production systems to the first industrial systems with a business logic based on principles that remain valid today. The invention of the steam engine allowed factories to introduce mechanization of production in the name of greater speed and power. Industry 2.0 (1870) represents the second generation of energy, and refers to the use of electricity first and of oil then, which allows to further increase the levels of mechanization and production and to arrive to the era of mass production. It was characterized, on the technological side, by a better implementation of the steam engine and the subsequent engineering of the internal combustion engines. Industry 3.0 (1970) summarizes the entry into the factory of information technology and electronics that further increases the levels of automation not only in production but also organizational, and improve the quality of production. Finally, Industry 4.0 (2011) includes a technological mix of robotics, sensors, connection and programming, and represents a new revolution in the way products are manufactured and work is organized. Industry 4.0 is based on the digitization of companies and more generally on Information Technology (IT). A common theme between Industry 3.0 and Industry 4.0 is that of environmental

Sustainable food processing and engineering challenges. http://dx.doi.org/10.1016/B978-0-12-822714-5.00001-2

sustainability with the introduction of the first restrictive regulations at the end of the twentieth century, which obviously also affect today's production systems. Although the implementation of the principles of Industry 4.0 is not yet complete, we are facing the fifth revolution, called Industry 5.0. Technological protagonists of Industry 5.0 could be collaborative robots (COBOT) equipped with highly evolved artificial intelligences, able to emulate human behavior. On the industrial side, we will find strongly interconnected and self-adaptive systems, extremely complex to manage, both for technical and social aspects. Fig. 1.1 summarizes these industrial revolutions and specifically indicates the period, the General Purpose Technologies (GPT), the main energy source, and the main industries involved by the different revolutions (Yin et al., 2018).

As for Industry 4.0, in the last years, more and more people are getting information on this topic, as shown by the number of results obtained searching on Google with the keyword "Industry 4.0": about 10,100,000 results only in the last year. Similarly, the same research on the Scopus database provides about 5,800 results in the same period, showing that academics are also increasingly paying attention on this topic. Also institutions have shown a growing interest toward Industry 4.0, thus contributing through public policies to its diffusion and adoption (Reischauer, 2018). Several nations are at the forefront internationally (Xu et al., 2018). The United States, for example, are characterized by a smaller presence of the central government and the dissemination of technology and expertise is entrusted to large private ICT groups, technology providers, universities and research centers, which have created networks of institutes and laboratories of excellence. The US model aims to improve activities that primarily concern services, with Internet of Things (IoT) as its reference technology, which includes the use of the cloud, sensors, and machine-to-machine connections, defining platforms designed to allow the interconnection of objects in an adaptable and open way. As far as Italy is concerned (Ardito et al., 2019), the potential of Industry 4.0 has been recognized to the Italian government, as demonstrated by the "National Industry Plan 4.0" aiming at encouraging companies to adapt and fully adhere to the fourth industrial revolution.

Notwithstanding its growing and recognized importance, no universal definition of "Industry 4.0" has been provided. Generally speaking, it can be said that the term implies a complex and flexible system of enabling technologies. Flexible because each entity/subject has its own vision and way of classifying technologies (Lu, 2017). These new digital technologies will impact on four different areas of development: the use of data and connectivity (such as big data, IoT, and cloud computing), analytics that allows to derive value from the data collected, the interaction man-to-machine (such as the increasingly widespread "touch" interfaces and augmented reality), and the transition from digital to "real" (including additive manufacturing, 3D printing, robotics, machine-to-machine interactions).

Researchers use to classify enabling technologies into several categories, as described later and schematically depicted in Fig. 1.2 (Ciffolilli and Muscio, 2018). Intelligent (or autonomous) robots are interconnected robots

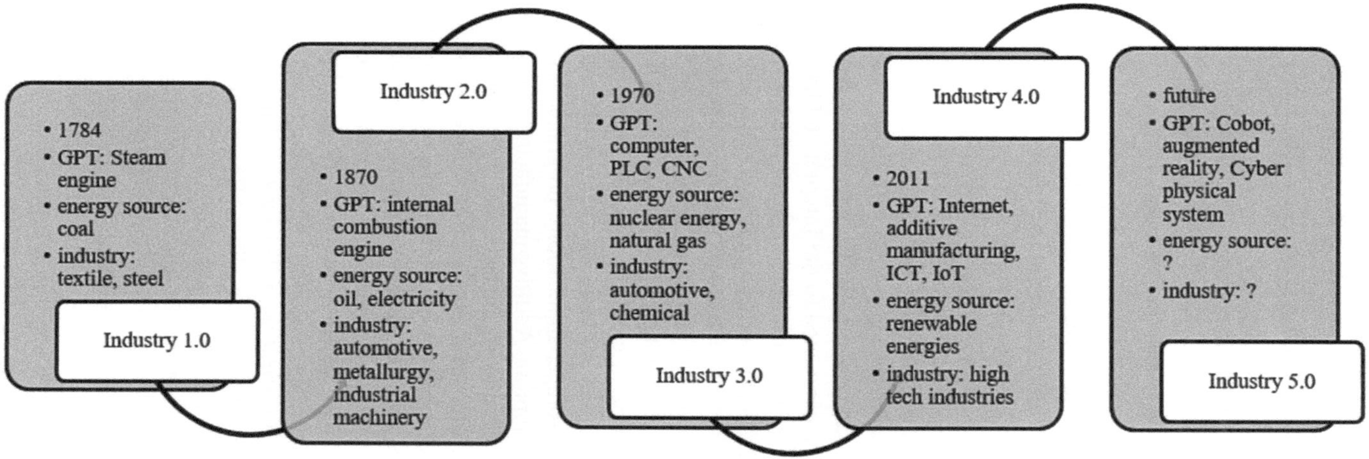

FIGURE 1.1 **The five industrial revolutions.** *(Author's own creation.)*

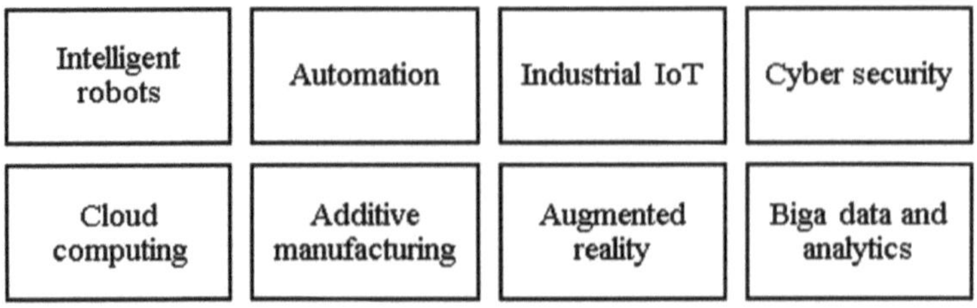

FIGURE 1.2 **The enabling technologies of Industry 4.0.** *(Author's own creation.)*

that allow greater flexibility and better performance. These robots work side by side with humans and learn from them.

The robots will cost less and less and have a greater range of capacities than those used to date. For example, companies like Kuka and ABB are launching new industrial robots that interact with each other and with humans. As far as automation is concerned, it refers to systems which will work alongside or replace their human co-workers. To date, only a few sensors and machines in manufacturing are connected to the Internet and have embedded computing systems. Thanks to the industrial IoT, more devices will be equipped with computing systems and connected to the Internet. This will allow devices to communicate and interact with each other. With the increase in connectivity and the use of standard protocols for communication, the need for security from cyberattacks grows. The latter are increasingly frequent and highly sophisticated. The future challenge will be to create services for prevention and defense. Cloud computing infrastructures are large data centers that allow the user to have the resources he needs (storage, applications, programs, services) with the formula of payment on consumption. Additive manufacturing is a new way of producing pieces no longer by removing material but by adding it. Companies have already started to use additive manufacturing technologies such as 3D printing to produce prototypes or individual components, reaching considerable technical advantages. For example, aerospace companies use additive manufacturing systems to produce aircraft components, thereby reducing overall weight. With the advent of "Industry 4.0," additive manufacturing methods will be widely used to produce small quantities of custom objects. High performance and decentralized systems will reduce transport distances and inventory, which in turn will reduce overall costs. Augmented reality (AR) systems bring actual reality back into a digital environment. These systems are still at a primordial stage but AR solutions are yet available (Smart assistance, app for warehouse management, app for visualization and system configuration). Finally, Big Data and analytics refer to the analysis of large amounts of data coming from different sources that can be, for example, the production lines or the customer. The collection of data and its complex analysis is and will increasingly become a standard of support for the decision-making process of a company.

This innovation, therefore, brings numerous benefits in many areas of people's daily lives, starting with work and food. In fact, according to the research "The Future of the Jobs," presented at the World Economic Forum 2016, it has emerged that technological factors will profoundly influence the evolution of the labor market (https://www.weforum.org/reports/the-future-of-jobs). The effect will be, on the one hand, a change in the skills and competencies, with problem-solving, critical thinking, and creativity becoming increasingly important, while on the other hand, the technological requalification and not technological substitution. This is due to the fact that the availability of sensors and automated production control and verification systems, for example, makes it possible to lift operators previously dedicated to repetitive and wear and tearful activities, without however excluding them from the production cycle. These operators acquire a different and more qualified role, becoming the ultimate supervisors of the work carried out by the technological equipment, bringing a contribution of human experience that technology cannot yet replace.

The benefits provided by these innovations can also be particularly effective in the food sector, where a large number of activities requiring meticulous attention and precision and above all a human assessment of detailed aspects, still have to be carried out manually. The availability of high-precision mechanisms and sensors can replace the intervention preserving and perhaps even better enhance the experience of individual operators called to supervise procedures.

2 Food industry and Industry 4.0

The food industry is a key sector in the Italian economy, both for its weight and for its advantages and potentials (Bigliardi & Galati, 2013). The industry is a multi-dimensional, complex, and challenging sector, including companies that produce food products, as well as companies that sell and serve food, beverages, and dietary supplements (Luque et al. 2017). In other words, it includes all the activities ranging from the manufacturing of raw materials to the production of semi-finished and finished products. Several studies exist stressing that the Industry 4.0 revolution is influencing almost all industries in several ways. In particular, it had a significant influence on the food industry, which is a mass-production type. In order to meet the nutritional needs in accordance with human health, food safety and food legislation, the food industry may benefit from technological resources. In such a sector, technology in the past has been used in several stages of food and beverage production (for instance, transportation, processing, packaging, and storage of food products) (e.g., Sun and Zheng, 2006). In more recent years, new food trends have been constantly emerging in the world and consumers are aware of these new trends through social media. This situation provides added pressure on the food industry that is forced to take advantage of using the aforementioned technological resources.

The available literature shows different papers highlighting the effect of Industry 4.0 on the food industry, as well as its technical and economic benefits.

As for the effects, Demir and Dincer (2020) highlighted digitalization (i.e., Big Data and cloud computing, virtualization and cyber security), interaction (i.e., IoT and Cyber-physical systems), and future of factories (i.e., additive manufacturing and automation) as the main effects of Industry 4.0 on the food sector. As for the technical and economic benefits, several authors (e.g., Luque et al., 2017) stressed that they include, among others, technological development, flexibility, products personalization, and improvement in consumer's satisfaction.

3 Objective and methodology

The objective of the chapter is twofold: first, to analyze what has been written in the literature on the topic Industry 4.0 in general and, second, to deepen the same topic in the food sector. The rationale behind the choice of this industry is its importance at the economical level. In particular, the chapter wants to analyze the effects that the new technological revolution has brought in the food sector: new production technologies, new systems to improve the efficiency and sustainability of the industry, new business models, and new characteristics of the working world. To reach this objective, a systematic review of the literature has been carried out.

This methodology evolved in the field of medicine intending to improve research performance and the quality of review processes then it was applied also to managerial studies (Tranfield et al., 2003). The systematic literature review was performed between 4 March and 4 April 2020, and it was articulated in three steps: (1) paper search and selection, (2) database construction containing all the characteristics of the selected papers, and (3) paper analysis (Fig. 1.3). The databases selected as source were considered as they are mainly used internationally in management studies (Abatecola et al., 2013), indeed they proved

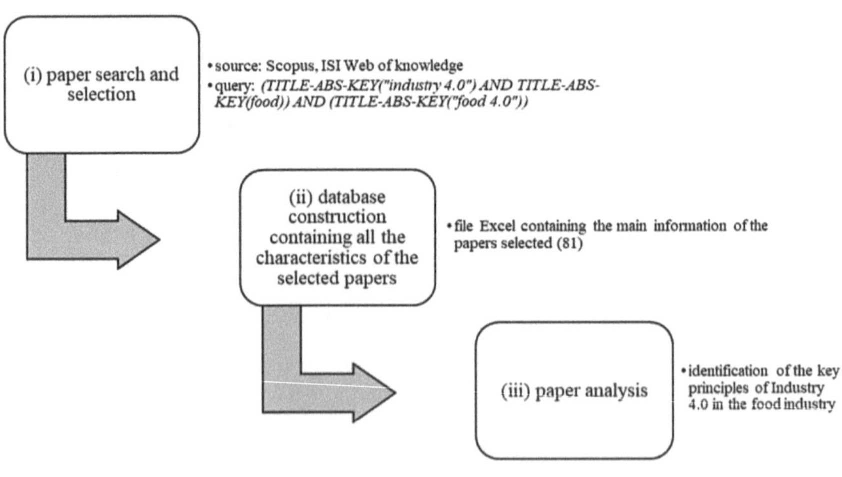

FIGURE 1.3 The three steps methodology. *(Author's own creation.)*

to be the most complete and to provide a great combination of variables, useful when conducting a systematic literature review. When required, the Google Scholar search engine was also used in addition to the first two sources, mainly in order to retrieve papers that were not available or downloadable from the first two. After the elimination of not pertinent, duplicate or unavailable papers, the final sample of papers included in the review comprised of 81 academic and conference papers. Each article was read and the main issues summarized on the initial Excel database. This final step leads to the identification of the key issues detailed in the next section.

4 Key principles of Industry 4.0 in the food industry

Industry 4.0 in the food context is often referred to the term Food 4.0, which means a radical transformation and innovation in manufacturing methods, such as, among others, new production methods (e.g., Saucedo-Martínez et al., 2018; Kiel et al., 2016; Yin et al., 2018; Kolberg et al., 2017), integration of external factors such as companies, suppliers, and customers (Burritt and Christ, 2016; Preuveneers and Ilie-Zudor, 2017; Xu et al., 2018), lean production (e.g., Kolberg and Zühlke, 2015), new qualification of workers (e.g., Weyer et al., 2015), or new managerial practices (Hozdić, 2015), and so on. Most of the papers included in the review investigated the application in the food context of the enabling technologies previously listed. Moreover, the analysis of the 81 papers shows that the research on Industry 4.0 topic has been conducted in the food sector at different levels: sector, supply chain in general, company and business area. Besides, the analysis conducted showed that particular attention has been paid to sustainability issues applied in Industry 4.0.

4.1 Enabling technologies in the food industry

The enabling technologies described in general terms in the first section find applications also in the food industry. As far as advanced robotic systems are concerned, they were initially designed to work only in certain areas of the food industry such as the palletizing of finished products. Today, however, they are used in many areas of production such as harvesting, cutting, processing and packaging of food products, meat processing, and automatic detection of the quality of the final baked goods. So the advantages of robots in the food industry can be traced back to kinematics, dynamics, control, hygiene, productivity, and worker safety (e.g., Masey et al., 2010).

Automated control systems play a key role in the production chain of the food industry in many areas. For example, da Silva Cotrim et al., (2020) described the application of a Computational Vision System (CVS) based on a Convolutional Neural Network (CNN) and operating with the Inception v3 module and a reduced number of convolutional layers. This system was used to classify the degree of browning of bread crust during baking. The system was able to correctly classify the images received from the test data set, extracting the colors present in the images already in the first convolutional layer, showing its potential use within food production control systems that require color changes. Similarly, Bowler et al. (2020) described how mixing is used to combine materials as well as to increase heat transfer uniformly in food products and to allow a change in material structure. For these reasons, sensors are required in critical processes such as mixing to provide automatic real-time data acquisition capabilities. There are several techniques available to monitor industrial mixing processes such as the use of ultrasound. Ultrasonic sensors are economical and applicable for characterizing opaque systems, allow real-time data acquisition, and using the low power, high-frequency sound waves in the ultrasonic range, they are used for characterization of the material without affecting its structure. The parameters used to characterize a system are the speed of sound, the attenuation of sound waves, and the acoustic impedance of the material (Hauptmann et al., 2002).

IoT allows the collection and use of data from different sources, and a greater digitization and automation of processes. This technological development, therefore, makes it possible to reduce the costs of food production and at the same time increase the efficiency of production times (Hong et al., 2014).

As far as the future of Food 4.0 is concerned, additive manufacturing with 3D molding, advanced robotic systems, and automated control systems are of great importance (e.g., Lipton et al., 2015). 3D printing is known as "additive manufacturing" or slow layer upon layer construction in the food industry: basically, food printing involves careful care, superimposing tiny, semi-liquid food particles on top of each other to create new food (Sun et al., 2015). The 3D molding technique in the food industry can be used both at the industrial or home level, and can be divided into three main categories: bio-driven (meat), bottom-up (uncommon sources such as algae and insects), and the mix of insect dust with other printable foods such as cheese. Despite these benefits, 3D food molding is still not widely used for two main reasons: the cost of machinery and the slowness of the process compared to other techniques currently in use.

Big Data allows the determination and storage of real and safe information, the prediction of malfunctions of machines used for production in the food industry, the control of maintenance, and the effectively improvement of the production process. A fundamental element for this technology is cloud computing (the "computer cloud") which allows to save Big Data, without the need of memory on the computer and without internet connection, and to share information quickly and securely with customers and workers (e.g., Bronson and Knezevic, 2016).

The technological revolution introduced with Industry 4.0 makes it possible to solve problems related to the serious shortage of labor and increased diversity of demand. Matsumoto et al. (2020) investigated these problems referring to the Japanese food industry, which is forced to improve its factories to be more productive and flexible to cope with future market expansion and with the diversity of products. The case study presented by Matsumoto et al. (2020) refers to a dairy plant where a horizontal five-level model with automation technologies is introduced to achieve high efficiency, rapid integration, and transfer of the production system. By comparing costs and working hours, it has been demonstrated that the five-level horizontal model has achieved better efficiency and production standards than in the previous case.

Also, Industry 4.0 also influences those activities that are not strictly productive but closely related to the food industry production chain such as the use of simulations as decision-making tools for balancing the production line or as the design of a new design for the workstation. Abd Rahman et al. (2020) studied the topic of analysis of production process data as a decision support method to reduce losses. They aimed to provide a model-driven decision support system (MD-DSS) framework in which simulation and data communication technologies are accompanied to improve the production process. OEE (Overall Equipment Effectiveness) data were acquired through the Internet network system and simulated to predict output improvement. To illustrate the applicability of MD-DSS, they investigated the food industry in Malaysia and the results show that it can be easily adopted in factories equipped with an Internet network to support decision making in the improvement plan. On the workplace side, the transition to smaller batches in production requires the design of highly flexible processes in which people and automation systems work together synergistically, safely, and efficiently. Papetti et al. (2019) discussed about the need to revise the human workstation with the advent of an automated technology. The new role of the operator requires researchers to study methods and tools able to evaluate workers' performance to maximize comfort and quality of work. To do this, it is, therefore, necessary to study methods to evaluate workers' well-being and, consequently, design operator-centered workstations. Therefore, the first objective is preparatory to the second in a perspective of continuous improvement.

Therefore, observing the numerous advantages and fields of application that the technological innovation of Industry 4.0 finds in the food industry, investments are needed to make these technologies more and more present within the sector, but at the same time, it is necessary to make skilled workers make the most of these technological innovations.

4.2 Industry-level

The food industry is a complex industry, with a wide range of processes, operations, and roles throughout the world. Plant or crop diseases are important elements in reducing quality and quantity in agriculture; therefore, the detection and diagnosis of these diseases are strongly needed. A solution to this problem, according to the literature, is the use of transfer learning, that is, the transfer of knowledge and skills learned in the past to solve future problems (e.g., da Silva et al., 2019). Computers are now everywhere in agriculture-related processes, from machinery to all food decision-making systems, and this includes the use of robots, sensors and technologies as support. Some papers (e.g., Panetto et al., 2020) highlight the new challenges, applications, and direction of the food sector. Currently, this sector uses modern machinery, tools, and emerging ICTs that take into account the capabilities of the IoT. Indeed, according to the article by Islam et al. (2020), one of the most promising technologies to address global challenges is the IoT based Trillion Sensor System (TSS). To support all the new technologies introduced with the advent of the fourth industrial revolution, the current IoT infrastructure needs to improve its spectrum from 1 to 10 trillion sensors. This new system would create a highly paid work base, allowing the creation of a new middle class and an abundant economy. Industry 4.0 promotes the automation of the food industry through IT systems and its goal is the Smart Factory. Lezama (2020), for instance, discussed about the implementation of a Business Intelligence (BI) system in a food industry system with large data masses (Big Data).

Demartini et al. (2018) stressed that "*the food industry has recently moved from a supply-based approach to a demand-based one, called inverse chain, in which consumers tell producers what they want to eat.*" Moreover, according to Lee and Lee (2020), the food industry, by using advanced digital technologies, has faced a change in consumer purchasing behavior from traditional retail to online channels. An example is the one provided by Lee and Lee (2020) describing the so-called "Untact" service and its use in different areas (e.g., ordering food from restaurants via digital devices) to facilitate customer meetings without direct contact with employees.

Industry 4.0 and its technological developments have to be considered an opportunity for the food industry for three main reasons. As first, they have lead to higher production efficiency and lower costs. As second, they have allowed generating products with higher added value. As third, they have ensured high standards of quality and food safety. Therefore, from a production point of view, we are increasingly moving toward higher product quality but with shorter production times and costs. Several papers exist investigating how the advent of Industry 4.0 has changed the food companies both from a productive and a professional and work point of view (e.g., Akyazi et al., 2020).

Industry 4.0 has been studied at an industry-level also in different countries. For example, Lezoche et al., (2020) analyzed how the Fourth Industrial Revolution has been carried out in the European and North America, while Fedotova et al. (2019) assessed the current state of Industry 4.0 in Russia. In addition, Luque et al. (2017) through a bibliographic review analyzed and described the state of the art of the Industry 4.0 on the Andalusian food industry. The article by Sharma et al. (2020) showed how the Indian food supply chain is still in full development and that it can represent a huge possibility of expansion and enrichment for all those sectors related to food. Similarly, the study by Rojas Fuentes et al. proposed the evaluation of the productivity and technical efficiency of the Chilean manufacturing sector. Their results showed that the country's manufacturing sector experienced a downward trend in technical efficiency during the study period but, at the same time, a period of productivity growth from 1986 to 1996, a slowdown since 1998 and only a slight increase in 2004. Similarly, within the sector, it is possible to see important changes in efficiency, especially as in the case of the food chain, where an efficiency range of 19%–91% is reported. Despite this, the food sector still represents the most important part of the Chilean manufacturing industry with 3.9% of the gross domestic product.

4.3 Supply chain-level

Papers dealing with the supply chain in the context of Food 4.0 deal with the main innovations introduced by the Fourth Industrial Revolution (e.g., Barata et al., 2018) as well as its impact on the companies belonging to the supply chain (e.g., Ivanov et al., 2016; Szozda, 2017). The studies by Sharma et al. (2020) and de la Fuente-Mella et al. (2020) both provided interesting comparative review of the food supply chain. The articles showed how the Indian and Chilean food supply chain are still in full development and that Industry 4.0 can represent a huge possibility of expansion and enrichment also for all those sectors related to food.

The food sector has been active in digital innovation for decades as evidenced above all by advances in precision farming, remote sensing, robots, farm management information systems, and decision support systems. This situation has thus paved the way for a broad digital transformation in the food sector. Demartini et al. (2018) have shown that recent technological developments (e.g., Cloud computing, IoT, Big Data, and so on) have allowed to the integration of food processes and to the digitization of the food industry as a whole. The main drivers for this digitization are usually increasing the flexibility and responsiveness of industrial and logistic systems and improving the robustness and resilience of the food supply chain. In this context, data become crucial and are the lifeblood of any company, and agricultural activity is not an exception but rather a reference point.

Industry 4.0 can be considered simultaneously as the biggest change but also a significant opportunity for the food industry. Indeed, all stages of the food supply chain can benefit from the efficiencies offered by connectivity, digitization, and data sharing. Companies need to create a digital roadmap for planning their food supply chain future needs in terms of technology, data, and employee training. Industry 4.0 creates jobs, but implies a transformation of the workforce with a new skill set. Producers who want to survive will have to hire, improve, and keep staff capable of the complicated business operations (Manavalan and Jayakrishna, 2019).

According to the supply chain-level, agriculture is the first step in the food supply chain that can benefit from Industry 4.0 because the whole process, from the farm to the distributor and the end customer, can be integrated. This leads to a reduction in delivery times, reduced product shelf life and therefore faster product arrival in the market. In addition to this, computerized data makes it possible to predict consumer demand and to act more efficiently by reducing overproduction and waste. Conversely, in the final phase of the food supply chain we find the retailer. For this stage, the literature collected is limited. The main technology applied in this phase resulted to be IoT (Nukala et al., 2016). Specifically, its main applications refer to the following:

- Loyalty card, that are NFC-based solutions used to offer discounts and loyalty card solutions, as well as to choose the right placement of food products in the retail space, or to improve consumers' engagement. The main objective of an IoT-based loyalty card program is to track and predict customer inflows and determine sales.
- On-shelf availability (OSA): in this case, an IoT architecture is proposed in order to determine shelf availability and to perform a demand forecasting to provide a real-time analysis.
- Real-time pricing mechanism: this area represents a challenge. Dynamic pricing is implemented in hotels, airline tickets, energy distribution, and the entertainment sector. A dynamic approach to pricing can be implemented via RFID/NFC by identifying food quality characteristics. The shelf life information printed on the labels is marked when the manufacturer packages the food. Perishable foods are sensitive to temperature and the conditions under which they are kept, and thus the dynamic price proposed for perishable products takes into account the dynamics during production, transport, and distribution and changes the price accordingly.
- Smart objects and environments in retail: smart shopping carts have always been of great interest in research in the commercial environment. Cambridge Consultants have developed an intelligent shopping cart that uses low energy consumption Bluetooth to allow monitoring and tracking of promotions and sales when customers move around the aisles in a smart retail environment. The generated mapping data can produce interesting results to enable smart pricing, product positioning, and reduced waiting times during checkout. The system based on RFID, barcode, and weight sensors is being researched to enable automated billing and checkout. Google has recently introduced the smart payment method via the "Hands-free" app that promises convenience and security during payments without taking the phone out of your pocket. The innovative approach does not require the use of a payment card (credit or debit) or physical user interaction. It uses location sensors such as low power Bluetooth, Wi-Fi, and Google location services to track the user's smartphone within retail space and the cashier checks using the document photo provided during registration.

In the food supply chain, transportation in cold chain logistics plays an important role in food and health products, to prevent the formation of harmful microorganisms and ensure that products remain intact (Buyukselcuk, 2019).

New technologies lead to an uncertainty reduction thus allowing to obtain accurate data in real-time and to increase efficiency and sustainability along the supply chain as a whole. Specifically, each technology will improve some of the following basic functions of the food supply chains: detection, monitoring, control, analysis, forecasting, decision making, and adaptive learning. Through detection, monitoring, control and analysis, it is possible to detect problems early and accurately and even to predict them before they occur, make appropriate decisions and improve the sustainability and resilience of food supply chains (Belaud et al., 2019).

Based on the articles analyzed, it is therefore possible to stress that the technological revolution brought by Industry 4.0 is influencing all the fields of the food supply chain.

4.4 Company-level

As previously stressed, the food industry is changing and recently it has to face rapid and constant change due to the current industrial revolution. Due to the emerging digitization, and specifically to the use of intelligent technologies, companies' production patterns are changing and leading to more and more efficient innovations and processes, products, and services (e.g., Manzouri et al., 2014).

Different studies have been collected by investigating companies of different size. For example, Garzoni et al. (2020) provides an example of how digital technologies have changed the business process of small and medium-sized enterprises (SMEs) operating in the agro-food industry of a regional contexts (specifically, in the Apulia region) where enterprises may access to limited knowledge and technologies.

According to the articles included in the review, the technological changes that are interesting the food industry are driven by market forces. Experts suggest creating a digital roadmap to meet future needs in terms of data, technology, and employee training. Companies must, therefore, assume what will add value in the future and thus offer the best chance to remain competitive: surviving and thriving in the era of Industry 4.0 will also mean investing in people, more than just investing in technology. History has shown that these production revolutions do not reduce work, but change it because even when more processes are automated, more jobs are created, requiring a transformed workforce with new skills to perform them (da Silva et al., 2019).

The intelligent and automated production systems characterizing Industry 4.0 will make the workers able to address complex issues and take accurate decisions in a shorter time, while simple and monotonous jobs will be carried out by robotic systems. At the same time, however, human intervention in machine maintenance and supervision operations will increase, so that industrial organizations will increasingly focus on teamwork while the classic top-down hierarchy will lose strength. In this way, teamwork between employees, as well as between workers and service systems, will become increasingly important: workers will, therefore, be asked to perform tasks in ever-widening areas and consequently to have greater knowledge and competence in different areas.

The main consequence of these technological changes is a growing demand for technological skills, including both basic and advanced skills, and programming skills. The consequence of this rapid technological development is an increasing demand for social and emotional skills, because machines are still a long way from learning emotions. Performing all these different tasks, therefore, requires mathematics, solid literacy, problem-solving skills, and ICT skills as well as transversal autonomy, collaboration, and coordination skills. Skills such as decision-making, critical thinking, and problemsolving are crucial especially for production operators and control technicians. Furthermore, the demand for managerial, communication and organizational skills will increase significantly. On the other hand, physical and manual skills will be redefined according to the degree of automation and in general, the demand for these skills will decrease. In conclusion, food companies require new professional skills able to adapt to digital transformations, changes in production processes and newly introduced work patterns.

Based on the documents analyzed, it is therefore possible to stress that, at a company-level, the technological revolution brought by Industry 4.0 is a difficult change but also an enormous possibility of development. The Fourth Industrial Revolution is approaching and, as any emerging opportunity, the first to benefit from it will be those able to respond to the new needs.

4.5 Business area-level

Several business areas have been investigated in the available literature, in particular: production, food processing, and logistics. As far as production is concerned, the main enabling technologies that have emerged at this stage are IoT and blockchain. Within IoT, several technological devices have been described (e.g., Nukala et al., 2016). These are:

- Radio-frequency identification (RFID), systems (tags) that use radio waves to transmit information about objects or people.
- Wireless Sensor Network (WSN), are sensors able to monitor the condition of products. They are low energy consumption and require minimal human intervention. Also, they allow data transmission using a wide range of wireless standards. Specifically, in the article mentioned earlier, the same authors said about the agricultural front: "*IoT devices provide precise information on a wide range of parameters that are required to improve cultivation methods and the production of fresh produce. These include environmental factors, growing conditions, cultivation equipment, greenhouse gas production, irrigation and the use of pesticides and fertilizers. Applications of WSN for real-time monitoring and management of these parameters are increasing. Several studies have proposed precise farming systems using WSNs for: crop cultivation, automation of irrigation processes, environmental and crop*

monitoring" (Nukala et al., 2016, p.2). On the livestock management front, the same authors say: "*RFID is widely known for applications in the field of traceability during the production phase. WSN is used to monitor animal behaviour and specifically for monitoring grazing livestock". Also, "RFID tags, wireless sensors and even the blockchain are three types of traceability technologies that start right at the first link in the chain: production*" (Nukala et al., 2016, p.2).

A second business area in the chain is the processing of food products. Although a digital food factory still seems to have a long way to be reached, there are many functions within the company that embrace the technologies of Industry 4.0. These are enterprise resource planning (ERP), manufacturing execution system (MES), food quality control, research and development (R&D), facility management, and many other departments (Hasnan and Yusoff, 2018). Demartini et al. (2018) summarized the technologies available in the various business areas of a food company, stressing that their results are applicable also to other sectors. In addition to the use of IoT, they highlighted, for example, the adoption of intelligent robots to enable smooth collaboration between R&D, laboratory and production, as well as to improve production performance, share best practices, allow a better use of resources, monitor the safety of devices and procedures, and send recipes and quality operative information on products. Cloud computing is useful to enable collaboration between R&D and production and implement data visibility at enterprise level. At the same time, additive manufacturing can be used in order to minimize garbage production and obtain capacity optimization. Finally, Big Data and analytics can help companies in automating data collection, reduce paperwork, monitoring job safety and employees wellbeing, and share best practices. Interestingly, they also highlighted the possibility to guarantee the efficiency of the operator and his safety through hologram technologies that can be both augmented reality and virtual reality. Much cited is the term CPPS (cyber physical production system) already introduced previously that in this it refers to cyber-physical systems in the productive field. Another technology that is often used in industry is Big Data Analytics, which is the enormous amount of data that is collected and stored thanks to sensors and IoT technology. However, only the collection of data is not enough, the part of statistical analysis that contributes enormously to the company's decision-making processes is fundamental.

As for the logistics business area, including activities such as transport and distribution, over the last few years, public awareness of food, particularly about the safety, freshness, and origin of products, has grown steadily. In this business area, the trends are traceability, safety, and sustainability. An important tool for traceability and food safety is the packaging, and in particular the so-called smart packaging, that is a combination of two types of packaging: active and intelligent (Dainelli et al., 2008). Active packaging is defined as the incorporation of certain additives into packaging systems to maintain or extend product quality and durability. Intelligent packaging is defined as a packaging system capable of performing intelligent functions (such as detection, traceability, registration, and communication) to facilitate decision-making, extend shelf life, improve quality, improve safety, provide information, and warn of potential problems. Consequently, smart packaging, being the union of the two described earlier, can be defined as a total packaging solution that on the one hand monitors changes in the product or its environment (intelligent) and on the other acts on these changes (active) (Dainelli et al., 2008). For traceability, other technologies that are needed may be, for example, QR code and barcode, that are codes allowing the identification of a product through contrasting black and white graphic bars (barcode), or the intersection of two barcodes (QR code) (Tarjan et al., 2014).

4.6 Sustainability and Industry 4.0

Finally, an important topic related to Industry 4.0 is that of sustainability. Sustainability is a broad concept addressing aspects that are not limited to the environmentalism, but also concerning economic and social issues (Beier et al., 2017; Choi and Ng, 2011; Ford and Despeisse, 2016). Environmental sustainability concerns maintaining the earth's environmental systems equilibrium (Glavič and Lukman, 2007). Economic sustainability refers to long-term economic growth and the preservation of environmental and social resources (Choi and Ng, 2011). Finally, the process of social sustainability consists in the recognition and management of both the positive and negative business, environmental, economic, and technological impacts on people. In other words, it aims at creating healthy and liveable communities where "*everyone is protected from discrimination and has access to universal human rights and basic amenities such as security or healthcare*" (Dempsey et al., 2011). While sustainability in general has a rich literature, the research on the sustainability impact of Industry 4.0 is in its nascence and requires further exploration (Ghobakhloo, 2020). However, the studies included in the review agree in stating that the sustainability impacts of Industry 4.0 has to be fully investigated by both academics and practitioners, because the previous industrial revolutions lead to unexpected economic, environmental, and social changes (Ghobakhloo, 2020).

In the Industry 4.0 context, the interconnected computers, smart materials, and intelligent machines communicate with one another, interact with the environment, and may make decisions with a minimal (if not any) human involvement (Gilchrist, 2016). The digitization, that from an Industry 4.0 point of view represents the true power of Industry 4.0, from the sustainability point of view it may lead to higher resource and energy consumption as well as elevated pollution concerns (e.g., Beier et al., 2017), disrupt the labor market (e.g., Brougham and Haar, 2018).

The industry has significant responsibility for pollution in general, depletion of fossil fuels and carbon dioxide emissions. Clairand et al. (2020) described those initiatives and technological innovations that have been implemented by the food industry to mitigate pollution problems. In the food context, food processes use significant amounts of work, technology, and energy to convert edible raw materials into more valuable food products. Although the amount of energy used is different in each country, in developing countries, the energy consumed by the food industry is generally very high. As the food industry is highly dependent on fossil fuels and demand for food is expected to grow, the development of energy efficiency strategies for this industry sector is crucial. In this context, a large amount of energy (heat in particular) is required, so clean energy production techniques and energy recovery are therefore of great importance. Heat recovery means using the wasted heat through energy generation technologies: by analyzing a milk production system it has been possible to show that there has been a 10% increase in energy efficiency by reprogramming the process and using heat exchangers. Another technique used in the food industry is heat storage. For energy production with a lower environmental impact, there are those energy sources related to waste management, that is, the use of biodiesel, biogas, biomass, and pyrolysis. In addition to these sources, there are renewable energy sources that would become increasingly important in the future of the food industry. Photovoltaic and wind power systems are considered a suitable alternative to conventional diesel generators and power distribution lines. In addition, solar energy could also be used to carry out thermal processes such as drying, thus reducing the energy consumption.

As stressed by Muñoz-Villamizar et al. (2020), environmental efficiency should not be treated independently of production efficiency: it has been shown that lean management can help improving the environmental performance of companies. By integrating existing methodologies and approaches from lean thinking, Industry 4.0 and mathematical optimization, an improvement in production and environmental performance can be achieved. This improvement of the industrial process in general, and more specifically for the food sector, can also be reflected in the processes of climate change adaptation and sustainable development of urban centers. Indeed, according to Balogun et al. (2020), digitization is a key factor for the sustainable development of the socio-economic dynamics of cities, with the power to promote climate-friendly urban environments and societies.

Based on the documents analyzed, it can, therefore, be concluded that the advent of the fourth revolution and its technological innovations has brought numerous advantages in terms of efficiency and sustainability to the food industry. However, this is not a point of arrival but only a starting point for a continuous search for new possibilities for improvement.

5 Food 4.0 framework

Based on the results of the systematic literature review, it is possible to obtain a framework for the implementation of the Industry 4.0 paradigm in the food industry. It is shown in Fig. 1.4, and it is possible to refer to it with the term “Food

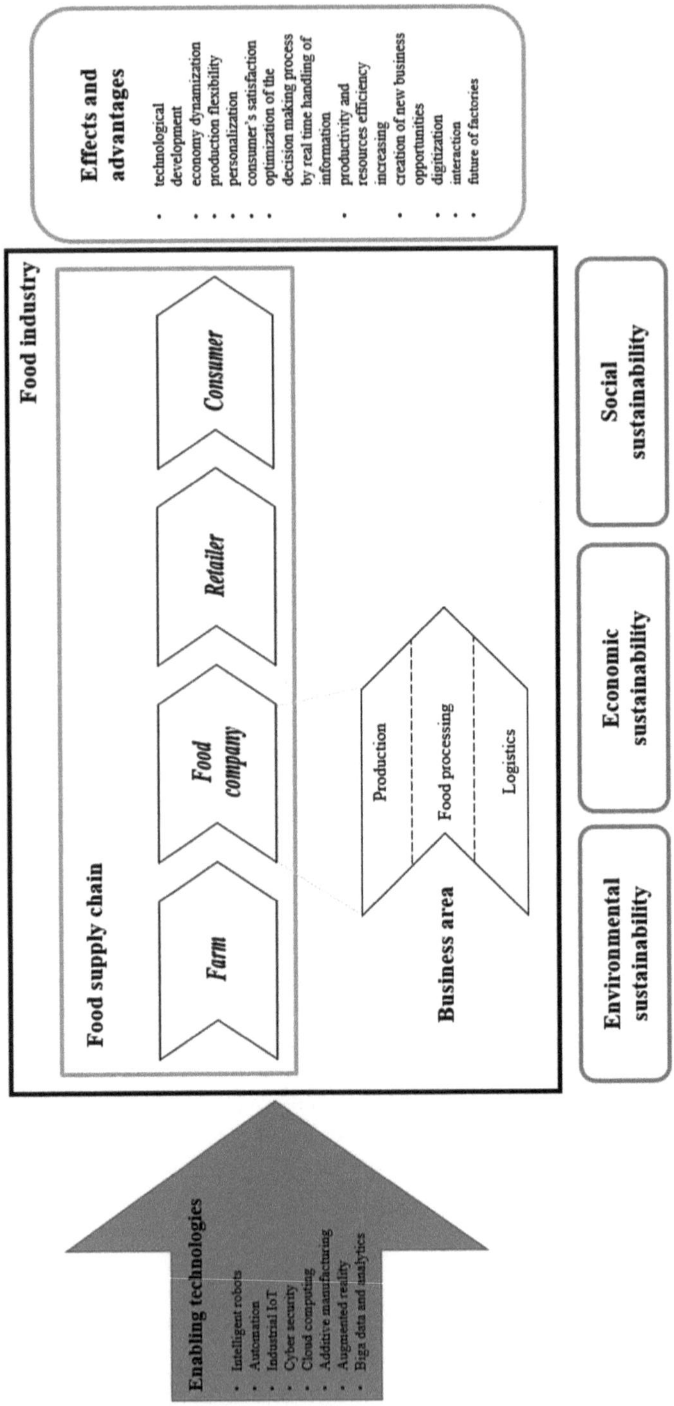

FIGURE 1.4 The Food 4.0 framework. *(Author's own creation.)*

4.0 framework." It summarizes the key issues described in the previous sections, specifically. These enabling technologies allow the adoption of the Industry 4.0 concepts, the main benefits and advantages obtained by companies operating in the food context, the different level of analysis (i.e., industry, supply chain, companies, business area) and the sustainability pillar on which Industry 4.0 is grounded.

6 Discussions and future research avenues

The papers analyzed in the chapter investigated the new available technologies and new food chain methods to understand the future paths of the food and agri-food sectors, which very often suffer from factors such as low efficiency and effectiveness, weather sensitivity and market disruption. In most cases, knowledge of agriculture transferred from generation to generation is culturally essential, but most of the time, it is unable to meet the needs or requirements of food supply chains. Therefore, Food 4.0 emerges to provide advanced technology to companies to meet the challenges of food production, to obtain cheaper prices for the open market and minimum cost for farmers. Food 4.0 is expected, on the one hand, to help meeting sustainable challenges and increasing stakeholders' revenues, and on the other hand to manage external factors such as time, market behavior, and policies (Miranda et al., 2019).

In reality, it is a much deeper revolution that starting from the technological tools already available must adapt the production methods and business to make them suitable for the technological tools of the future. It could be defined as a fluent industrial revolution, that is, generating a flow of continuous and transversal change. It is continuous because it is a revolution that potentially feeds itself and triggers further novelties on the technological level but also on the economic, financial, productive, social, political ones. It is transversal because of the number of consequences it may have on companies' business models and people lifestyles.

Even if the industry has shown a positive trend toward digitization, but introducing it efficiently will be the new challenge.

References

Abatecola, G., Mandarelli, G., & Poggesi, S. (2013). The personality factor: how top management teams make decisions. A literature review. *Journal of Management & Governance*, *17*(4), 1073–1100.

Abd Rahman, M. S., Mohamad, E., & Rahman, A. A. (2020). Enhancement of overall equipment effectiveness (OEE) data by using simulation as decision making tools for line balancing. *Indonesian Journal of Electrical Engineering and Computer Science*, *18*(2), 1040–1047.

Akyazi, T., Goti, A., Oyarbide, A., Alberdi, E., & Bayon, F. (2020). A guide for the food industry to meet the future skills requirements emerging with Industry 4.0. *Foods*, *9*(4), 492.

Ardito, L., Petruzzelli, A. M., Panniello, U., & Garavelli, A. C. (2019). Towards Industry 4.0. *Business Process Management Journal*, *25*(2), 323–346.

Balogun, A. L., Marks, D., Sharma, R., Shekhar, H., Balmes, C., Maheng, D., et al. (2020). Assessing the potentials of digitalization as a tool for climate change adaptation and sustainable development in urban centres. *Sustainable Cities and Society*, *53*, 101888.

Barata, J., Da Cunha, P. R., & Stal, J. (2018). Mobile supply chain management in the Industry 4.0 era: An annotated bibliography and guide for future research. *Journal of Enterprise Information Management*, *31*(1), 173–192.

Beier, G., Niehoff, S., Ziems, T., & Xue, B. (2017). Sustainability aspects of a digitalized industry–A comparative study from China and Germany. *International Journal of Precision Engineering and Manufacturing-Green Technology*, *4*(2), 227–234.

Belaud, J. P., Prioux, N., Vialle, C., & Sablayrolles, C. (2019). Big data for agri-food 4.0: Application to sustainability management for by-products supply chain. *Computers in Industry*, *111*, 41–50.

Bigliardi, B., & Galati, F. (2013). Models of adoption of open innovation within the food industry. *Trends in Food Science & Technology*, *30*(1), 16–26.

Bowler, A. L., Bakalis, S., & Watson, N. J. (2020). Monitoring mixing processes using ultrasonic sensors and machine learning. *Sensors*, *20*(7), 1813.

Bronson, K., & Knezevic, I. (2016). Big Data in food and agriculture. *Big Data & Society*, *3*(1), . 2053951716648174.

Brougham, D., & Haar, J. (2018). Smart technology, artificial intelligence, robotics, and algorithms (STARA): Employees' perceptions of our future workplace. *Journal of Management & Organization*, *24*(2), 239–257.

Burritt, R., & Christ, K. (2016). Industry 4.0 and environmental accounting: a new revolution? *Asian Journal of Sustainability and Social Responsibility*, *1*(1), 23–38.

Buyukselcuk, E. C. (2019). Cold chain logistics firm selection by using AHP-VIKOR integrated method and a case study in food industry. In *Proceedings of the international symposium for production research 2019* (pp. 403–415). Springer, Cham.

Choi, S., & Ng, A. (2011). Environmental and economic dimensions of sustainability and price effects on consumer responses. *Journal of Business Ethics*, *104*(2), 269–282.

Ciffolilli, A., & Muscio, A. (2018). Industry 4.0: national and regional comparative advantages in key enabling technologies. *European Planning Studies*, *26*(12), 2323–2343.

Clairand, J. M., Briceño-León, M., Escrivá-Escrivá, G., & Pantaleo, A. M. (2020). Review of energy efficiency technologies in the food industry: Trends, barriers, and opportunities. *IEEE Access*, *8*, 48015–48029.

da Silva Cotrim, W., Minim, V. P. R., Felix, L. B., & Minim, L. A. (2020). Short convolutional neural networks applied to the recognition of the browning stages of bread crust. *Journal of Food Engineering*, *277*, 109916.

da Silva, V. L., Kovaleski, J. L., & Pagani, R. N. (2019). Technology transfer in the supply chain oriented to industry 4.0: a literature review. *Technology Analysis & Strategic Management*, *31*(5), 546–562.

Dainelli, D., Gontard, N., Spyropoulos, D., Zondervan-van den Beuken, E., & Tobback, P. (2008). Active and intelligent food packaging: legal aspects and safety concerns. *Trends in Food Science & Technology*, *19*, S103–S112.

de la Fuente-Mella, H., Fuentes, J. L. R., & Leiva, V. (2020). Econometric modeling of productivity and technical efficiency in the Chilean manufacturing industry. *Computers & Industrial Engineering*, *139*, 105793.

Demartini, M., Pinna, C., Tonelli, F., Terzi, S., Sansone, C., & Testa, C. (2018). Food industry digitalization: from challenges and trends to opportunities and solutions. *IFAC-PapersOnLine*, *51*(11), 1371–1378.

Demir, Y., & Testa, F. I. (2020). The effects of Industry 4.0 on the food and beverage industry. *Journal of Tourismology*, *6*(1), 133–145.

Dempsey, N., Bramley, G., Power, S., & Brown, C. (2011). The social dimension of sustainable development: Defining urban social sustainability. *Sustainable Development*, *19*(5), 289–300.

Fedotova, G. V., Gorlov, I. F., Glushchenko, A. V., Slozhenkina, M. I., & Natyrov, A. K. (2019). Trends of scientific and technical development of agriculture in Russia. In *Institute of Scientific Communications Conference* (pp. 193–200). Springer, Cham.

Ford, S., & Despeisse, M. (2016). Additive manufacturing and sustainability: an exploratory study of the advantages and challenges. *Journal of Cleaner Production*, *137*, 1573–1587.

Galati, F., & Bigliardi, B. (2019). Industry 4.0: Emerging themes and future research avenues using a text mining approach. *Computers in Industry*, *109*, 100–113.

Garzoni, A., De Turi, I., Secundo, G., & Del Vecchio, P. (2020). Fostering digital transformation of SMEs: a four levels approach. *Management Decision*.

Ghobakhloo, M. (2020). Industry 4.0, digitization, and opportunities for sustainability. *Journal of Cleaner Production*, *252*, 119869.

Gilchrist, A. (2016). *Industry 4.0: the industrial internet of things*. Apress.

Glavič, P., & Lukman, R. (2007). Review of sustainability terms and their definitions. *Journal of Cleaner Production*, *15*(18), 1875–1885.

Hasnan, N. Z. N., & Yusoff, Y. M. (2018). Short review: Application areas of industry 4.0 technologies in food processing sector. In *2018 IEEE Student Conference on Research and Development (SCOReD)* (pp. 1–6). IEEE.

Hauptmann, P., Hoppe, N., & Püttmer, A. (2002). Application of ultrasonic sensors in the process industry. *Measurement Science and Technology*, *13*(8), R73.

Hong, I., Park, S., Lee, B., Lee, J., Jeong, D., & Park, S. (2014). IoT-based smart garbage system for efficient food waste management. *The Scientific World Journal*, *2014*, 1–14.

Hozdić, E. (2015). Smart factory for Industry 4.0: A review. *International Journal of Modern Manufacturing Technologies*, *7*(1), 28–35. https://www.weforum.org/reports/the-future-of-jobs.

Islam, N., Marinakis, Y., Majadillas, M. A., Fink, M., & Walsh, S. T. (2020). Here there be dragons, a pre-roadmap construct for IoT service infrastructure. *Technological Forecasting and Social Change*, *155*, 119073.

Ivanov, D., Dolgui, A., Sokolov, B., Werner, F., & Ivanova, M. (2016). A dynamic model and an algorithm for short-term supply chain scheduling in the smart factory industry 4.0. *International Journal of Production Research*, *54*(2), 386–402.

Kiel, D., Arnold, C., Collisi, M., & Voigt, K. I. (2016). The impact of the industrial internet of things on established business models. In *Proceedings of the 25th international association for management of technology (IAMOT) conference* (pp. 673–695).

Kolberg, D., & Zühlke, D. (2015). Lean automation enabled by Industry 4.0 technologies. *IFAC-PapersOnLine*, *48*(3), 1870–1875.

Kolberg, D., Knobloch, J., & Zühlke, D. (2017). Towards a lean automation interface for workstations. *International Journal of Production Research*, *55*(10), 2845–2856.

Lee, S. M., & Lee, D. (2020). "Untact": a new customer service strategy in the digital age. *Service Business*, *14*(1), 1–22.

Lezama, O. B. P. (2020). Applying a business intelligence system in a Big Data context: Production companies. *Computational Vision and Bio-Inspired Computing: ICCVBIC 2019*, *1108*, 268.

Lezoche, M., Hernandez, J. E., Díaz, M. D. M. E. A., Panetto, H., & Kacprzyk, J. (2020). Agri-food 4.0: a survey of the supply chains and technologies for the future agriculture. *Computers in Industry*, *117*, 103187.

Lipton, J. I., Cutler, M., Nigl, F., Cohen, D., & Lipson, H. (2015). Additive manufacturing for the food industry. *Trends in Food Science & Technology*, *43*(1), 114–123.

Lu, Y. (2017). Industry 4.0: A survey on technologies, applications and open research issues. *Journal of Industrial Information Integration*, *6*, 1–10.

Lukač, D. (2015). The fourth ICT-based industrial revolution "Industry 4.0"—HMI and the case of CAE/CAD innovation with EPLAN P8. In *2015 23rd telecommunications forum telfor (TELFOR)* (pp. 835–838). IEEE.

Luque, A., Peralta, M. E., De Las Heras, A., & Córdoba, A. (2017). State of the Industry 4.0 in the Andalusian food sector. *Procedia Manufacturing*, *13*, 1199–1205.

Manavalan, E., & Jayakrishna, K. (2019). A review of Internet of Things (IoT) embedded sustainable supply chain for industry 4.0 requirements. *Computers & Industrial Engineering*, *127*, 925–953.

Manzouri, M., Ab-Rahman, M. N., Zain, C. R. C. M., & Jamsari, E. A. (2014). Increasing production and eliminating waste through lean tools and techniques for halal food companies. *Sustainability*, *6*(12), 9179–9204.

Masey, R. J. M., Gray, J. O., Dodd, T. J., & Caldwell, D. G. (2010). Guidelines for the design of low-cost robots for the food industry. *Industrial Robot: An International Journal*, *37*(6), 509–517.

Matsumoto, T., Chen, Y., Nakatsuka, A., & Wang, Q. (2020). Research on horizontal system model for food factories: A case study of process cheese manufacturer. *International Journal of Production Economics*, 107616.

Miranda, J., Ponce, P., Molina, A., & Wright, P. (2019). Sensing, smart and sustainable technologies for Agri-Food 4.0. *Computers in Industry*, *108*, 21–36.

Muñoz-Villamizar, A., Santos, J., Grau, P., & Viles, E. (2020). Toolkit for simultaneously improving production and environmental efficiencies. *Central European Journal of Operations Research*, 1–12.

Nukala, R., Panduru, K., Shields, A., Riordan, D., Doody, P., Walsh, J. (2016). Internet of Things: a review from 'farm to fork'. In *27th Irish conference on Signals and Systems Conference (ISSC)*, pp. 1–6. IEEE, (2016).

Panetto, H., Lezoche, M., Hernandez, J., Diaz, M. D. M. E. A., & Kacprzyk, J. (2020). Special issue on Agri-Food 4.0 and digitalization in agriculture supply chains-New directions, challenges and applications. *Computers in Industry*, *116*, 103188.

Papetti, A., Scafà, M., Brunzini, A., & Mandolini, M. (2019). Multiperspective ergonomic assessment approach for human centered workplace design. In *International conference on design, simulation, manufacturing: The innovation exchange* (pp. 675–685). Springer, Cham.

Preuveneers, D., & Ilie-Zudor, E. (2017). The intelligent industry of the future: A survey on emerging trends, research challenges and opportunities in Industry 4.0. *Journal of Ambient Intelligence and Smart Environments*, *9*(3), 287–298.

Reischauer, G. (2018). Industry 4.0 as policy-driven discourse to institutionalize innovation systems in manufacturing. *Technological Forecasting and Social Change*, *132*, 26–33.

Saucedo-Martínez, J. A., Pérez-Lara, M., Marmolejo-Saucedo, J. A., Salais-Fierro, T. E., & Vasant, P. (2018). Industry 4.0 framework for management and operations: a review. *Journal of Ambient Intelligence and Humanized Computing*, *9*(3), 789–801.

Sharma, J., Tyagi, M., & Bhardwaj, A. (2020). Parametric review of food supply chain performance implications under different aspects. *Journal of Advances in Management Research*, *17*(3), 421–453.

Sun, D. W., & Zheng, L. (2006). Vacuum cooling technology for the agri-food industry: Past, present and future. *Journal of Food Engineering*, *77*(2), 203–214.

Sun, J., Zhou, W., Huang, D., Fuh, J. Y., & Hong, G. S. (2015). An overview of 3D printing technologies for food fabrication. *Food and Bioprocess Technology*, *8*(8), 1605–1615.

Szozda, N. (2017). Industry 4.0 and its impact on the functioning of supply chains. *Logforum*, *13*, 401–414.

Tarjan, L., Šenk, I., Tegeltija, S., Stankovski, S., & Ostojic, G. (2014). A readability analysis for QR code application in a traceability system. *Computers and Electronics in Agriculture*, *109*, 1–11.

Tranfield, D., Denyer, D., & Smart, P. (2003). Towards a methodology for developing evidence-informed management knowledge by means of systematic review. *British Journal of Management*, *14*(3), 207–222.

Weyer, S., Schmitt, M., Ohmer, M., & Gorecky, D. (2015). Towards Industry 4.0-Standardization as the crucial challenge for highly modular, multi-vendor production systems. *IFAC-PapersOnLine*, *48*(3), 579–584.

Xu, L. D., Xu, E. L., & Li, L. (2018). Industry 4.0: state of the art and future trends. *International Journal of Production Research*, *56*(8), 2941–2962.

Yin, Y., Stecke, K. E., & Li, D. (2018). The evolution of production systems from Industry 2.0 through Industry 4. 0. International Journal of Production Research, 56(1–2), 848–861.

Chapter 2

Pasteurization of Juices with Non-Thermal Technologies

Sara Aghajanzadeh and Aman Mohammad Ziaiifar*
Department of Food Process Engineering, Gorgan University of Agricultural Sciences and Natural Resources, Basij Square, Gorgan, Iran
**Corresponding author*

1 Introduction

Fruit and vegetable are popular and vastly consumed due to their low-cost, high nutritional value, fantastic color, pleasant odor, and taste. Fruit and vegetable have a limited shelf life; in order to minimize the losses, different products such as juice, jam, marmalade, puree, and paste are produced. These products, easily transported to different parts of the world, make the seasonal fruit and vegetable available throughout the year and satisfy the consumer's tastes.

The juice is one of the most important fruits and vegetable-based products with almost the same nutritional values and health benefits. It is generally preferred over the full product, especially by children, elderly, and patients. It mainly contains vitamins (thiamin, riboflavin and niacin, vitamin C, A, and E) Zou et al., 2016, minerals (calcium, phosphorus, sulfur, folate, zinc, selenium, potassium, and magnesium) (Zhou, 2012), sugars (mainly sucrose, fructose, and glucose), organic acids, and phytochemicals with small quantities of proteins and no fat (Rampersaud and Valim, 2017). The juice is also known as a good source of bioactive and antioxidant compounds such as phenolic compounds, carotenoids, anthocyanins, and flavonoids. The consumption of juice could be useful in reducing the risk of chronic diseases such as cardiovascular diseases and different types of cancer.

The juice is defined as "the unfermented but fermentable liquid which obtained by mechanical extraction from the edible part of appropriately mature and fresh fruit under the provisions of the Codex Alimentarius Commission" (di Caracalla, 2000). As the endocarp cells are ruptured during the mechanical juice extraction, the membrane and high molecular weight compounds (proteins, hesperidin, cellulose, hemicellulose, and pectin) are suspended in the serum phase.

Sustainable food processing and engineering challenges. http://dx.doi.org/10.1016/B978-0-12-822714-5.00002-4

These colloidal suspensions are responsible for the cloudiness and turbid appearance of the juice (Kimball, 1999). Clarified and cloudy juices are produced because of entirely or partially separation of the pulp and pectin, respectively. In general, juice can be categorized in different types: (1) fresh juice, (2) single strength, produced by reconstituting the concentrated juice up to a soluble solids content equal to the fresh juice, (3) nectar, obtained by adding water and sugar to clarified or cloudy juice containing at least 25%–50% juice or puree, and (4) squashes and cordials, prepared by reconstituting the juice or puree with other ingredients such as sweeteners.

There are different steps in juice production, which include cleaning, sorting, peeling, juice extraction, filtration, pasteurization, filling, sealing, and storage, depending on the type of fruit and vegetable. Pasteurization is one of the most important steps as it defines the safety, quality, and shelf life of the juice. A minimum 5-log reduction in target pathogens of concern in the pasteurized juice is introduced as a performance standard (FDA, 2001). Microorganisms are usually considered as pasteurization index in controlling the safety of juice; however, the microbial studies are generally expensive and time-consuming. Some enzymes (such as pectin methylesterase) show higher thermal resistance than the spoilage microorganisms in acidic foods, their inactivation can be therefore considered as a juice pasteurization index (Kimball, 1999; Polydera et al., 2004). Thermal treatment is the most common method to pasteurize the juice (Aghajanzadeh et al., 2016). The intensity of the thermal process (temperature and time) is, therefore, essential to control the stability, shelf life, and nutritional values. Based on the pH, food products can be categorized into two groups of low-acids (pH > 4.6) and high-acids (pH < 4.6) (Aghajanzadeh and Ziaiifar, 2018). In general, low-acid foods (pH > 4.6) are heated at a higher temperature than 100°C; while, high-acidic foods (pH < 4.6) require the heat treatment below 100°C. The juice is generally known as a high-acid product, which is a suitable medium for the growth of spoilage yeasts, molds, and some acid-tolerant bacteria (Turtoi, 2014). As these kinds of microorganisms are relatively thermal sensitive, applying a mild thermal treatment (below 100°C) is sufficient for their inactivation (McGlynn, 2003). However, thermal processing shows some disadvantages, mainly nutritional value loss due to the degradation of chemical compounds such as pigments and vitamins. Chemical changes affect the sensorial quality and physical aspects of the juice and finally reduce its acceptance; for instance, the color of the juice is affected by the non-enzymatic browning via thermal degradation of ascorbic acid (Burdurlu et al., 2006). Different non-thermal food processing methods have been recently developed to decrease the undesirable effects of thermal treatments. These methods change the biological and chemical structures of microorganisms and enzymes and inactivate them (Table 2.1). In these methods, a negligible rise in juice temperature occurs which minimizes the undesirable effects on the juice.

TABLE 2.1 The inactivation mechanisms of microorganisms and enzyme during pasteurization of juice using different non-thermal methods.

	Inactivation mechanisms of	
Method	Microorganisms	Enzymes
Pulsed electric field	Interfering in the membrane permeability via electroporation, increase in the temperature by ohmic heating	Protein denaturation during electroporation, increase in the temperature by ohmic heating, pH shifting, oxidation of amino acid residues
High-pressure	Permeabilization of the cell membrane, changes in phase transition fluidity of cell membrane, ribosome breakdown, inactivation of the vital enzymes	Changes in the native structure of the enzyme by folding/unfolding and formation/disruption of interactions
High-pressure carbon dioxide	Cell rupturing during pressurizing and depressurizing of CO_2, leakage out of the vital cell components via CO_2 diffusion into the cell membrane, disturbing the vital biological procedures by dropping the intracellular pH, altering in the osmotic balance	Conformational change in the enzyme protein bonds at low pH
Sonication	Cavitation phenomenon inactivates the microorganisms by interfering the membrane permeability and thinning the membrane, expansion/compression phenomena causes the cell rupturing, generating the localized heating by absorbing the acoustic energy	The shear force resulting in the denaturation of the protein structure of the enzyme, the formed free radicals affects the stability, substrate binding, catalytic functions of the enzyme
Ultraviolet	DNA and RNA structure are damaged by absorbing the UV light	No considerable effect
Ozone	Oxidative properties and altering the cell permeability, damaging the DNA structure	Oxidation of protein structures of the enzymes causes the covalent modification

2 Pulsed electric field treatment

2.1 Definition of pulsed electric field process

Pulsed electric field (PEF) is applied in the processing of homogenous juice containing no large solid particles and air bubbles. In the PEF process, the juice is subjected to specific controlled high voltage electric pulses (up to 70 kV/cm) for a few microseconds to inactivate target microorganisms and enzymes (Buckow et al., 2011). In batch and continuous PEF systems, the juice is placed between or passed through two electrodes, respectively. The juice temperature can be controlled and maintained almost constant at ambient temperature or near ambient temperature. Therefore, the PEF treated juice become microbiologically safe with almost similar sensorial aspects, physical properties, and nutritional values to the fresh juice. The quality of the treated juice is influenced by a different parameter, including treatment conditions and juice properties.

2.2 Pulsed electric field system

The PEF system comprises a high-voltage generator, one or more energy storage capacitor(s), a switch, two electrodes (treatment chamber), pump (in continuous PEF design), monitoring and controlling devices (such as oscilloscope and temperature sensors). High-voltage generator supplies pulses with specific intensity, shape, and duration. The generator could be a source of direct current (DC) or also a capacitor charging power supply with high-frequency AC inputs (Zhang et al., 1996). A capacitor is required to store the generated high voltage energy from the pulse generator and transfer this energy during a short period. This energy is discharged to the electrodes via activation of a switch. There are two types of switches, including ON (having lower cost and shorter lifetime) and ON/OFF. While an ON switch provides full capacitor discharging and it turns off at the end of complete discharge, an ON/OFF switch provides the partial or complete discharge of the capacitors.

There are different electrode configurations, including parallel plates, wire-cylinder, rod–rod, rod–plate, and coaxial (concentric) cylinders. The parallel configuration is frequently used because of producing a more uniform distribution of PEF intensity and having a simple design. Uniform juice flow can also be obtained when the coaxial electrodes are applied (Jeyamkondan et al., 1999). Stainless steel is frequently used by designing the electrode in the PEF system. Non-conductive materials are used to separate two electrodes and prevent the electrical flow from one to the other. Multiple treatment chambers (in series) can be used to prevent the over-processing of the juice; as the high rise in the outlet temperature of the juice will be prevented by dividing the required total PEF treatment time to several periods (depending on the number of chambers) and inserting cooling sections between each chamber (Buckow et al., 2013).

2.3 Critical parameters in pulsed electric field process

PEF efficiency is affected by PEF system characteristics. Electric field intensity and treatment time are known as the most critical factors affecting the performance of PEF treatment. Electric field intensity is calculated by dividing the applied peak electric voltage by the gap distance between the electrodes (Agcam et al., 2014). This factor has a critical role in the effect of PEF on juice properties. Applying higher PEF intensity causes more rise in the juice temperature and a better inactivation of the microorganisms and enzymes. "PEF treatment time" is defined as the sufficient time during which the juice is exposed under the PEF treatment. The more prolonged PEF treatment may bring adverse effects on the juice quality by causing a considerable rise in the juice temperature. Considering the applied PEF intensity, shorter treatment time is required when the higher intensity PEF is applied.

Different shapes of the pulse, such as exponential decaying, square and oscillatory, are used. The square wave pulse shows the highest energy efficiency in the inactivation of microorganisms. The application of oscillatory shape waves shows a slight effect on the irreversible inactivation of the microorganism due to the discontinuous exposure of the cell to the PEF intensity (Jeyamkondan et al., 1999). Pulse is categorized as monopolar (only positive pulses) or bipolar (alternating positive and negative pulses). In comparison to the monopolar pulse, the bipolar form is more efficient due to the re-orientation of charged molecules in the cell membrane. The application of bipolar pulse reduces energy consumption, juice electrolysis, and solids deposition on the electrode surface (Barbosa-Canovas et al., 1999). The pulse width is the duration of the peak field being constant as in square wave pulses or the time until decay to 37% in exponential decay pulses. Frequency defines the number of pulses per second. The efficiency of the PEF process and the juice temperature are considerably affected by pulse width and frequency (Jeyamkondan et al., 1999).

PEF efficiency is also affected by the physicochemical properties of the juice. During PEF treatment, the electrical conductivity (ability to conduct the electric current) is related to the ionic strength of the juice (Jeyamkondan et al., 1999). The pH of juice is another factor affecting the PEF effectiveness as an additional stressing factor during PEF treatment (Góngora-Nieto et al., 2002).

2.4 Electroporation phenomena

The microorganism cell membrane naturally has a particular dielectric strength. The cytoplasm, containing free charges (ions), shows a higher dielectric constant than the cell membrane and medium (juice). This difference brings a transmembrane potential (~10 mV) between the inside (cytoplasm) and outside (membrane and juice) of the cell. When PEF is applied, the ions move along the electrical field and their accumulation at the membrane results in compression, which decreases the membrane thickness and increases the transmembrane potential. When the intensity of the applied electrical filed rises more than the critical threshold of transmembrane potential (1 V), pores with different sizes are formed in the membrane (Buckow et al., 2013). It interferes with the natural permeability of the membrane to enter the small molecules and brings swelling and rapturing of the cell. This event is called electroporation, known as a non-thermal phenomenon during PEF treatment. The irreversible effect of electroporation on cell membrane structure and the increase in its permeability causes the death of the cell. At a lower and controlled PEF intensity, this effect can be reversible; it means that the cell pores are considerably small about the membrane surface in which the cell can repair and reseal its membrane after removing the PEF exposure (Kaur and Singh, 2016).

2.5 Ohmic heating effect during pulsed electric field

The difference between the PEF treatment and ohmic heating method is mainly related to the intensity and duration of the treatment. In PEF, a 100-fold or higher field strengths than ohmic heating are usually used (kilovolt/cemtimeter vs. tens of Volt/centimeter) in microseconds or milliseconds (μs or ms) while it is based on seconds to minutes in ohmic heating treatment (Kulshrestha and Sastry, 2006). Nevertheless, the temperature of the juice could increase during PEF treatment due to the "ohmic heating effect," resulted from exposure of the juice to the electrical current (Kaur and Singh, 2016). The alteration in the juice temperature is directly related to the process conditions (PEF intensity, treatment time, pulse characteristics, and electrodes properties) and the product properties (electrical conductivity, particle size, and ionic concentration) (Kaur and Singh, 2016). For instance, applying a higher PEF intensity induces higher temperatures in the juice. Besides, foods containing sufficient ionic concentration can be quickly heated due to a reduction in electrical conductivity (Demirdöven and Baysal, 2014).

2.6 Susceptibility of microorganisms to pulsed electric fields

Electroporation and ohmic heating effects are responsible for the inactivation of the microorganism during the PEF treatment of juice. The effect of ohmic heating is related to the temperature dependence of the membrane fluidity and permeability properties. At higher temperatures, the cell membrane stability reduces due to alteration in the structure of the phospholipid of the cell membrane from the crystalline–crystalline state to liquid–crystalline (Stanley and Parkin, 1991). The osmotic imbalance is another mechanism involving in the inactivation of the bacteria. This mechanism is related to the loss in functionality of the cell membrane channels by denaturing the protein structure of these channels and forced them to open (Buckow et al., 2013). Treatment of mango juice at 35 kV/cm for 800 μs ensured a 5-log reduction in *Listeria monocytogenes* (Salinas-Roca et al., 2017). *Listeria monocytogenes* was inactivated

(3-log) by PEF treatment of sour cherry juice at 27 kV/cm for 131 μs (Altuntas et al., 2010). Total aerobic plate count, yeasts and molds of the tomato juice decreased from 10^6 cfu/mL to <10 cfu/mL after PEF treatment at 40 kV/cm for 57 μs (Min et al., 2003). After treatment of orange-carrot juice (25 kV/cm for 280 ms) and pomegranate juice (38 kV/cm for 281 μs) significant growth inhibition of total aerobic bacteria, yeast and mold were reported (Rodrigo et al., 2003; Guo et al., 2014).

The PEF treatment of orange juice at 35 kV/cm for 59 μs showed 7-log reduction in total aerobic plate count, yeasts and molds. While applying lower PEF intensity (up to 30 kV/cm) caused less inactivation of these microorganisms. Also, less reduction (3.3-log) in total aerobic plate count, yeasts and molds were achieved as a shorter PEF process was applied (35 kV/cm for both 39 and 49 μs) (Yeom et al., 2000). After PEF processing (20 kV/cm) of apple juice with 20°C, 30°C, and 40°C inlet temperature, 0.5-, 1.5-, and 2.8-log decrease in *Escherichia coli* was observed, respectively (Saldana et al., 2011). This fact shows the importance of juice temperature.

Type of bacteria is another factor influencing the PEF efficiency. While the only 2.5-log decrease in *S. cerevisiae ascospores* was achieved in orange juice processed at 30 kV/cm and 50°C, it was 5-log in *L. mesenteroides, E. coli, and L. innocua* at the same PEF condition (McDonald et al., 2000). The growth stage of microorganisms influences the effectiveness of its inactivation by the PEF process. Gram-positive bacteria show higher resistance to the PEF treatment rather than the Gram-negative ones. It was found that PEF treatment (32 kV/cm for the 90 s) was more efficient in the inactivation of yeast than *E. coli,* which reflects the higher sensitivity of the yeast to the PEF process due to their larger size than the bacteria (Zhang et al., 2010). It was found that the PEF treatment is only capable of inactivation of vegetative bacteria with less effect on the bacteria in lag and stationary phases. PEF is ineffective in the inactivation of spores because they are small, round-shaped and protected by an energetic shield. These factors limit the application of PEF to the processing of high acid juices, which are distributed in the refrigerated chain (Mastwijk and Bartels, 2005).

2.7 Pulsed electric field effects on enzymes

During PEF treatment, the enzyme is re-directed due to the presence of the net charges in its structure and dipole moments in an aqueous medium (juice). Hence, the enzyme inactivation possibly takes place due to the alteration in its secondary and tertiary protein structures (Samaranayake and Sastry, 2016). Modification of specific molecular linkages in the active site of the globular protein chain and the loss of α-helix and β-sheet in the enzyme structure could be the reasons for its inactivation. Also, electrolysis and free radical formation can bring pH shifts (near the electrodes) in juices as well as oxidation of amino acid residues, which impact the stability of enzymes (Meneses et al., 2011).

Pectin methylesterase is naturally found in fruit and vegetable with vital roles in fruit ripening. This enzyme is responsible for the de-esterification of methoxylated pectin and the formation of insoluble calcium pectate. During these phenomena, precipitation of the pectin occurs. Inactivation of pectin methylesterase influences cloud stability, viscosity, color, mouthfeel, and flavor of the juice (Tiwari et al., 2009). Pectin methylesterase in tomato juice was inactivated up to 93.8% after PEF treatment (400 pulses, 0.02 ms pulse width, 24 kV/cm) in a condition in which the temperature of the sample never be allowed to exceed 15°C (Giner et al., 2000). Eighty percent reduction in pectin methylesterase activity was observed after treating orange juice at 35 kV/cm for 1500 μs (<37.5°C) (Elez-Martinez et al., 2007).

Polyphenol oxidase and peroxidase enzymes can involve in anthocyanins degradation, causing colorless juice or development of undesirable brown-colored compounds in the juice (CAO et al., 2011). Lipoxygenase plays a vital role in the formation of the off-flavor compounds and loss in nutritional quality via oxidation of unsaturated fatty acids and carotenoid (Min et al., 2003). During PEF treatment of mango juice at 35 kV/cm for 1800 μs, polyphenol oxidase, lipoxygenase, and residual peroxidase activities were reduced to 70%, 53%, and 44%, respectively (Salinas-Roca et al., 2017). Fifty-four percent irreversible reduction in the lipoxygenase activity was obtained after the PEF treatment of tomato juice at 40 kV/cm for 57 μs (Min et al., 2003).

The effect of PEF on enzyme inactivation depends on the intensity of PEF treatment and juice properties (Buckow et al., 2011). Pectin methylesterase activity in orange juice treated by PEF (35 kV/cm for 184 ms) decreased by 83.2% with a final juice temperature of 61.9°C. Just 24.1% of pectin methylesterase was inactivated when this juice was lonely heated in an oil bath at 62.9°C (Yeom et al., 2002). A 90% reduction in pectin methylesterase activity was observed during PEF treatment of the orange juice at 25 kV/cm for 250 ms at 50°C; while, during the same PEF treatment condition, only 36.3% enzyme inactivation was achieved when the temperature was set at 10°C (Yeom et al., 2002). When the PEF is applied, the enzyme surface and its environment are modified by ionization and distribution of solution components. It can be concluded that the combination of ohmic heating and non-thermal effects of PEF processing is responsible for the inactivation of the enzymes (Buckow et al., 2011; Yeom et al., 2002).

Longer treatment time shows positive effects on inactivation of enzymes; as 5% rise in pectin methylesterase inactivation was observed when the orange-carrot juice was treated at 25 kV/cm for 330 μs (70°C) in comparison to the processing for 280 μs (68°C) (Rivas et al., 2006). After the treatment of orange juice at 35 kV/cm (<40°C) for 1000 and 1500 μs, 93% and 100% inactivation in peroxides were observed, respectively (Elez-Martínez et al., 2006). The application of bipolar improves the stability of the juice. For instance, bipolar and monopolar pulses caused respectively 73% and 52.2% decrease in the peroxidase activity in PEF processed carrot juice (35 kV/cm for 1000 μs at <35°C) (Quitão-Teixeira et al., 2008).

2.8 Impact of pulsed electric field treatment on nutrients

One of the essential benefits of PEF is no or lower nutrients loss in juices. Cortés, Esteve (Cortés et al., 2006) found that the carotenoids content remained constant or reduced at 25 kV/cm while their content increased at 35 and 40 kV/cm. Vitamin A content was better maintained when higher PEF intensity for the shorter time was applied, as 11.1% and 9.9% decrease in vitamin A content was observed as the juice treated at 25 kV/cm for 340 μs and 40 kV/cm for 130 μs, respectively. The PEF treatment had no significant effect on the ascorbic acid content of the orange (Cortés et al., 2006), tomato (Min et al., 2003), watermelon (Oms-Oliu et al., 2009), and longan juices (Zhang et al., 2010). In other research, about a 6.2% reduction in ascorbic acid content was observed after treatment of different citrus juices (grapefruit, orange, lemon, & tangerine) at 28 kV/cm for 100 μs (Cserhalmi et al., 2006). Applying more intense PEF (20–40 kV/cm for 100–360 μs) brought more decrease in (26%) ascorbic acid content of mixed papaya-mango juice (Carbonell-Capella et al., 2017). PEF treatment (40 kV/cm for 57 μs) showed no significant effect on lycopene content in the tomato juice (Min et al., 2003). No changes in the content of total phenolic compounds and anthocyanin were reported during the processing of pomegranate juice at 35 and 38 kV/cm for 281 μs (55°C) (Guo et al., 2014). Decrease of 18% and 14.5% in total phenol content was found after the PEF treatment of longan juice (32 kV/cm) (Zhang et al., 2010). PEF process affected the odor–flavor attribute of the apple juice by causing decrease in ethyl acetate (77.5%), ethyl butyrate (60.2%), acetic acid (39.8%), methyl butyrate (30%), butyl hexanoate (18.1%), 1-hexanal (14%), Hexyl acetate (8.4%), and hexanal (7%) (Aguilar-Rosas et al., 2007). After applying 35 kV/cm for 200 ms, limonene and ethyl butyrate, the essential flavor components in orange juice, were decreased by 15% and 26%, respectively (Jia et al., 1996). Also, the electroporation effect can bring more release of components into the juice. As 13%–28% increase in total polyphenolic content in fresh grape juice was reported after the PEF process at 0.5–2.4 kV/cm with 50 pulses (Balasa et al., 2006). Also, the content of total carotenoids increased (107.3%) by PEF processing of mixed papaya-mango juice at 20 kV/cm for 360 μs (Carbonell-Capella et al., 2017).

2.9 Physicochemical properties of the pulsed electric field treated juice

During juice processing, preservation of chemical compounds such as ascorbic acid, antioxidants and pigments are considered as a significant challenge, because they are highly sensitive to processing conditions (Lee and Kader, 2000). The change in these compounds appears as a physical alteration of the juice. For example, hydroxymethylfurfural is produced during the degradation of ascorbic acid and its reaction with amino groups yields to brown pigments and finally brings off-taste and off-color development. After PEF treatment, hydroxymethylfurfural content of the orange-carrot juice (25 kV/cm for 80 μs) and different citrus juices (28 kV/cm for 100 μs) were constant. Lightness and chroma of orange-carrot juice were constant and the hue index shifted toward the yellowish during PEF treatment (25 kV/cm for 80 μs) (Rivas et al., 2006). PEF process (28 kV/cm for 100 μs) showed no considerable effect on the browning index of different juices (grapefruit, orange, lemon, & tangerine). However, a*-value of all samples, L*and b* values of the grapefruit and tangerine juices reduced which affect the appearance of the juices as the total color difference of all juices slightly increased (grapefruit (0.45), orange (0.47), lemon (0.59), and tangerine (2.44)) (Cserhalmi et al., 2006). The total color difference after the PEF treatment of longan juice was 0.28 reflecting the effectiveness of the treatment (32 kV/cm for the 90 s) on color preservation (Zhang et al., 2010). Color analysis of the treated pomegranate juice showed lower L* values, higher a* values, and constant b* value (35 and 38 kV/cm for 281 μs at 55°C) (Guo et al., 2014). More prolonged PEF treatment at 20 kV/cm showed no significant alteration in lightness and a* value of the juice; while, the b*-value increased between 32.5 (100 μs) and 33 (360 μs). Raise in the PEF intensity showed more change in the color of the juice as the total color differences of the juices were 8.72 and 10.05 at 20 and 40 kV/cm for 360 μs, respectively (Carbonell-Capella et al., 2017).

During the PEF processes at 25 kV/cm for 80 μs, the turbidity of the fresh orange-carrot juice increased from 0.64 to 1.06 (Rivas et al., 2006). The more turbid appearance of the juice can be related to the higher pectin methylesterase inactivation and more release of pectin into the juice due to electroporation. Turbidity or cloudy appearance positively influences the color and sensorial properties of the juices (Aghajanzadeh et al., 2017).

Sensory evaluations revealed that the PEF treated tomato juice (40 kV/cm for 57 μs) (Min et al., 2003), mango juice (35 kV/cm for 1800 μs) (Salinas-Roca et al., 2017), and pomegranate juice (35 and 38 kV/cm for 281 μs) (Guo et al., 2014) had an almost similar flavor and overall acceptability to the fresh samples. It was found that the color of the PEF processed mango juice obtained a higher score by the panelist, which can be due to the release of more pigments and reduction of oxidative enzyme activity (Salinas-Roca et al., 2017). These results showed the desirable effect of the PEF process in producing the fresh-like product without alternating the pH, conductivity, acidity, total soluble solids, and viscosity of the juice (Salinas-Roca et al., 2017; Min et al., 2003; Zhang et al., 2010).

2.10 Benefits and drawbacks

There are several limitations to the application of PEF on an industrial scale. The primary limitations are associated with the cost, non-uniform treatment, electrode metal emission, and juice properties. PEF is a high cost in electrical energy consumption and initial investment in equipment. However, these costs can be justified considering its capacity in producing the fresh-like juices in large-scale.

Juice directly contacts electrodes and emission of electrode metal, leading to the formation of electrolytic products. Besides, the possible release of metal within the juice, even if at low concentrations, brings high toxicity for the consumers (Mastwijk and Bartels, 2005). The amount of released metal depends on the PEF system (such as the design of electrodes and pulse properties) and the product properties. The non-uniformity results from the intensity and temperature distributions relating to juice flow rate, pulse shape, electrode insulator configuration, the design of the electrode and chamber (Buckow et al., 2011; Mastwijk and Bartels, 2005).

PEF cannot be used in the processing of the juices containing gases (air bubbles) at an intensity higher 20 kV/cm. The presence of bubbles leads to a non-uniform process. On the other hand, exceeding the electric field than the dielectric strength of the bubbles causes partial discharges as sparks. These events cause liquid evaporation, emission of electrode metal, and connecting the two electrodes (Morris et al., 2007). These events can be almost controlled by using a vacuum degassing or pressurizing systems to minimize the air bubbles content of the juice before pumping it through the PEF chamber. The other factor that should be considered is related to the homogeneity of the juice. The PEF system is only practical in the processing of the juice containing suspended particles smaller than the electrodes gape.

On the other hand, there are various advantages in the application of PEF technology in juice production. PEF is useful in the inactivation of different microorganisms and enzymes in various juices. PEF treatment can enhance the spore germination and make it sensitive to thermal inactivation; hence, based on the hurdle concepts, its combination with mild heating can be useful in the processing of low acidic juices. Also, as the juice is treated at low or medium intensity during a short time, electroporation enhances the extraction of chemical compounds and also the yield of juice from cells (Rawson et al., 2011).

3 High-pressure processing

3.1 Principle of high-pressure processing

High-pressure processing (HPP) is known as one of the most innovative technologies in which the juice is rapidly and uniformly subjected to very high pressures (100–700 MPa) to inactivate the microorganisms and enzymes (Welti-Chanes et al., 2009). Depending on the food ingredients, the temperature can increase by approximately 3°C per 100 MPa (Rastogi et al., 2007). If the food contains a high amount of fat, the more change in the temperature would be observed (9°C per 100 MPa). Also, this method impacts only the covalent bonds (presence in high molecular weight components such as proteins) with no influence on the low molecular weight juice components (flavoring agents, pigments, vitamins) (Kadam et al., 2012). This fact results in better preservation of the sensorial and color aspects of the treated juice (Chakraborty et al., 2014).

3.2 Design of high-pressure system

The HPP system usually consists of a high-pressure vessel, means to close the vessel off, a pressure medium (usually water containing a small amount of soluble oil) and a controlling system for monitoring the temperature and pressure.

HPP processes the juices as pre-packed and bulk. The pre-packed juices in a flexible container are transferred into the pressure vessel. Using a pump, the pressure is then established in the vessel (Kadam et al., 2012). The stability and flexibility of the packaging materials are critical to producing a safe juice with attractive package shape. Different suitable materials are using in juice packaging, including plastic stomacher bags, sterile tubes, polyester tubes, polyethylene pouches, and nylon cast polypropylene pouches (Kadam et al., 2012). In HPP treatment of bulk juice, it is pumped into the pressure vessel and free-floating piston transfers the pressure of the pressurizing fluid to the juice. After depressurizing, the treated juice is aseptically transferred in the container.

3.3 Microbial inactivation by high-pressure application

Vegetative bacteria, yeasts and molds are inactivated under pressure ranged from 300 to 600 MPa due to the permeabilization of the cell membrane (Rastogi et al., 2007). During HPP, inactivation of microorganisms takes place due to change in phase transition fluidity of cell membrane, ribosome breakdown (damaging the cell genetic), and changes in the activity of the enzyme (being vital for survival and reproduction of the cells) (Kadam et al., 2012; Roobab et al., 2018). Gram-negative bacteria and bacteria in the exponential growth phase are usually more pressure-sensitive than the Gram-positive ones and bacteria in the stationary phase (Kadam et al., 2012). It was reported that HPP (500 MPa for 10 min) was capable of inactivating viable microorganisms more than 5-log in broccoli juice (Houška et al., 2006). Pomegranate juice treatment at 350 MPa for 30 s resulted in 1.8- and 2.1-log reductions in aerobic mesophilic bacteria and counts of molds and yeasts, respectively (Varela-Santos et al., 2012). Counts of aerobic bacteria, yeasts and molds in treated orange juice (600 MPa, 1 min, 20°C) were less than the minimum level of detection (<10 cfu/mL) (Bull et al., 2004). The count of microbial colonies in fresh blueberry juice (10^4 cfu/mL) reduced to less than 10 cfu/mL by treating at 200–600 MPa for 5–15 min (Barba et al., 2013). Reduction of 1- and 5-log in the total count of *E. coli* O157 were reported after treating orange and apple juice at 500 MPa at 20°C for 5 min, respectively (Jordan et al., 2001).

The effect of high-pressure processing also depends on the microorganism strain as *E. coli* K12 strain showed more resistance than *E. coli* O157: H7 during processing at a constant condition (Torres et al., 2015). It was found that the processing of orange juice (pH = 3.4–4.5) at 550 MPa for 5 min (20°C) brought 6-log reduction in *E.coli* O157: H7 while treating the juice with pH of 5 was not such active until increasing the temperature to 30°C (Linton et al., 1999). The pressure treatment causes the cells to be more vulnerable to acid conditions and unable to repair their injuries (Bayındırlı et al., 2006). The higher inactivation of pathogens and spoilage microorganisms is applicable by a combination of high-pressure processing with a mild (45°C–50°C) heating treatment (Rastogi et al., 2007). As the processing of different juices at 350 MPa for 5 min revealed no detectable survive foodborne pathogens by 10°C rise in the juice temperature (from 30°C to 40°C) (Bayındırlı et al., 2006).

The spores are pressure-resistant because of protecting the effect of the thick outer cell membrane, small acid-soluble proteins shielding DNA, being small and having spherical shape (Roobab et al., 2018). Two steps pressurizing or combination of HPP with other method is required to initiate the spore germination and then its inactivation (Rastogi et al., 2007). Case in point, the combination of high pressure (500–700 MPa) with heating (90°C–110°C) is suggested to inactivate the spores of *Clostridium botulinum* (Kadam et al., 2012).

3.4 Effect of high-pressure process on enzyme structure

The effect of HPP on the enzyme structure is complex as it can bring inactivation and reactivation of enzyme depending on the applied pressure and the juice properties. Several events, such as formation or disruption of interactions and changes in the native structure of the protein by folding or unfolding involved in the inactivation of enzymes (Chakraborty et al., 2014). The volume of protein-solvent decreases upon unfolding that is related to the existence of voids and internal cavities in the folded state and this phenomenon changes the structure of the enzyme during HPP (Alexandrakis et al., 2014). The unfolding of protein chains takes place by affecting the non-covalent bonds such as hydrogen, ionic, and hydrophobic bonds. The changes in the active site and protein denaturation lead to the inactivation of the enzyme. However, HPP shows a low impact on the primary protein structure of the enzyme due to the presence of covalent bonds. During HPP, the increase in hydrogen bond formation causes less impact on the secondary structure (Alexandrakis et al., 2014). However, this structure can be irreversibly unfolded under pressures higher than 700 MPa. The decrease in enzyme activity using high-pressure treatment is almost related to alteration in electrostatic, hydrophobic interactions, and the hydrogen bonds in tertiary and quaternary structures (Chakraborty et al., 2014).

A 90% reduction in polygalacturonase activity was reported during tomato juice treatment at 400 MPa for 10 min (Hsu, 2008). Application of HPP (600 MPa for 60 min) in watermelon juice decreased the activity of polyphenol oxidase (88%), peroxidase (42%), and pectin methylesterase (77%) (Liu et al., 2013). Rise in the applied pressure and time induced a higher enzyme inactivation; for instance, polyphenol oxidase inactivation increased from 35% to

83% when the fruit smoothies (homogenate mixture of apple juice concentrate, apple, strawberry, banana, and orange) were treated at 450 MPa for 5 min and 600 MPa for 10 min, respectively (Keenan et al., 2012).

Studies represent the effect of juice properties in the inactivation of enzymes during HPP. It was found that lowering the juice pH increased the pressure sensitivity of the α-amylase (Riahi and Ramaswamy, 2004). During treatment at 100–400 MPa, the D-value of pectin methylesterase inactivation in orange juice (pH = 3.7) varied from 8250 to 260 min. The calculated D-value during HP processing (270–400 MPa) of the orange juice with the lower pH (3.2) was ranged 145–264 min (Basak and Ramaswamy, 1996). Lower pectin methylesterase inactivation rate was reported when the juice with higher soluble reliable content treated using HPP. This result can be related to the protective effect of higher concentration levels on pressure inactivation of the enzyme (Basak and Ramaswamy, 1996; Basak et al., 2001).

Considering the protein structure of the enzymes, a combination of mild heating with HPP improve is effective to increase the juice stability (Alexandrakis et al., 2014). Pectin methylesterase activity was reduced to 50.4%, 49.4%, and 37.8% during high-pressure treatment (250 MPa) of orange juice with initial temperatures of 22°C, 35°C, and 45°C, respectively (Welti-Chanes et al., 2009). During treatment (400 MPa) of the orange juice at 25°C and 50°C, D-values of pectin methylesterase inactivation reduced from 116.9 to 45.6 min, respectively (Nienaber and Shellhammer, 2001). The pectin methylesterase activity in treated watermelon juice (Zhang et al., 2011) and orange juice (Torres et al., 2015) decreased 65% (900 MPa, 60 min, 60°C) and 90% (500 MPa, 15 min, 50°C), respectively. Peroxidase was inactivated (30%) when the kiwifruit juice was treated at 600 MPa for 30 min at 50°C (Fang et al., 2008). A decrease of 91% in polyphenol oxidase activity in apple juice was obtained after processing at 450 MPa for 60 min at 50°C (Bayındırlı et al., 2006).

In comparison to the fresh juice, the activity of pectin methylesterase showed an increase when the tomato juice was treated using HPP in the range of 300–500 MPa (Hsu, 2008). An increase of 3 and 2.5 folds in polyphenol oxidase activity was observed after processing the grape juice at 400 and 550 MPa, respectively (Barba et al., 2013). An increase in pectin methylesterase activity (2.6%) was reported during the processing of peach juice at 400 MPa for 15 min at ambient temperature (Rao et al., 2014). An increase in the enzyme activity can be related to the alteration in the properties of the substrates and juice. The other reason can be explained by the enzyme conformation reversibility or its active site rearrangement during high-pressure treatment. Also, the release of the enzyme can take place during processing and damaging the cell membrane (Chakraborty et al., 2014).

Inactivation of the enzymes could be irreversible or reversible (Rastogi et al., 2007). HPP was applied to treat the orange and grapefruit juices at 600–900 MPa for 1–30 s. Applying higher pressures than 600 MPa was effective in the inactivation of the heat-labile isoform of pectin methylesterase, with no effect on the heat-stable isoform (Goodner et al., 1998). Due to the existence of the stable isoforms of pectin methylesterase, the enzyme's activity increases during the storage of the treated juice (Welti-Chanes et al., 2009).

3.5 Effect of high-pressure processing on juice properties

HPP showed no effect on total soluble solids, pH and acidity of blueberry, pomegranate, watermelon, orange, and guava juices (Varela-Santos et al., 2012; Bull et al., 2004; Barba et al., 2013; Liu et al., 2013; Yen and Lin, 1998).

HPP (450 MPa for 30 min) at ambient temperature has no considerable effect on the ascorbic acid content of orange juices; however, a 7% loss in this vitamin content was detected by 10°C rise in the juice temperature (Bayındırlı et al., 2006). The ascorbic acid content of blueberry juice was constant during processing at 200 MPa for 15 min; however, processing at higher pressure resulted in a 5.8% drop in the content of this vitamin (600 MPa for 15 min). Total phenols were relatively resistant to high-pressure treatment as its content increased by 14% and 24% when blueberry juice was, respectively, treated at 200 MPa and 400 MPa for 15 min. The total anthocyanin content increased (7%–16%) in the same treatment condition (Barba et al., 2013). An increase of 11% and a decrease of 7% in total phenol content was observed after processing of fruit smoothies at 450 MPa for 5 min and 600 MPa for 10 min, respectively. Processing at 450 MPa for 5 min increased (17%) the total anthocyanins content of fruit smoothies; however, the application of higher pressure (600 MPa for 10 min) brought a considerable reduction. Additionally, pressure treatment enhanced the content of ascorbic acid in comparison to the fresh fruit smoothies (Keenan et al., 2012). An increase of 62% and 56% in the contents of carotenoids and lycopene was observed after pressure imposing the tomato juice (300–500 MPa for 10 min at 5°C and 25°C). This fact was related to the effect of the high-pressure application on modifying the structures of macromolecular due to the release of tightly bound carotenoids to protein and membrane lipids (Hsu, 2008). No change in lycopene and total phenols content of watermelon juice was observed during HPP (200–600 MPa for 60 min). After HPP, all-*trans* lycopene might transform to *cis*-isomers which is better absorbed in the body (Liu et al., 2013). There was no significant difference between DPPH scavenging ability of fresh and high pressure processed (400 MPa, 40°C, 10 min) orange juice (Sánchez-Moreno et al., 2005).

3.6 Effect of HPP on physical properties of the juice

The results of the sensorial evaluations represented that the color of the HPP processed orange juice (600 MPa for 4 min) was the same as the fresh one (Polydera et al., 2004). a* and b* values were constant after watermelon juice processing at 200–600 MPa for 60 min due to the stability of the lycopene content. However, the lightness of the juice increased by reducing the browning index. The total color difference between treated juices was less than two, which

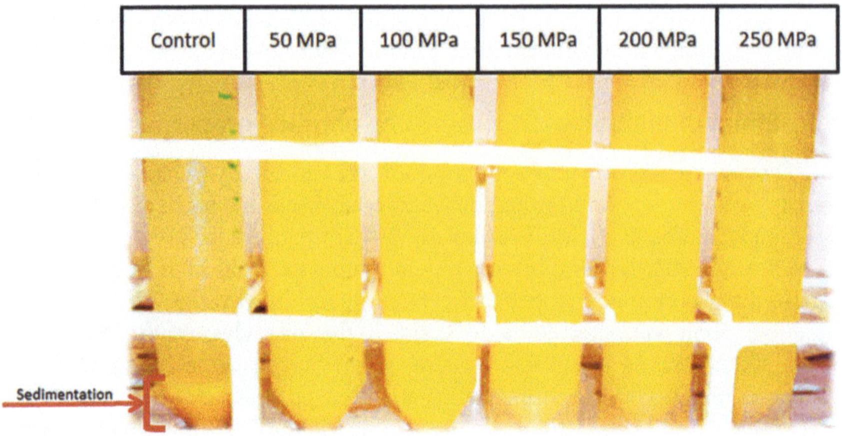

FIGURE 2.1 Effect of high pressure processing on the cloud stability of orange juice during storage. *(Welti-Chanes et al., 2009).*

reflects the effectiveness of this method in preserving the color of fresh juice (Liu et al., 2013). HPP of blueberry juice at 200–600 MPa for 5–15 min caused an increase in chroma, decrease in L* and b* values with no effect on a*-value. Total color differences were in the range of 0.83–1.44 (Barba et al., 2013). The color of pomegranate juice became brighter and more yellowish with less redness as higher pressures (350–550 MPa) was applied (Varela-Santos et al., 2012). Browning index of the fresh and HPP treated (600 Mpa, 1 min) orange juices were 0.05 and 0.08, respectively. No significant difference between the total color difference of these samples was distinguished (Bull et al., 2004). After processing at 500 MPa for 10 min at ambient temperature, a/b values (red–yellow ratio) of tomato juice increased from 3.62 to 3.84. The juice viscosity of the juice increased as higher pressure levels were imposed (up to 500 MPa) at various temperatures (4°C, 25°C, and 50°C) because of more release of pectin into the juice and more inactivation of polygalacturonase (Hsu, 2008).

The cloud stability of the orange juice (Fig. 2.1) and watermelon juice increased after HPP processing (Welti-Chanes et al., 2009; Bull et al., 2004; Liu et al., 2013). This fact can be related to the more release of pectin and other cloud particles during damaging the cell walls as well as more inactivation of pectin methylesterase (Aghajanzadeh et al., 2017). In addition, applying high-pressure treatment causes a reduction in the size of pectin and other suspended particles in the juice and prevents their aggregation and sedimentation from bringing further juice clarification (Lacroix et al., 2005).

3.7 Advantages and limitations of high-pressure processing

The stability of covalent bonds to HPP brings the better preservation of the nutritional values (such as vitamins and pigments), sensorial and quality aspects of the juice (Polydera et al., 2004; Welti-Chanes et al., 2009). However, the high cost of installation and operation of HPP limit its application in industrial scale (Kaur and Singh, 2016). It was mentioned that the application of 600 MPa would be inappropriate for commercial use due to rising in the operating expenses; hence, a combination of the HPP with mild thermal treatment can be a wise way to overcome this problem (Nienaber and Shellhammer, 2001).

Unlike the PEF, this method can be used before or after the packaging of a wide range of food products. The high-pressure system can be operated for different packaging sizes and geometry (Rastogi et al., 2007). In HPP, the least amount of headspace should be considered in pre-packed juice since air and other gases are compressed to zero volume and cause the deformation of the package (Kadam et al., 2012).

4 High-pressure carbon dioxide processing

4.1 Description of high-pressure carbon dioxide

High-pressure carbon dioxide (CO_2) or dense phase CO2 is a cold pasteurization method running at moderate pressure (<50 MPa) at low temperatures (Ferrentino and Spilimbergo, 2011). The application of CO_2, a non-toxic, non-flammable, and inexpensive gas, shows no undesirable effect on juice safety (Truong et al., 2002; Damar and Balaban, 2006). In food packaging, CO_2 at atmospheric pressure is used to control the microbial growth and respiration of the agricultural products. In juice treatment, CO_2 under pressure is introduced to the juice. In this case, carbonic acid is produced by dissolving the CO_2 in the juice, which causes a drop in pH. The change in pH is temporary up to removing the pressure (Bertoloni et al., 2006). This method is operative in destroying the microorganisms and enzymes as well as preserving juice's nutritional values.

4.2 High-pressure carbon dioxide process system

Different types of high-pressure CO_2 are batch, semi, and continuous systems. These systems usually contain a gas cylinder, a CO_2 release valve, a pressure vessel, a heater, devices for controlling the temperature and pressure. In the batch system, the CO_2 and juice are stationary in a container. The CO_2 is introduced into the pressure vessel (juice chamber) up to achieve the desired pressure. The temperature is set to wanted value. After some time of the juice exposure to CO_2, the gas is released to reduce the pressure and finalize the process (Damar and Balaban, 2006). Ishikawa, Shimoda (Ishikawa et al., 1995) designed a semi-continuous system equipped with a cylindrical filter made of porous stainless steel for feeding the CO_2. When the pressure has reached the desired level, the CO_2 micro-bubbles are fed into the full pressure vessel with the sample. The process is then stopped and the vessel is depressurized by opening the pressure-regulating valve. A continuous system was designed to continuously mix and pressurize a stream of juice with a CO_2. The gas pressure is adjusted with a regulator valve to keep it in the liquid phase. A reciprocating intensifier pump is applied to have a defined pressure of the CO_2-juice mixture. After a static mixing section, the mixture is transferred to a holding tube to maintain the pressure and dissolved CO_2 in the juice at a controlled temperature during a particular time. CO_2 depressurization and its vacuum separation from the juice are finally performed. In general, the difference in this system is related to its effectiveness. Continuous type in which better CO_2 and juice mixing occurs provides a higher efficiency due to a sudden release of pressure causing microbial cell rupture.

4.3 Mechanism of carbon dioxide bactericidal action

During high-pressure CO_2 processing, two different stages are usually observed in linear curves of microorganism inactivation. The first stage, known as the controlling step, is characterized by a slow inactivation rate, in which CO_2 penetrates the cells. During the second stage, the CO_2 pressurizing profoundly affects the microorganism and results in fast inactivation (Erkmen, 2000).

Several mechanisms are involved in microorganisms inactivation by this method (Zhang et al., 2006). The first one is the rupturing and death of the cells due to swelling by pressurizing and depressurizing of CO_2. The second mechanism is related to CO_2 diffusion into the cell membrane. The CO_2 accumulation in the lipophilic inner layer of the membrane influences the membrane lipid chain, which increases its fluidity. These events cause the membrane permeability and leakage out of the vital cell components (Damar and Balaban, 2006). The third mechanism is related to CO_2 breaking down to form bicarbonate, carbonate, and H^+ (Eqs. 2.1–2.3) as it is dissolved in an aqueous medium (Damar and Balaban, 2006). During these reactions, CO_2 penetrates through the cell membrane, causing a drop in the intracellular pH. The cytoplasm of most microorganisms has a high buffering capacity to prevent the variations in internal pH. If too much-dissolved CO_2 enters the cytoplasm, the cells may be unable to expel all the protons and pH will start to decrease. Therefore, a pH gradient is created between the internal (cytoplasm) and external environment (juice) by pumping H^+ ions out of the cell. These events disturb the vital biological procedures, including enzyme activity, glycolysis, amino acid and peptide transport, active transport of ions and proton translocation (Damar and Balaban, 2006; Garcia-Gonzalez et al., 2007).

$$CO_2 + H_2O \leftrightarrow H_2CO_3 \tag{2.1}$$

$$H_2CO_3 \leftrightarrow H^+ + HCO_3^- \tag{2.2}$$

$$HCO_3^- \leftrightarrow H^+ + CO_3^{-2} \tag{2.3}$$

Intracellular inorganic electrolytes (Ca^{2+}, Mg^{2+}, and similar ions) in cytoplasm and cell membranes can participate in microorganisms inactivation via conversion of CO_2 to CO_3^{2-}. These electrolytes play an essential role in the osmotic balance between the cell and its environment. This fact increases the lethality by affecting the volume of the cell (Garcia-Gonzalez et al., 2007).

Vegetative bacteria, yeasts and molds are sensitive to the high-pressure CO_2 processing. The Gram-positive bacteria species show a higher resistance than the Gram-negative ones due to the presence of a thick peptidoglycan layer in the cell wall, which results in lower permeability (Zhang et al., 2006). The spores are resistant to the high-pressure CO_2 treatment condition; as, the multilayer spore coat acts as a barrier again to the chemicals and increases the spore stability to be survived during juice processing (Zhang et al., 2006).

Several factors, particularly temperature and pressure, involved in the inactivation of microorganisms during high-pressure CO_2 processing. Temperature and pressure are determinant factors in CO_2 sate, being subcritical (liquid or gaseous) or supercritical. CO_2 is in its supercritical state at temperature and pressure higher than 31°C and 7.34 MPa, correspondingly (Gunes et al., 2005). This state is characterized by gas-like diffusivity and liquid-like density, which bring the quick diffusion and high extraction power of the CO_2, respectively (Zhang et al., 2006). Applying pressure results in damaging the cell wall, improving CO_2 solubilization and more CO_2 contact with the cells (Garcia-Gonzalez et al., 2007).

Higher temperature brings more inactivation of microorganisms by affecting the fluidity of cell membranes, protein denaturation, and improving the CO_2 diffusivity. However, this process should be operated at a near-critical temperature of CO_2 since its dissolving capacity in the juice will decrease as more rise in the temperature (Garcia-Gonzalez et al., 2007). The effect of different factors including temperature (25°C and 35°C), pressure (6.9, 27.6, and 48.3 MPa), and CO_2 concentration (0, 85, and 170 g of CO_2/kg of juice) was investigated during grape juice processing. The inactivation of the yeasts increased by rising in the applied temperature, pressure, and CO_2 concentration. The CO_2 concentration was introduced as the most crucial factor in inactivating the yeast cells (*S. cerevisiae, C. stellate, and K. apiculata*). CO_2 in the supercritical state showed more effectiveness as the yeast inactivation decreased at 25°C and 6.9 MPa. The yeast populations decreased to about an undetectable level (<10 cfu/mL) at 27.6 and 48.3 MPa along with 85 and 170 g/kg at 35°C (Gunes et al., 2005). The total microbial count of the fresh melon juice (5-log) decreased as higher pressure and temperature were applied for a longer duration. The total microbial count was zero in the processed juice at 35 MPa and 65°C for 5–60 min (luan Chen et al., 2010). Both *L. monocytogenes* and *Y. enterocolitica* cells were completely inactivated after processing the juices of carrot, orange, peach, and apple for 720, 480, 240, and 480 min at 45°C under 6 MPa CO_2 pressure (Erkmen, 2000, 2001). Reduction of 4-log was observed in the population of aerobic organisms by processing the carrot juice under 4.9 MPa CO_2 pressure for 10 min at 5°C (Park et al., 2002).

Other factors are influencing the process efficiency, including applying a higher rate of gas penetration, faster pressurization and depressurization of CO_2, and repeated release-compression of CO_2 (pressure cycling), which improve the inactivation of the microorganism via causing the cell rupture and increasing the mass transfer (Zhang et al., 2006). CO_2 agitation is another factor that can improve the gas solubilization, which increases the CO_2 contact with cells and bring a higher cellular penetration (Garcia-Gonzalez et al., 2007).

Also, the juice acidity shows a synergistic effect with high-pressure CO_2 processing in microorganisms inactivation as it affects the stability of microorganisms by increasing the cell permeability, altering the enzyme activity, disturbing the homeostasis and proton force (Zhang et al., 2006). The water content of the treatment medium and the vegetative cells affects the CO_2 solubility and shows a positive effect on microbial inactivation (Damar and Balaban, 2006).

4.4 Inactivation mechanism of enzymes

Enzymes are inactivated during high-pressure CO_2 treatment due to two mechanisms. First, CO_2 dissolves in the juice and reduces its pH due to the carbonic acid formation. At low pH, the enzyme protein bonds can interact with CO_2 and form a bicarbonate complex that causes a conformational change, especially in the secondary structure of the enzyme (Bertoloni et al., 2006). Second, the precipitation of the Ca^{+2} sensitive proteins occurs as a result of the formation of carbonate-Ca^{+2} complexes, depending on the binding site of the ion and chemical structure of the enzyme (Damar and Balaban, 2006).

Pectin methylesterase was completely inactivated during supercritical CO_2 treatment (29 MPa, 50°C, 4 h) of orange juice. However, the enzyme activity regenerated gradually during 15 days of storage at 4.4°C (Arreola et al., 1991). Reduction of 46.3% in pectin methylesterase activity was observed when the orange juice was treated using 107 MPa for 10 min at ambient temperature (0.43 w_{CO2}/w_{juice}). In this study, no reversible activity in the enzyme was found during 4 weeks of storage at 1.7°C (Kincal et al., 2006). Inactivation of 64.2% and 85.2% pectin methylesterase were achieved in non-carbonated and carbonated orange juice (600 MPa, 20°C, 5 min), respectively. It was also found that carbonating the juice had no considerable effect on the total percentage of pectin methylesterase activity, but it influenced the rate of its inactivation (Truong et al., 2002). During high-pressure CO_2 treatment, no change in the SDS-PAGE electrophoretic behavior of pectin methylesterase was found, while the secondary and tertiary structures were altered, which affect the enzyme inactivation (Zhou et al., 2009).

Applying a higher CO_2 pressure results in the more enzyme inactivation, as polyphenol oxidase activity of the carrot juice reduced to less than 3% after processing at 10–30 MPa for 15 min at ambient temperature (Zhou et al., 2009). It was reported that activities of polyphenol oxidase, lipoxygenase and pectin methylesterase decreased as the carrot juiced was treated under high CO_2

pressure (4.90 MPa) (Park et al., 2002). Melon juice processing at 35 Mpa pressure of CO_2 induced a 74.7%, 61.5%, and 99.9% decrease in activities of polyphenol oxidase, lipoxygenase, and peroxidase, respectively(luan Chen et al., 2010).

4.5 Effect on the physicochemical attributes of juice

During high-pressure CO_2 processing, the nutritional values and quality of the fresh juice are almost preserved due to low alteration in the juice temperature. As ascorbic acid content of melon juice was completely preserved during processing at 35 MPa for 60 min at 55°C (luan Chen et al., 2010). Retention of 71%–98% in the ascorbic acid content of orange juice was reported during high-pressure CO_2 processing at 7–34 MPa (35°C–60°C). This high retention was explained as the ascorbic acid is not soluble in supercritical CO_2, as well as the lack of oxygen during treatment (Arreola et al., 1991). No change in carotenoid content of the fresh carrot juice was observed after high-pressure CO_2 processing at 10–30 MPa for 15–60 min (Zhou et al., 2009). The anthocyanin and phenol contents of the fresh grape juice and processed juice by 34.5 MPa pressure of CO_2 were similar (Del Pozo-Insfran et al., 2006). Analysis of volatile compounds of melon juice showed that high-pressure CO_2 brought no alteration in ester compositions (ethyl acetate, ethyl propanoate, ethyl butyrate, butyl acetate, and ethyl-2-methyl butyrate). However, slight changes in alcohols and aldehydes, such as (Z)-nonel-3-ol, (E,Z)-2,6-nonadien, (Z)-nonel-6-ol, and nonanol, hexanal, (Z)-6-nonenal, nonanal, and 2-nonenal were observed (luan Chen et al., 2010).

There were no significant differences between the nutritional values (contents of ascorbic acid and folic acid), aroma profile, pH, total soluble solids, and titratable acidity of the fresh and CO_2 treated orange juice (Ho, 2003). The acidity of orange juice, carrot juice, and mandarin juice increased after processing, while there was no difference between the pH of fresh and treated juices (Kincal et al., 2006; Zhou et al., 2009; Yagiz et al., 2005). Under CO_2 pressure, the pH of the juice could reduce, but it returns to its initial value by depressurization of CO_2 (Arreola et al., 1991).

The cloudiness of fresh carrot juice increased 159%, 145%, and 126% after 10 min CO_2 processing at 10, 20, and 30 Mpa, respectively (Zhou et al., 2009). It was found that the gas/juice ratio had a more impact on cloud improvement in comparison to the applied pressure. The increase in the cloud value could be related to diminishing in the suspended particle size during the depressurizing step (Kincal et al., 2006).

The lightness and yellowness of the treated mandarin, orange, and carrot juices were enhanced, but the redness of their color was reduced (Park et al., 2002; Kincal et al., 2006; Yagiz et al., 2005). High-pressure CO_2 treatment (10–30 MPa, 15–60 min, 25°C) had no impact on the b*-value of the carrot juice color, but its lightness and a*-value increased after treatment

(Zhou et al., 2009). The total color difference of the treated juices at 10 MPa for 45 and 60 min and 20 or 30 MPa for 60 min was higher than 2, which reflects the undesirable effect of the long processing on the juice appearance. Different process conditions had no considerable effect on the viscosity of the carrot juice and the processed juice followed the Newtonian flow behavior as same as the fresh one (Zhou et al., 2009).

The results of the panel test revealed that CO_2 treatment improved the color and cloudiness of the juice with no impact on flavor, aroma, and overall acceptability (Arreola et al., 1991). Processed coconut water, at 34 MPa with 13% w/w (CO_2/juice ratio) for 6 min at ambient temperature, had a similar overall acceptance with the fresh one (Damar and Balaban, 2005). There was no difference between the sensorial aspects of the fresh and treated grape juice (170 g of CO_2/kg of juice, 35°C and 48.3 MPa); however, in some cases, a more acidic or carbonated taste of treated juice was detected which has no negative impact on the juice acceptance by the panelists (Gunes et al., 2005). Grape juice processing (34.5 MPa, 6.25 min, 30°C, 8% and 16% CO_2) caused no change in the sensorial properties of the juice, including aroma, flavor, color, and total acceptance (Del Pozo-Insfran et al., 2006).

4.6 Potentials and limitations of high-pressure carbon dioxide technology

CO_2 gas shows several advantages such as no trace, inert in most situations, no reaction with polymers, recyclable and having low critical temperature and pressure (Zhang et al., 2006). Low viscosity in its supercritical state ($3–7 \times 10^{-5}$Ns/m^2) and zero surface tension of this gas enhance its fast penetration in different products even in compound structures materials (Zhang et al., 2006). During this process, there is no need to apply high pressure and temperature to inactivate the enzymes and microorganisms, resulting in producing a desirable juice having an appropriate nutritional value, appearance, and sensorial properties.

The main hazard of CO_2 is related to its asphyxiation effect in high concentrations and unsafe conditions (Zhang et al., 2006). In order to inactivate the spores, long process time, high temperatures, and pressure are required, which limits the application of this method as showing an undesirable effect on the quality of the juice. In order to overcome this problem, pressure cycling is suggested, which is effective in reducing the treatment time at the same temperature and pressure conditions (Zhang et al., 2006).

5 Sonication treatment

5.1 Definition of sonication

Sonication or ultrasound processing is one of the alternative technologies in juice processing. Ultrasound is categorized as vibrational energy, which produces sound energy from electrical or mechanical energy at a frequency higher than the human hearing threshold (Majid et al., 2015). In the food industry, sonication is generally applied in two main areas, including food properties measurement and food processing. Low energy ultrasound (below 1 W/cm^2) with frequencies higher than 100 kHz is used to measure some food properties. High-energy ultrasound (higher than 1 W/cm^2) with frequencies between 20 and 100 kHz is used in food processing, for example, in the inactivation of the enzymes and microorganisms.

During sonication, enzymes and microorganisms inactivation occurs via different phenomena such as mechanical cavitation and free radicals formation (Majid et al., 2015; Terefe et al., 2009). During the juice sonication, cavitation bubbles are created as a result of alteration in pressure. The bubbles collapse in consequence of compression cycles of the ultrasonic waves (Tiwari et al., 2008). The growth and collapse of the bubbles define the amount of released energy during cavitation. This energy is affected by the interfacial bubble tension and vapor pressure. High bubble interfacial tension found in hydrating foods such as juices increases the energy (Knorr et al., 2004).

5.2 Design of sonication system

Batch and continuous types of sonication systems are available in processing the food products (Fig. 2.2). The system comprises a controlling section (temperature, pressure, frequency, time, and power), jacketed vessel, horn, ultrasonic probe, ultrasound generator (producing vibration amplitudes), an impedance adaptation box, and an automatic generation unit. Automatic generation unit including a power amplifier and part for resonance frequency monitoring to control the applied power to transducers (Zinoviadou et al., 2015). The ultrasonic transducers convert electrical or mechanical energy to sound energy. There different types of transducers, including liquid-driven, magnetostrictive and piezoelectric (the most common) transducers (O'donnell et al., 2010). The piezoelectric one is the most common, which works based on electric oscillations of a particular frequency (Zinoviadou et al., 2015; O'donnell et al., 2010). Application of a non-contact or coated transducer made of non-reactive materials and erosion resistance is generally suggested in the design of the probe for food applications (O'donnell et al., 2010).

There are different methods in applying ultrasound, including directly to the product, coupling with the device, and submerging in an ultrasonic bath. In the ultrasonic bath system, low power is applied to prevent the cavitational damage to the tank walls and the large volume of the juice brings a low acoustic energy density (O'donnell et al., 2010).

5.3 Mechanism of cavitation

The effect of sonication in the inactivation of microorganisms and enzymes is related to several physical and chemical mechanisms (O'donnell et al., 2010).

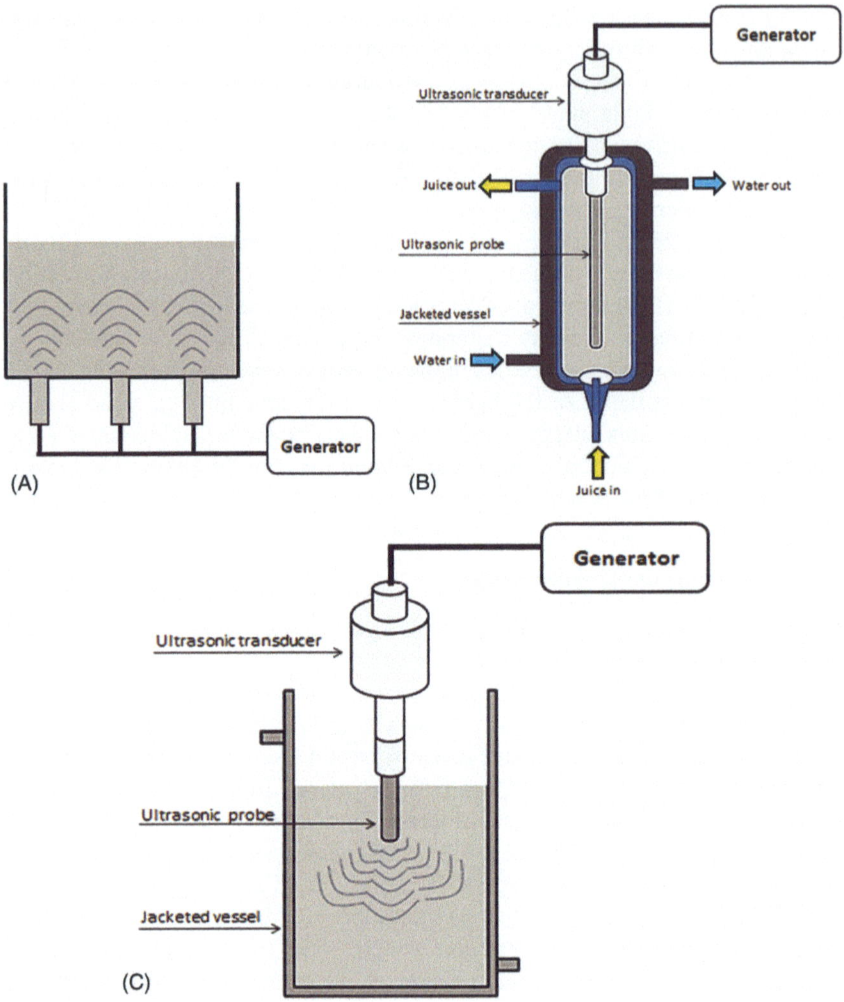

FIGURE 2.2 Scheme of different types of sonication systems including ultrasonic bath (A), batch type probe (B), and continuous probe (C) according to Zinoviadou, Galanakis (Zinoviadou et al., 2015).

The physical mechanisms are associated with the cavitation phenomena referring to the formation, growth, and implosion of bubbles (Tiwari et al., 2009). Ultrasound produces a stable cavitating bubble vibration, which creates "shock waves" and causes strong shear and microstreaming in the juice (Terefe et al., 2009). These microbubbles are generated due to the pressure changes as high power ultrasound is propagated through the juice. These bubbles grow with the compression and expansion cycles of the ultrasonic wave that results in expanding and eventually collapse when reaching a critical size. These events show chemical (free radicals formation), mechanical (collapse pressure, turbulences, and shearing), and thermal effects (Majid et al., 2015). Instantly (about a nanosecond), the temperature and pressure of the gas in the bubble will significantly increase (Tiwari et al., 2009). Cavitation effect has a role in microorganisms' and enzymes' inactivation as well as the extraction of the cell compounds into the juice.

5.4 Effect of sonication on microorganisms

Different mechanisms are involved in the inactivation of microorganisms during the ultrasonic process. Cavitation phenomenon inactivates the microorganisms by rising the membrane permeability, reducing the membrane selectivity, thinning the membrane, and causing the chemical reaction within the cell (Majid et al., 2015; O'donnell et al., 2010). It was observed that applying a low-intensity sonication may produce steady cavitation and reparable cell injuries, while high-intensity one mostly results in irreversible inactivation of the microorganisms (Roobab et al., 2018). During sonication, alternation of positive and negative pressures (expansion or compression phenomena) causes the cell rupturing (Majid et al., 2015). Additionally, the absorption of acoustic energy, particularly by the cell membrane, generates localized heating, which can bring the inactivation of some microorganisms (Piyasena et al., 2003).

The microorganism's sensitivity to sonication is associated with process condition (amplitude of the ultrasonic waves, treatment time, and temperature), juice properties, type, and characteristics of the microorganism (shape, structure, growth stage, and Gram negativity or positivity) (Roobab et al., 2018; Knorr et al., 2004; Piyasena et al., 2003). Generally, spores are relatively more resistant to this treatment in comparison to the vegetative forms (Zinoviadou et al., 2015).

The sonication process at different powers (20%–60%) and pulses (2–6 s) during the various times (2–6 min) were used to process different juices (strawberry, orange, apple, pineapple, and red-fruits juices). The inactivation of the microorganism was affected by the applied power and treatment time. The highest reduction in yeasts (*S. cerevisiae*, *P. membranifaciens*, *W. anomalus*, *Z. bailii*, *Z. rouxii*, *C. norvegica*) was observed at the most intensive process condition (60%, 4 min, 6s) (Bevilacqua et al., 2014). Processing of blueberry juice brought a considerable reduction in aerobic plate count, total coliforms, and yeast counts as a higher juice flow rate (24 and 93.5 mL/min) and amplitude level (40%, 80%, and 100%) were applied. The effect of flow rate on the inactivation of microorganisms was possibly related to the more juice mixing, which brings more exposure of the juice (close to the ultrasonic probe) to a higher intensity field (Mohideen et al., 2015). The total plate count of the fresh lime juice (3.7-log) was reduced to 1.9-log after 60 min processing at 25 kHz, while lower treatment time of 30 min had no significant inactivation impact. The counts of yeasts and molds in the fresh juice (2.9-log) decreased to 2.5- and 2.4-log after juice processing for 30 and 60 min, respectively (Bhat et al., 2011). After sonication of carrot juice (20kHz, 70% amplitude,15°C, 5 min), about 1.5- and 1.8-log decrease in total plate count and population of the yeasts and molds were achieved, respectively (Jabbar et al., 2014). Reduction of 5-log in yeast population (*P. fermentans*) was observed during the processing of the tomato juice at an amplitude level of 61 μm for 7.5 min (Adekunte et al., 2010).

Only a 1-log decrease in the mesophilic aerobic population was reported after sonication of the calcium-added orange juice at 20 kHz (75% wave amplitude, 10°C) for 10 min (Gómez-López et al., 2010). Lower than 1-log decrease in the total plate count and total yeasts and molds were reported after sonication of the strawberry juice at 35 kHz for 30 min (Cheng et al., 2007). Total plate count and total yeasts and molds counts in sonicated strawberry juice (30 min at 20°C, 25 kHz) showed a non-significant difference with the fresh juice. Applying forced water rinsing, juice treatment with mild temperature water or electrolyzed water as pretreatments are suggested to increase the effectiveness of the juice sonication process (Bhat and Goh, 2017).

5.5 Ultrasonic application in enzymes inactivation

A sudden increase in pressure and temperature brings hydrolysis of water inside the oscillating bubbles and therefore form the free radicals of H^+ and OH^-. The formed free radicals can participate in some chemical reactions such as scavenging with some amino acid residues in the active sites of the enzyme structure. This result affects the stability, substrate binding, and catalytic functions of the enzyme, damaging its biological activity (Aghajanzadeh and Ziaiifar, 2018). Also, the formed bubbles in cavitation are continuously changed in shape and size, making a stream adjacent to the bubble. The formation or collapse of the bubbles causes severe shear force resulting in the protein denaturation by affecting the secondary and tertiary structures of protein via breaking up of hydrogen bonds and Van der Waals interactions in the polypeptide chains of the enzyme (Terefe et al., 2009).

The effect of ultrasound treatment is dependent on the enzyme concentration, juice properties, and process conditions. For instance, the effect of cavitation on the enzyme could be altered as a result of changing the volume of the sample and the probe properties and location at a constant ultrasound intensity. Generally, a higher enzyme inactivation is reported when the probe type systems are applied (O'donnell et al., 2010).

Using sonication lonely may not be sufficient in almost complete inactivation of the enzymes (Tiwari et al., 2009; Jabbar et al., 2014). The residual activities of polyphenol oxidase, peroxidase, pectin methylesterase, and lipoxygenase in carrot juice were higher than 99.9% when treated by sonication (20 kHz, 70% amplitude,15°C, 5 min) (Jabbar et al., 2014). During sonication, the inactivation of peroxidase happens by altering the conformational protein structure and splitting the prosthetic group from haloenzyme (O'donnell et al., 2010). No change in polyphenol oxidase activity in strawberry juice was shown during 15 min of the sonication treatment (20°C, 25kHz), while sonication for longer time (30 min) showed a 23% increase (Bhat and Goh, 2017). However, this process could enhance the extraction of the compounds from the cell into the juice, as the sonicated guava juice (35 kHz for 30 min) had relatively higher polyphenol oxidase activity (18 units) than the control juice (10 unit) (Cheng et al., 2007).

A synergistic effect found between sonication and mild heating or pressurizing treatment to achieve a higher enzyme inactivation. However, thermosensation at higher temperatures was less effective in enzyme inactivation since rising in the vapor pressure inside of the bubbles resulting in cushioning effect and lowering the effective bubble collapse (O'donnell et al., 2010).

5.6 Effect of sonication on physicochemical attributes of juice

Ultrasonic processing showed no impact on pH, acidity, and total soluble solids of different juices, including grapefruit (Aadil et al., 2013), tomato (Adekunte et al., 2010), lime (Bhat et al., 2011), strawberry (Bhat and Goh, 2017), guava (Cheng et al., 2007), and blueberry juices (Mohideen et al., 2015).

After sonication of the lime juice at 25 kHz and 20°C for 60 min, 6.8%, 27.4%, 42.3%, and 127.4% raise in the contents of ascorbic acid, total phenols, total flavonoids, and total flavonols were observed (Bhat et al., 2011). Total phenol content and antioxidant activity of the blueberry juice increased when higher juice flow rate (24 and 93.5 mL/min) and higher amplitude level (40%, 80%, and 100%) were applied; as the highest increase in the amount of phenol content (32%) and antioxidant activity (20%) were observed in the sonicated juice with 93.5 mL/min flow rate (Mohideen et al., 2015). The enhancement in the total phenols is related to the cell walls rupture and facilitating the release of the bound phenols into the juice. This increase could also result from the addition of hydroxyl radicals to the aromatic ring of the phenols, enhancing their antioxidant activity (Bhat et al., 2011). The cavitation phenomena and the reduction in oxygen content of the grapefruit juice caused an increased phenol, flavonoids, flavonols content (Aadil et al., 2013). After sonication (20 kHz, 70% amplitude, 15°C, 5 min), the contents of ascorbic acid, carotenoids, β-carotene, and lycopene were increased in carrot juice (Jabbar et al., 2014). Strawberry juice processing at 20°C and 25 kHz brought an increase in the content of ascorbic acid, total phenols, and total anthocyanins (Bhat and Goh, 2017). The residuals of ascorbic acid content were 93.7%, 80.3%, and 54.5% after ultrasonic processing (20kHz, 75% wave amplitude, 10°C) of calcium-added orange juice for 6, 8, and 10 min, respectively (Gómez-López et al., 2010). Ultrasonic processing of the grapefruit juice (28 kHz, 90 min) and guava juice (35 kHz, 30 min) enhanced the ascorbic acid about 28.4% (Aadil et al., 2015) and 8.2% (Cheng et al., 2007), respectively. While, 32.4% reduction in the content of ascorbic acid of the treated tomato juice (amplitude level of 61 μm, 20 kHz, 10 min) was observed. This vitamin degradation was explained by occurring of the sonochemical reactions such as the formation of hydroxyl radicals during cavitation (Adekunte et al., 2010).

Over 200% rise in the cloud value of processed orange juice at 8.61 W/cm^2 for 2 min was reported (Tiwari et al., 2008). Application of ultrasound (28 kHz, 600 W) for 30, 60, and 90 min enhanced the cloud value of fresh grapefruit juice from 0.42 to 1.04, 1.09, and 1.11, respectively (Aadil et al., 2015). Grapefruit juice sonication (28 kHz, 420 W, 20°C, 30 min), increased the cloud value of the juice (from 0.11 to 1.18) (Aadil et al., 2015). It could be related to the formation of the weaker network due to the breakage of the linear pectin. The damages of pectin structure and also other cloud particles are related to microjets of liquid, which is generated by the asymmetrical collapse of cavitation bubbles (Tiwari et al., 2008, 2009). These changes in the structure of the pectin also decrease the accessibility of the pectin molecule to pectin methylesterase (Tiwari et al., 2009). Also, after sonication treatment, the cloud stability of juice usually enhances due to mechanical damage of the pectin methylesterase protein structure (Knorr et al., 2004). The sonication process decreased the viscosity of the grapefruit juice by breaking the pectin, which reduces the molecular weight (Aadil et al., 2015).

In a study, different levels of ultrasound intensity ranging from 8.61 to 22.79 W/cm^2 were applied to process the orange juice for 2–10 min. In comparison to the fresh sample (59.7), the color of the processed orange juice became lighter (62.03–63.54) after treatment for 2 min. However, longer process time slightly reduced the L*-value due to the partial participation of unstable suspended particles and the occurrence of oxidation reactions. Unlike the b*-value, the a*-value was reduced as higher intensities are applied for longer treatment times. As the more intense process was applied, the more increase in the browning index was observed; for example, a 261% increase in this value was reported during treatment at 8.61 W/cm^2 for 2 min. The increase in browning index was resulted from interaction between carotenoids and free radicals such as hydroxyl radicals. Besides, during cavitation contact between the pigments and dissolved gasses like water vapor and O_2 could be enhanced. The total color difference of the processed juice at 17.17 and 22.79 W/cm^2 for 10 min was higher than 3, representing the considerable alteration in the juice color (Tiwari et al., 2008). During sonication of orange juice (28 Hz, 600 W, 90 min), the breakage of pigments structure brought raise in the browning index from 0.221 (fresh juice) to 0.315. Decreased L*, a*, and b* values of the fresh juice were obtained after sonication, which was not recognizable by the naked eyes (Aadil et al., 2015). The L*, a*, and b* of the treated tomato juice (amplitude level of 61 μm, 20 kHz, 10 min) showed 3.7%, 8.1%, and 12.4% increase in comparison to the fresh juice (Adekunte et al., 2010). The processed lime juice at 25 kHz and 20°C for 60 min showed a lower lightness, redness and higher yellowness rather than the fresh juice. The change in the color of sonicated juice was attributed to the accelerated carotenoid isomerization and occurrence of the oxidation reactions resulting from free radicals formation (Bhat et al., 2011). The color analysis showed an increase in a* and b* values as well as a decrease in lightness of the sonicated guava juice (35 kHz, 30 min) resulting in 0.84 total color difference (Cheng et al., 2007). Neither the applied flow rate (24 and 93.5 mL/min) nor the applied amplitude (40%, 80%, and 100%) affected the lightness and a* value of the blueberry juice color, while the hue angle and chroma of the fresh juice were higher than the treated samples resulting from alteration in b* value (Mohideen et al., 2015). Subjecting the strawberry juice to ultrasonic waves (30 min at 20°C, 25 kHz) did not affect color aspects and browning degree (Bhat and Goh, 2017). The sensorial analysis represented that sonication did not cause significant differences in color, aroma, and flavor of the calcium-added orange juice (Gómez-López et al., 2010).

5.7 Advantages and limitations of ultrasonication

The acoustic energy transferring throughout the food products is instantaneous, which decreases the processing time, resulting in the minimal undesirable effects on the quality of fruit juices (Tiwari et al., 2008). Also, processing time reduction results in energy consumption and increasing production throughput (Gómez-López et al., 2010). These waves are non-toxic and safe for application in the food industry. Besides increasing the safety and stability of the juice, this method can enhance the extraction of some nutrients and homogenize the juice (Majid et al., 2015).

However, the high required energy and production of free radicals with high reaction tendencies are the significant limitations in the ultrasound application, especially on a commercial scale. This process is usually capable of reducing the microbial and enzyme activities to the acceptable level. In order to increase the process efficiency, this method is applied in combination with thermal treatment, HPP and also both of them called thermosonication, manosonication, and manothermosonication, respectively.

6 Ultraviolet treatment

6.1 Definition and description of ultraviolet treatment

Ultraviolet radiation (UV) is classified in the electromagnetic spectrum ranging from 100 to 400 nm, which can be divided into UV-A (320–400 nm), UV-B (280–320 nm), and UV-C (200–280 nm). In general, UV-C with 254 nm is suggested to be applied in juice treatment (Keyser et al., 2008; Kaya et al., 2015). In addition, UV radiation is used in air, water surfaces, and packaging disinfection in some industries such as soft drinks and cheese production (Tran and Farid, 2004).

6.2 Ultraviolet equipment

In the design of a UV treatment system, type of system (continuous or batch), the number of juice passes through the reactor (in the continuous system), lamp properties (such as power, number, and wavelength) are the essential aspects (Koutchma et al., 2016). In the food industry, continuous UV, pulsed UV, and excimer lamps are generally used (Koutchma, 2008). In a UV reactor, the juice flow pattern determines the applied UV dose, exposure, and residence time. In addition, the juice properties can affect the process efficiency; as, low UV-transmittance of juice containing a high amount of color pigments, organic solutes, and suspended particles brings some difficulties in the UV processing of the juices. Although these compounds increase the UV absorption, they can induce the reflection and scattering of the light (Koutchma, 2008).

Tran and Farid (Tran and Farid, 2004) designed a vertically falling film UV reactor in which a thin uniform film of the juice is exposed to UV as shown in Fig. 2.3. In this system, a low-pressure UV lamp is fixed at the axis of the glass tube. It is enclosed by a quartz tube (high UV transmitted) to prevent direct contact of the juice and lamp. The cooling of the lamp is performed by pumping air through the space surrounding the lamp. The flow rate of the juice is controlled using a peristaltic pump to modify penetration of the UV light the juice film thickness.

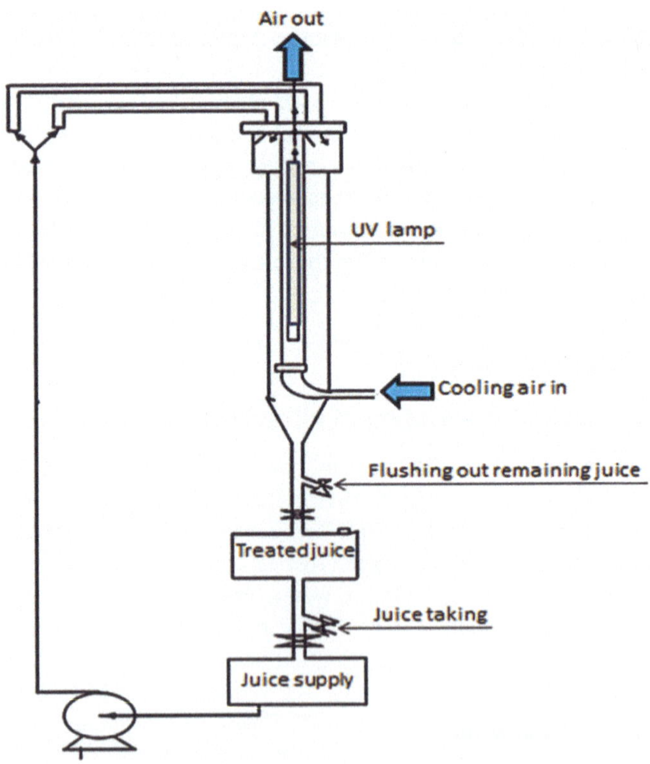

FIGURE 2.3 Flow diagram of vertically falling film UV reactor designed for juice processing. *(Reproduced from Tran and Farid, 2004.)*

This reactor consisted of either a helically wound polytetrafluoroethylene (PTFE) or Teflon tube wrapped around a quartz glass tube with a UV lamp to ensure complete passing the juice through the UV zone, providing effective disinfection. In this system, freshwater circulation is applied to avoid the juice temperature increase. The application of high flow rates to produce a turbulent homogeny flow improves the juice UV exposure. Very high flow rates may result in the pressure drops across the reactor, which decreases the juice residence time (Koutchma, 2008). To ensure UV exposure, a rotating drum or screw conveyor to lift and tumble the product can be used (Koutchma, 2008).

6.3 Sensitivity of microorganisms to ultraviolet

UV destroys different microorganisms such as bacteria, viruses, protozoa, yeasts, and molds. The UV light is absorbed by the microbial DNA and RNA structure causing their inactivation. Absorbed radiant energy can affect several structural bonds (O–H, C–C, C–H, H–N, and S–S). This energy prompts the crosslinks between neighboring pyrimidine nucleotide bases, which blocks the transcription or replication of microbial DNA or RNA, resulting in the death of the cell (La Cava and Sgroppo, 2015). The environmental factors (temperature, nutrients, etc.), type and growth stage of the microorganisms affect the UV efficiency in inactivation of microorganisms. Organic media, rich in peptones and sera, shows a protective effect for the microorganisms because they provide a shield for the DNA against the UV by increasing the number of ribosomes within the cell. At low temperatures ($< 25°C$), the natural state of single-stranded DNA is a stacked structure, which increases the rate of dimer formation between adjacent thymine nucleotides in single-stranded DNA. The cells show higher UV resistance in the lag phase than cells in the logarithmic and stationary phase (Tran and Farid, 2004).

Orange juice was processed at up to 0.12 J/m^2 UV dosage. The predicted necessary UV dosages to inactivate 90% of viable microorganisms were 0.087 J/cm^2 for aerobic plate counts and 0.12 J/cm^2 for yeasts and molds. Yeasts and molds are more resistant due to their large size, low pyrimidine base in their DNA structure and thick cell wall (Tran and Farid, 2004). Reduction of 3.5- and 3-log in aerobic plate count and yeast and mold were respectively observed after apple juice processing at 0.23 J/mL. After the juice treatment at 1.377 J/mL, a 7.42-log reduction in the population of the *E. coli* was obtained. Apple juice was clarified. Hence, it easily penetrated the UV light. In comparison to the apple juice, the guava juice was more turbid, so a higher UV dosage was necessary for its processing. In processed guava-and-pineapple juice (1.38 J/mL), 3.31 and 4.48-log reduction in aerobic plate count and yeast and mold were respectively, reported. The mango nectar showed zero aerobic plate count, yeast and mold at a UV dosage of 1.38 J/mL. In the strawberry nectar, 1.32- and 2.45-log reduction in aerobic plate count, yeast and mold were respectively reported after a high dose of UV treatment (2.06 J/mL) (Keyser et al., 2008). The population of yeast and mold, total aerobic bacteria and coliform in the fresh watermelon juice were 3.47-, 2.73-, and 1.62-log, respectively. After juice treatment at 37.5 J/mL, total aerobic bacteria (1.26-log) and yeast and mold (2.48-log) were reduced significantly, but there was no detectable coliform (Feng et al., 2013).

Processing of starfruit juice at 2.158 J/m^{2}for 60 min showed no aerobic plate count with no impact on yeast and molds (Bhat et al., 2011). *L. innocua* and *E. coli* counts were below the limit of detection after exposing apple juice to UV treatment (10.62–53.10 J/cm^2) (Caminiti et al., 2012). Reductions of 2.8-, 0.34-, and 5.72-log in aerobic plate count, yeast and mold count and *E. coli* ATCC 25922 were, respectively, observed after UV treatment of orange juice at 36.09 J/mL (Pala and Toklucu, 2013). The population of *E. coli K-12* was decreased by 5.34-log after grape juice exposure to UV dosage at 9.92 J/cm^2. In addition, no yeast, lactic acid bacteria, and total aerobic bacteria were detected in the UV treated juice (Unluturk and Atilgan, 2015).

Applying the lowest juice depth and the highest level of UV dosage brought more than a 5-log reduction in the *E. coli* O157: H7 population during UV treatment of apple juice (Ngadi et al., 2003). During the processing of melon-lemon juice, it was found that applying a higher flow rate produced a less residence time and less inactivation in *E. coli* K12. The rise in the pump frequency reduces the residence time and increases in flow rates, numbers of Reynolds. Applying flow rates ranging from 7.8–121 mL/min produced a laminar flow (Reynolds < 2100). After processing pineapple juice at the frequency of 40, 35, and 30 Hz, 3.89-log (10.10 mJ/cm^2), 3.99-log (10.37 mJ/cm^2), and 5-log (13.75 mJ/cm^2) reduction in *S. typhimurium* were observed, respectively (Mansor et al., 2014).

6.4 Effects on nutritional aspects

In most cases, the soluble solids, titratable acidity and pH of the juices were not affected by UV treatment (Kaya et al., 2015; Feng et al., 2013; Unluturk and Atilgan, 2015; Shamsudin et al., 2014). After UV processing at 9.92 J/cm^2, a remarkable decrease in the ascorbic acid content of fresh grape juice was observed (from 0.56 to 0.01 mg/L) (Unluturk and Atilgan, 2015). The contents of the ascorbic acid, naringin, and total phenol content in the fresh grapefruit juice were 41.0–56.9 mg/100 mL, 17.5–27.8 mg/100 mL, and 68.9–86.5 mg/100 mL, respectively. Ascorbic acid content reduced up to 17%, 29%, and 35% after UV treatment at 1.83, 2.84, and 3.94 J/cm^2, respectively. However, the level of naringin and total phenol content remained constant after UV treatment (La Cava and Sgroppo, 2015). After 2 cycles processing of pineapple juice at 0.011 J/cm^2 (per cycle), 13% and 20% decrease in total phenol and ascorbic acid contents were, respectively reported (Shamsudin et al., 2014). Applying 0.053 J/cm^2 UV dosage caused a 17% decrease in total phenolic content in pineapple juice (Chia et al., 2012). The total phenol content of apple juice was not affected by UV treatment. However, an 11% decrease in antioxidant activity was observed after juice processing at 53.10 J/cm^2 (Caminiti et al., 2012). After UV treatment (36.09 J/mL), no alterations in the content of ascorbic acid, organic acids, antioxidant capacity, and phenols were reported in orange juice (Pala and Toklucu, 2013). In another study, the application of different UV dosages (2.7 and 37.5 J/mL) showed no effect on the contents of total phenolic compounds and lycopene in watermelon juice (Feng et al., 2013). After UV treatment of mango juice at 3.5 J/m^2 for 15 and 60 min, an increase in the content of carotenoids (6%), polyphenols (31%), and flavonoids (3%) were observed (Santhirasegaram et al., 2015). High flow rates and treatment times decrease the betalain content of pitaya juice due to the formation of photons by organic molecules such as betalains to possess conjugated bonds and aromatic rings responsible for color. The antioxidant activity of the juice decreased when the juice was processed for longer treatment times and higher flow rates (Ochoa-Velasco and Guerrero-Beltrán, 2012). UV process (2.158 J/m^2) had no significant effect on total soluble solids and pH of the starfruit juice while the acidity of fresh juice reduced from 6.73% to 6.54% and 6.24% after processing for 30 and 50 min, respectively. Increase of 6%, 3%, 6%, and 17% in the contents of total phenols, antioxidant capacity, total flavonols, and total flavonoids were detected after juice treatment for 60min. This increase could be related to the accumulation of phenolic compounds and their products as a means of defense against UV treatments or enhanced activities of phenylalanine ammonia-lyase and the polyphenol oxidase enzyme. But, the ascorbic acid content of the fresh juice (27.31 mg/100 mL) was reduced to 21.77 mg/100 mL after juice treatment at 2.15 J/m^2 for 60 min (Bhat et al., 2011).

Apple juices (Golden, Starking, Fuji, and King David varieties) were irradiated in incident energy of 3.88×10^{-7} E/min for 120 min. UV process had no effect on total phenol and sugars contents. Ascorbic acid degradation in juices obtained from Golden, Starking, Fuji, and King David were 5.7%, 5.6%, 4%, and 70%, respectively. The difference between the losses in the content of this vitamin in juices of various varieties could be resulted from the lack of pigmentation of King David apple juice due to the pigments absorb the radiation energy and protect other juice molecules. Initial activities of peroxidase in fresh juices were small, which were inactivated after treatment for 15 min. (Falguera et al., 2011). Only a 5% reduction in the activity of pectin methylesterase was observed after the UV processing of orange juice (0.074 J/cm^2) (Tran and Farid, 2004). The residual activity of pectin methylesterase was 82% after processing (10.62 J/cm^2, 1 min) the orange-carrot juice. Total phenols content and antioxidant activity were similar in the fresh and treated carrot juices (Caminiti et al., 2012).

6.5 Impact on juice appearance

There was no significant difference between the color properties of the fresh and processed pineapple juice, orange-carrot juice, and orange juice (Tran and

Farid, 2004; Caminiti et al., 2012; Chia et al., 2012). Lightness and a*-value of the processed watermelon juice were similar to the fresh sample at lower UV dosage. At higher UV dosage, the b*-value of the juice was reduced and the total color difference was increased to 0.31 and 0.90 after juice processing at 2.7 and 37.5 J/mL, respectively (Feng et al., 2013). Generally, the effect of the UV process on the color aspects depends on the juice properties and varieties. For instance, the lightness of apple juices (Golden, Starking, Fuji, and King David varieties) increased during processing (250 and 740 nm for 120 min). Unlike King David apple juice, a*-value was decreased during radiation. While the b*-value of King David apple juice remained almost constant, it was decreased in the golden sample. However, a slight increase in the b*-value of the samples from Starking and Fuji was observed (Falguera et al., 2011).

The turbidity of the fresh lemon-melon juice (278.5 NTU) was increased up to 292.75 NTU after processing for (124s, 2.46 J/mL, 8 cycles). Degradation of the carotenoids, besides UV phytochemical destruction effect of polymeric compounds to change b*-value from 2.48 (fresh juice) to 1.88 and brought total color difference of 0.63 (Kaya et al., 2015). The turbidity of the grape juice was increased (21%) after juice treatment at 9.92 J/cm^2. A juice with a darker color and higher a* and b* values were obtained after UV treating. These results indicate the photo-degradation of juice pigments as a result of long UV process time (32.5 min). About 76% increase in the browning index of the juice was observed after UV treatment (Unluturk and Atilgan, 2015). In another study, the turbidity of the fresh lime juice decreased by 50% after exposure of the UV lamp at 0.044 J/cm^2 (Mohd-Hanif et al., 2016). It was explained by the inactivation of the yeasts and molds besides the less sedimentation of the fragmented pectin. After juice processing, the lightness (61.47), b*-value (8.54) and chroma (9.09) of the fresh juice were reduced to 55.28, 1.64, and 3.03, respectively. a*-value and hue angle of fresh juice were −3.12 and 110.04, respectively. Applying the more intense treatment (0.023–0.044 J/cm^2) raised a*-value (−2.98 to −2.55), hue angle (113.58–147.28), and total color difference (1.78–9.30). An increase in the lightness of the starfruit juice was reported corresponding to increase the UV process time (30 and 60 min) at 2.158 J/m^2, which could be related to the degradation of the colored compounds and making the juice more transparent. However, a* and b* values decreased during UV treatment. Total color differences of the processed juice for 30 and 60 min were 0.43 and 1.98, respectively (Bhat et al., 2011). The color analysis of the UV treated apple juice represented an increase in the juice lightening, a decrease in redness and a non-enzymatic browning index with no significant difference in b*-vale. Based on the sensorial evaluation (color, odor, flavor, and overall acceptability), the apple juice samples exposed to 5.31 and 10.62 J/cm^2 energy dosages were more acceptable due to having similar attributes to the fresh juice. However, the panelists disliked the color of processed juice at 53.10 J/cm^2 (Caminiti et al., 2012).

6.6 Advantages and disadvantages

UV application is a low-cost operation, representing good investment potential, and a shorter payback period (Shamsudin et al., 2014). Besides low energy consumption in comparison to the conventional thermal pasteurization, the use of UV-C showed no critical adverse effects such as toxicity or producing the toxic by-products, off taste and odor (Keyser et al., 2008; La Cava and Sgroppo, 2015). It becomes popular due to showing consumer's satisfaction as well as being environment-friendly.

One of the most limiting factors in the UV-C application is its low penetration capacities, as it is about 1 mm for the absorption of 90% of the light in the juices (Keyser et al., 2008). Producing a turbulent juice flow or a thin layer of the juice can be an effective way to make sure of an optimum juice exposure to the light (Keyser et al., 2008). It should be noticed that UV could show an unwanted effect on the furan development in fruit juice; hence, it is necessary to utilize the appropriate dose and duration of juice exposure to UV (Shah et al., 2016).

7 Ozone processing

7.1 Ozone overview

Ozone (O_3), a bluish gas with a strong odor, is a triatomic form of oxygen (Kim et al., 1999). It shows a strong disinfection effect due to its high potential to diffuse through biological membranes and showing high oxidizing reactive properties to inactivate microorganisms at a low concentration during short pasteurization time. Therefore, ozone is used in the washing, storage, and processing of different food products in the form of gases and liquids (Sukarminah et al., 2017).

7.2 Methods of ozone generation

Combination of single "O" with O_2 results in the formation of the very reactive O_3. However, splitting the O_2 (oxygen) molecule to form a single "O" needs a high energy input. In nature, UV radiation (185 nm) from the sun prepares this energy. There are different commercial methods to produce O_3, such as UV radiation, corona discharge, an electrochemical method (Kim et al., 1999; Sukarminah et al., 2017).

In a UV-based generator, ambient air (containing about 20% of O_2) or oxygen-enriched air is passed through the UV light source (typically 140–190 nm). This system is low-cost with more limited output (Sukarminah et al., 2017; Patil, 2010). In order to produce high amounts of ozone, the corona discharge method is generally used. In this method, dry O_2 enriched air or highly purified oxygen is passed through a high electric voltage-current (>5000 V) or corona (Sukarminah et al., 2017). In the generator, the oxygen should be free from

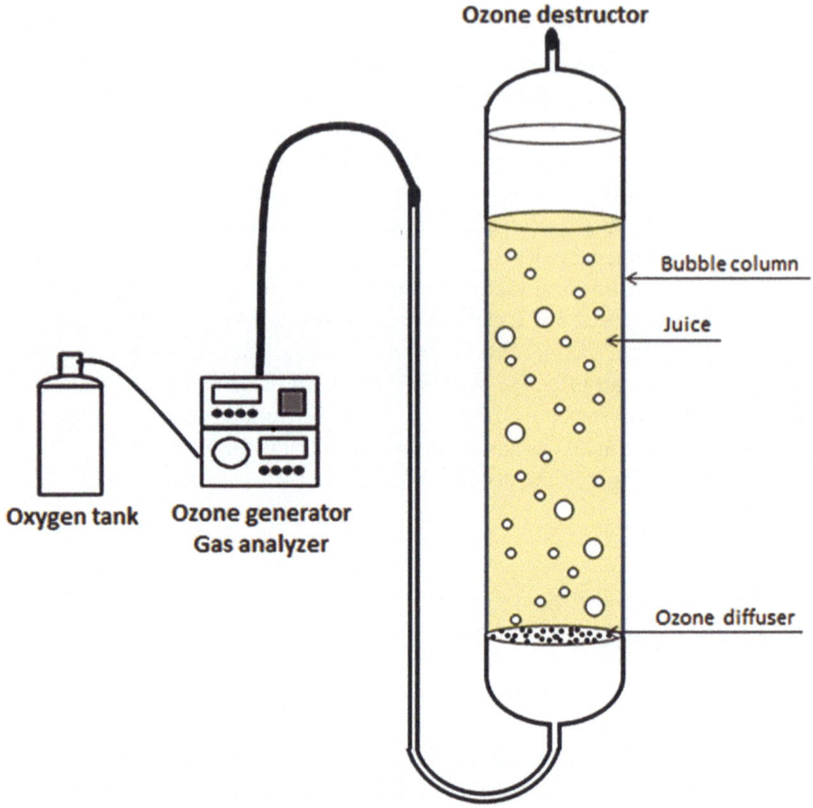

FIGURE 2.4 **Scheme of ozone processing system.** *(Reproduced from Tiwari et al., 2009.)*

moisture, which produces nitric acid deposits and brings corrosion of metal surfaces (Kim et al., 1999). Application of high-voltage alternating current provides oxygen electrons excites and splitting of O_2, which results in ozone production at concentrations up to 4% by weight (Kim et al., 1999). In an electrochemical method, water is split into hydrogen and oxygen atoms by electrolysis between an anode and cathode in an electrolytic solution containing water and a solution of highly electronegative anions. After separating the hydrogen, water mixture and the oxygen atoms are combined to form ozone and diatomic oxygen (Fig. 2.4). In comparison to the applying corona discharge generator, this method shows a higher efficiency in ozone production, needs lower voltage DC and no need to feed gas preparation (Kim et al., 1999; Patil, 2010).

7.3 Factors affecting the efficacy of ozone treatment

Ozone is very unstable in water and decomposes to oxygen during a short time. Different factors are influencing the efficiency of this gas to inactivate the microorganisms. However, solubility/stability and reactivity of the ozone are highly affected by the process conditions; hence, it is difficult to predict the effect of only one specific factor on the ozone effectiveness (Patil, 2010). For example, lowering the temperatures brings the higher ozone solubility in juice and longer gas half-life. These provide a more inactivation chance by increasing the concentration of dissolved ozone and improving the contact of ozone and juice (Sukarminah et al., 2017). On the other hand, the ozone decomposition (being unstable) will be accelerated by increasing the temperature. Acidic pH can increase the ozone retention time and so its lethality effect (Sukarminah et al., 2017). While the rapid decomposition of ozone would happen in an aqueous medium with a high pH (Kim et al., 1999). In addition, the effect of ozone on microorganism inactivation is increased at higher relative humidity (Kim et al., 1999).

The concentration of ozone is one of the most critical parameters in food processing affecting the safety of the products as well as the security of the workers. To inactivate microorganisms in juices, ozone concentration about 0.5–4.0 mg/L is recommended (Sukarminah et al., 2017). Long exposure of ozone at concentrations higher than 4 ppm is highly corrosive to equipment and lethal to humans. Exposure of ozone concentration at 1–2 ppm for 3–6 h shows strong displeasing odor and irritating effect on eyes and throat. Ozone level above 50 ppm could be fatal for the human. Besides, oxidation of chemical compounds, undesirable change in color and odors of the juice could happen under excessive use of ozone (Rawson et al., 2011). Therefore, calculating the amount of required disinfectant effect on the microorganism within the juice is essential, based on the ozone concentration and process time (Sukarminah et al., 2017). The other important factor is related to the conditions of contacting the juice with the ozone. The produced ozone can introduce to liquid food products based on the gas bubbling method, in which ozone reacts chemically with the food in a bubble ozone gas column (Sukarminah et al., 2017). Depending on the applied flow rate, different sizes of bubbles are produced. It was found that the smaller bubbles enhance its disinfection effect by providing higher solubilization of ozone and interfacial contact area (Patil, 2010).

7.4 Antimicrobial action of ozone

Ozone is known as a powerful agent against the activity of different microorganisms, including bacteria, fungi, viruses, and protozoa (Tiwari et al., 2009). In general, bacteria are more sensitive than fungi. Gram-positive bacteria are more ozone sensitive than the Gram-negative ones (Patil, 2010). It was found that vegetative cells show less ozone resistance in comparison to the spores. Also, molds are more resistant to ozone treatments rather than the yeasts (Kim et al., 1999).

Ozone shows high oxidative properties and the ability to diffuse through a biological membrane. These properties bring it attacks directly to surface layers microorganisms to breakdown the cell permeability defense and paralysis it. This high lethality of ozone is related to affecting the unsaturated lipids, oxidizing the sulfhydryl groups in cysteine residues of the enzymes, protein, and peptidoglycans (in coatings of spores and virus capsids) and oxidizing the lipoprotein and the lipopolysaccharide (in the inner layer of the Gram-negative bacteria) (Sukarminah et al., 2017; Patil, 2010). In addition, inactivation of microorganisms can be caused by damaging the genetic material during ozone treatment (Kim et al., 1999). It could affect purines and pyrimidines in nucleic acid producing single-strand breaks in DNA and altering the closed circular DNA to open structure (Patil, 2010).

Inactivation of *E. coli* O157: H7 and *Salmonella* in apple cider (180 min) and orange juice (240 min) were achieved during ozone treatment (0.9 g/h, 4°C & 50°C) (Williams et al., 2004). Treatment of apple cider and orange juice at 50°C resulted in 4.8- and 5.4-log reduction in *E. coli*, respectively. Processing at lower temperatures (4°C) reduced the *Salmonella* by 4.5-log (apple cider) and 4.2-log (orange juice). Ozone treatment of the juices, in combination with mild heating, enhances the microorganism inactivation. *E. coli* O157: H7 populations were undetectable after processing of apple cider and orange juice at 50°C for 45 and 75 min, respectively. After 15 min of processing at 50°C, *Salmonella* was undetectable in orange juice and a 4.8-log reduction in its population was found in the apple cider.

Different types of orange juice (fresh unfiltered juice, without pulp and filtered through 500 μm or 1 mm sieves) were processed using ozone at 75–78 μg /mL for the different duration (0–18 min) (Patil et al., 2009). *E.coli* inactivation in low pulp contained juice took a shorter time (6 min) in comparison to the unfiltered sample (15–18 min). The higher amount of organic compounds in the unfiltered juice (such as sugars, fibers, and ascorbic acid) brings a negative impact on the ozone disinfection rate by influencing the ozone dissolution and diminishes its availability for inactivation of the bacteria.

During apple juice treatment at 22°C for 40 min, the initial level of *A. acidoterrestris* spores is decreased by 1.8- and 2.4-log at ozone concentrations of 2.8 and 5.3 mg/L, respectively. However, 40 min juice treatments at a lower temperature (4°C) caused more reduction in the spores at the same ozone concentrations of 2.8 mg/L (2.2-log) and 5.3 mg/L (2.8-log) (Torlak, 2014). In another study, 0.73- and 2.22-log reduction in *A. acidoterrestris* spores were observed after 30 and 60 min melon juice ozone processing at 7.0 g/L gas concentration (Fundo et al., 2018).

7.5 Effect on juice quality

It was found that the juice processing with ozone had no considerable impact on pH, total soluble solids, and titratable acidity (Tiwari et al., 2008, 2009; Jaramillo-Sánchez et al., 2018). However, the higher rate in peroxidase and polyphenol oxidase inactivation was found as higher ozone concentration

(0.11 and 0.20 mg/min.mL) and process time (up to 12 min) was applied in the processing of the peach juice (Jaramillo Sanchez et al., 2018). After juice treatment for 12 min, less than 1% in peroxidase activity and 2.7%–6% polyphenol oxidase activity was observed. Via direct or indirect ozone reactions, oxidation of protein structures of the enzymes causes the covalent modification. In this phenomenon, amino acid side chains and protein backbones involve in protein fragmentation or protein–protein cross-linkages.

The vigorous oxidizing activity of ozone causes a loss in bioactive antioxidant compounds. Degradation of ascorbic acid in ozone-treated products is mainly related to the activation of ascorbate oxidase. Under stress conditions, this enzyme will be activated and produce dehydroascorbic acid via the degradation of ascorbic acid (Rawson et al., 2011). A significant reduction in the content of ascorbic acid (30.53%) was observed after orange juice treatment at 7.8% ozone concentration (Tiwari et al., 2008). About 23.8% (1.6% w/w, 3 min) and 85.8% (7.8% w/w, 10 min) decrease in the content of ascorbic acid were reported after the strawberry juice ozone processing. Pelargonidin-3-glucoside, the major anthocyanin in the strawberry juice, degraded 39.9% and 98.2% after treatment at 1.6% w/w for 3 min and 7.8% w/w for 10 min, respectively (Tiwari et al., 2009). During ozonation at 1.6% w/w concentration for 10 min, 78%, 95%, and 99% reduction in the content of three significant anthocyanins of the grape juice, including of cyanidin-3-O-glucoside, delphinidin-3-O-glucoside, and malvidin-3-O-glucoside were respectively observed (Tiwari et al., 2009). In ozone treatment of juice, degradation of chemical compounds such as carotenoids and anthocyanins could happen through a direct reaction of the dissolved substances with ozone molecules or indirect reactions via ozone decomposition to form secondary oxidants (free radical). The indirect reactions based on rapid ozone decomposition produce radicals such as hydroxyl (•OH), hydroperoxyl (•HO_2), and superoxide $\left(O_2^-\right)$. These radicals can lead the electrophilic and nucleophilic reactions with aromatic compounds, which are substituted with an electron donor with high electron density on the carbon compounds in ortho and para positions (Tiwari et al., 2009).

During ozonation of the apple juice at 4.8% w/w for 10 min, a considerable decrease in chlorogenic acid (99%), caffeic acid (97%), cinnamic acid (99%), and total phenol content (50%) were reported (Torres et al., 2011). The phenolic content and total antioxidant capacity of the processed juice were the same as the control orange juice (Almeida et al., 2015). Decrease of 65.67% in the total phenol content of the apple juice was observed after processing (5.3 mg/L, 40 min) (Torlak, 2014). In another study, it was found that the total phenol content of fresh melon juice (164.70 μg/mL) was increased to 198.63 and 266.71 μg/mL after processing for 30 and 60 min, respectively. This increase could be related to oxidative stress resulting in the activation of pre-existing enzymes, such as phenylalanine ammonia-lyase, with a critical role in the biosynthesis of the phenolic compounds. After juice treatment in either 30 or

60 min, 19% degradation in the content of ascorbic acid was reported. Residual carotenoid content after juice ozonation for 30 and 60 min were about 24% and 17%, respectively (Fundo et al., 2018). Unlike the total soluble solids, the pH of the melon juice reduced during more extended ozone processing (7.0 g/L) (Fundo et al., 2018).

The effect of process variables including gas flow rate (0–0.25 L/min), ozone concentration (1.2%–4.8% w/w), and treatment time (2–10 min) was investigated on the color properties of the orange juice (Tiwari et al., 2008). Unlike a* and b* values, an increase in the lightness and hue angle of the juice was observed in higher ozone flow rate, ozone concentration, and treatment time. Treatment time showed the dominant effect on the total color difference. In comparison to the fresh melon, the processed juice was lighter with lower redness and yellowness, causing a high total color difference of 19.51 and 22.70 after treatment at 30 and 60 min, respectively. Reduction of 60% and 70% in chroma and 4% and 9% reduction in hue angle were observed after juice treatment for 30 and 60 min, respectively (Fundo et al., 2018). A reduction in a*-value and increase in L* and b* values in the color of the apple juice were reported with an increase in processing time (10 min) and ozone concentration (1%–4.8% w/w) (Torres et al., 2011). The total color difference of the processed juice at the most intensive process condition (4.8% w/w, 10 min) was about 30.84. Ozone concentration in the range of 3%–10%w/w of oxygen was used to process the orange juice. The results showed that the cloud value and browning index of the fresh and ozonated juice were similar. While the processed juice showed lower redness and yellowness, the lightness and total color difference of the juice were increased at higher ozone concentration (Tiwari et al., 2008). By the rise in ozone concentration (up to 7.8% w/w) and treatment time (0–10 min), the L*-value of the strawberry juice was increased while a* and b* values were decreased (Tiwari et al., 2009). For instance, after processing at 7.8% w/w for 10 min, L*, a*, and b* values of the fresh juice altered 28.0–53.4, 44.8–11.1, and 33.7–17.2, respectively. During ozonation of blackberry juice at higher ozone concentration (0%–7.8% w/w) and longer treatment time (0–10 min), increase in lightness as well as decreases in redness and yellowness were observed (Tiwari et al., 2009). During the processing of the grape juice at a concentration of 1.6%–7.8% w/w for 10 min, a*-value was decreased from 9.32 to 6.40 and L* and b*-values were increased from 22.09 to 26.55 and 8.85 to 5.83, respectively (Tiwari et al., 2009). Lightness and chroma of fresh and the treated (0.057, 0.128, and 0.230 mg/ mL) prebiotic orange juice were similar (Almeida et al., 2015). However, the hue angle of the ozonated juice showed a slight reduction (less than 2 degrees) by a rise in the ozone concentration. After processing (0.06–2.48 g/L) of peach juice, lightness was slightly decreased at first and then remained constant, while a*-value and browning index were increased (Jaramillo-Sánchez et al., 2018). The study of the rheological properties revealed that both fresh and the processed juices showed pseudo-plastic flow behavior.

7.6 Advantages and limitations

FDA approved the application of ozone as a safe (GRAS) additive in processing the food products (Rawson et al., 2011). It is known as an eco-friendly gas with no hazardous toxic residues in the juice (Sukarminah et al., 2017).

Ozone is very unstable and it is difficult to predict the mechanisms of ozone decomposition and its reaction with the compounds. It could not be stored and should be produced at the point of application. Mechanisms of decomposition of ozone are complex depending on several factors, such as the types of formed radicals in the solution and the present organic compounds in the medium that controls the radical chain reaction. These make the challenge to define precisely the needed ozone concentration to disinfection of the juice (Kim et al., 1999). In the food industry, it is essential to consider the systems of ozone detection, destruction, and respirators to guarantee the safety of workers (Kim et al., 1999). In addition, the excess of ozone not dispersing in the food could be captured and destroyed to prevent from corrosion of the equipment (Sukarminah et al., 2017).

8 Conclusion

Pasteurization is one of the most critical steps in juice production to satisfy its safety, quality, and shelf life. Different non-thermal food processing methods, including pulsed electric field, high-pressure processing, high-pressure carbon dioxide, sonication, ultraviolet, and ozone have been introduced to prevent the undesirable effects of thermal processing on juice nutritional values and physical aspects. During these processing methods, no considerable change in the juice temperature occurs and the inactivation of the microorganisms and enzymes takes place via different mechanisms. The chemical properties of the juice can alter during these processes, which influence the physical and sensorial aspects of the juice. In comparison to the thermal pasteurization, the application of non-thermal processing methods results in producing fresh-like or high-quality juice. However, there is some limitation in their industrial applications such as the high-cost requirement for installation and operation. Besides, considering the protein structure of the enzymes and resistance of some microorganisms, a combination of these methods with the mild heating process will enhance the effectiveness of the juice pasteurization based on the hurdle concept.

References

Aadil, R. M., et al. (2013). Effects of ultrasound treatments on quality of grapefruit juice. *Food Chemistry*, *141*(3), 3201–3206.

Aadil, R. M., et al. (2015). Thermosonication: a potential technique that influences the quality of grapefruit juice. *International Journal of Food Science & Technology*, *50*(5), 1275–1282.

Adekunte, A., et al. (2010). Effect of sonication on colour, ascorbic acid and yeast inactivation in tomato juice. *Food Chemistry*, *122*(3), 500–507.

Agcam, E., Akyıldız, A., & Evrendilek, G. A. (2014). Effects of PEF and heat pasteurization on PME activity in orange juice with regard to a new inactivation kinetic model. *Food Chemistry*, *165*, 70–76.

Aghajanzadeh, S., & Ziaiifar, A. M. (2018). A review of pectin methylesterase inactivation in citrus juice during pasteurization. *Trends in Food Science & Technology*, *71*, 1–12.

Aghajanzadeh, S., et al. (2016). Thermal inactivation kinetic of pectin methylesterase and cloud stability in sour orange juice. *Journal of Food Engineering*, *185*, 72–77.

Aghajanzadeh, S., Kashaninejad, M., & Ziaiifar, A. M. (2017). Cloud stability of sour orange juice as affected by pectin methylesterase during come up time: Approached through fractal dimension. *International Journal of Food Properties*, *20*(Suppl. 3), S2508–S2519.

Aguilar-Rosas, S., et al. (2007). Thermal and pulsed electric fields pasteurization of apple juice: Effects on physicochemical properties and flavour compounds. *Journal of Food Engineering*, *83*(1), 41–46.

Alexandrakis, Z., et al. (2014). Comparative structural changes and inactivation kinetics of pectin methylesterases from different orange cultivars processed by high pressure. *Food and Bioprocess Technology*, *7*(3), 853–867.

Almeida, F. D. L., et al. (2015). Effects of atmospheric cold plasma and ozone on prebiotic orange juice. *Innovative Food Science & Emerging Technologies*, *32*, 127–135.

Altuntas, J., et al. (2010). Effects of pulsed electric field processing on the quality and microbial inactivation of sour cherry juice. *International Journal of Food Science & Technology*, *45*(5), 899–905.

Arreola, A., et al. (1991). Supercritical carbon dioxide effects on some quality attributes of single strength orange juice. *Journal of Food Science*, *56*(4), 1030–1033.

Balasa, A., Toepfl S., & Knorr, D. (2006). P9s. In Food factory of the future 3. Gothenburgh, Sweden.

Barba, F. J., Esteve, M. J., & Frigola, A. (2013). Physicochemical and nutritional characteristics of blueberry juice after high pressure processing. *Food Research International*, *50*(2), 545–549.

Barbosa-Canovas, G. V., et al. (1999). In *Preservation of foods with pulsed electric fields*. Academic Press.

Basak, S., & Ramaswamy, H. (1996). Ultra high pressure treatment of orange juice: a kinetic study on inactivation of pectin methyl esterase. *Food Research International*, *29*(7), 601–607.

Basak, S., Ramaswamy, H., & Simpson, B. (2001). High pressure inactivation of pectin methyl esterase in orange juice using combination treatments. *Journal of Food Biochemistry*, *25*(6), 509–526.

Bayındırlı, A., et al. (2006). Efficiency of high pressure treatment on inactivation of pathogenic microorganisms and enzymes in apple, orange, apricot and sour cherry juices. *Food Control*, *17*(1), 52–58.

Bertoloni, G., et al. (2006). A study on the inactivation of micro-organisms and enzymes by high pressure CO2. *Biotechnology and Bioengineering*, *95*(1), 155–160.

Bevilacqua, A., et al. (2014). Inactivation of spoiling yeasts of fruit juices by pulsed ultrasound. *Food and Bioprocess Technology*, *7*(8), 2189–2197.

Bhat, R., & Goh, K. M. (2017). Sonication treatment convalesce the overall quality of hand-pressed strawberry juice. *Food Chemistry*, *215*, 470–476.

Bhat, R., et al. (2011). Sonication improves kasturi lime (*Citrus microcarpa*) juice quality. *Ultrasonics Sonochemistry*, *18*(6), 1295–1300.

Buckow, R., et al. (2011). Effect of dimensions and geometry of co-field and co-linear pulsed electric field treatment chambers on electric field strength and energy utilisation. *Journal of Food Engineering*, *105*(3), 545–556.

Buckow, R., Ng, S., & Toepfl, S. (2013). Pulsed electric field processing of orange juice: a review on microbial, enzymatic, nutritional and sensory quality and stability. *Comprehensive Reviews in Food Science and Food Safety*, *12*(5), 455–467.

Bull, M. K., et al. (2004). The effect of high pressure processing on the microbial, physical and chemical properties of Valencia and Navel orange juice. *Innovative Food Science & Emerging Technologies*, *5*(2), 135–149.

Burdurlu, H. S., Koca, N., & Karadeniz, F. (2006). Degradation of vitamin C in citrus juice concentrates during storage. *Journal of Food Engineering*, *74*(2), 211–216.

Caminiti, I. M., et al. (2012). The effect of pulsed electric fields, ultraviolet light or high intensity light pulses in combination with manothermosonication on selected physico-chemical and sensory attributes of an orange and carrot juice blend. *Food and Bioproducts Processing*, *90*(3), 442–448.

CAO, S. -q., Liang, L., & PAN, S. -y. (2011). Thermal degradation kinetics of anthocyanins and visual color of blood orange juice. *Agricultural Sciences in China*, *10*(12), 1992–1997.

Carbonell-Capella, J., et al. (2017). Influence of pulsed electric field processing on the quality of fruit juice beverages sweetened with *Stevia rebaudiana*. *Food and Bioproducts Processing*, *101*, 214–222.

Chakraborty, S., et al. (2014). High-pressure inactivation of enzymes: A review on its recent applications on fruit purees and juices. *Comprehensive Reviews in Food Science and Food Safety*, *13*(4), 578–596.

Cheng, L., et al. (2007). Effects of sonication and carbonation on guava juice quality. *Food Chemistry*, *104*(4), 1396–1401.

Chia, S., et al. (2012). The effect of storage on the quality attributes of ultraviolet-irradiated and thermally pasteurised pineapple juices. *International Food Research Journal*, *19*(3), 1001–1010.

Cortés, C., et al. (2006). Changes of colour and carotenoids contents during high intensity pulsed electric field treatment in orange juices. *Food and Chemical Toxicology*, *44*(11), 1932–1939.

Cserhalmi, Z., et al. (2006). Study of pulsed electric field treated citrus juices. *Innovative Food Science & Emerging Technologies*, *7*(1–2), 49–54.

Damar, S. & Balabanm, M. (2005). Cold pasteurization of coconut water with a dense phase CO2 system. In IFT annual meeting book of abstracts. Institute of Food Technologists New Orleans, La.

Damar, S., & Balaban, M. O. (2006). Review of dense phase CO_2 technology: microbial and enzyme inactivation and effects on food quality. *Journal of Food Science*, *71*(1).

Del Pozo-Insfran, D., Balaban, M. O., & Talcott, S. T. (2006). Enhancing the retention of phytochemicals and organoleptic attributes in muscadine grape juice through a combined approach between dense phase CO2 processing and copigmentation. *Journal of Agricultural and Food Chemistry*, *54*(18), 6705–6712.

Demirdöven, A., & Baysal, T. (2014). Optimization of ohmic heating applications for pectin methylesterase inactivation in orange juice. *Journal of Food Science and Technology*, *51*(9), 1817–1826.

di Caracalla, V.D.T. (2000). Subject: Request for comments on the proposed draft codex general standard for fruit juices and nectars. Codex.

Elez-Martinez, P., Suarez-Recio, M., & Martin-Belloso, O. (2007). Modeling the reduction of pectin methyl esterase activity in orange juice by high intensity pulsed electric fields. *Journal of Food Engineering*, *78*(1), 184–193.

Elez-Martínez, P., Aguiló-Aguayo, I., & Martín-Belloso, O. (2006). Inactivation of orange juice peroxidase by high-intensity pulsed electric fields as influenced by process parameters. *Journal of the Science of Food and Agriculture*, *86*(1), 71–81.

Erkmen, O. (2000). Effect of carbon dioxide pressure on *Listeria monocytogenes* in physiological saline and foods. *Food Microbiology*, *17*(6), 589–596.

Erkmen, O. (2001). Note. Antimicrobial effect of pressurized carbon dioxide on *Yersinia enterocolitica* in broth and foods. *Food Science and Technology International*, *7*(3), 245–250.

Falguera, V., Pagán, J., & Ibarz, A. (2011). Effect of UV irradiation on enzymatic activities and physicochemical properties of apple juices from different varieties. *LWT-Food Science and Technology*, *44*(1), 115–119.

Fang, L., Jiang, B., & Zhang, T. (2008). Effect of combined high pressure and thermal treatment on kiwifruit peroxidase. *Food Chemistry*, *109*(4), 802–807.

Food and Drug Administration, U.S. (2001). Hazard analysis and critical point (HACCP); procedures for the safe and sanitary processing and importing of juice; final rule. *Federal Register*, *66*, 6137–6202.

Feng, M., et al. (2013). Effects of ultraviolet-C treatment in Teflon®-coil on microbial populations and physico-chemical characteristics of watermelon juice. *Innovative Food Science & Emerging Technologies*, *19*, 133–139.

Ferrentino, G., & Spilimbergo, S. (2011). High pressure carbon dioxide pasteurization of solid foods: current knowledge and future outlooks. *Trends in Food Science & Technology*, *22*(8), 427–441.

Fundo, J. F., et al. (2018). Quality assessment of Cantaloupe melon juice under ozone processing. *Innovative Food Science & Emerging Technologies*, *47*, 461–466.

Garcia-Gonzalez, L., et al. (2007). High pressure carbon dioxide inactivation of microorganisms in foods: the past, the present and the future. *International Journal of Food Microbiology*, *117*(1), 1–28.

Giner, J., et al. (2000). Inhibition of tomato (*Licopersicon esculentum* Mill.) pectin methylesterase by pulsed electric fields. *Innovative Food Science & Emerging Technologies*, *1*(1), 57–67.

Gómez-López, V., et al. (2010). Microbiological and sensory quality of sonicated calcium-added orange juice. *LWT-Food Science and Technology*, *43*(5), 808–813.

Góngora-Nieto, M., et al. (2002). Food processing by pulsed electric fields: treatment delivery, inactivation level and regulatory aspects. *LWT-Food Science and Technology*, *35*(5), 375–388.

Goodner, J., Braddock, R., & Parish, M. (1998). Inactivation of pectinesterase in orange and grapefruit juices by high pressure. *Journal of Agricultural and Food Chemistry*, *46*(5), 1997–2000.

Gunes, G., Blum, L. K., & Hotchkiss, J. H. (2005). Inactivation of yeasts in grape juice using a continuous dense phase carbon dioxide processing system. *Journal of the Science of Food and Agriculture*, *85*(14), 2362–2368.

Guo, M., et al. (2014). Evaluation of microbial stability, bioactive compounds, physicochemical properties and consumer acceptance of pomegranate juice processed in a commercial scale pulsed electric field system. *Food and Bioprocess Technology*, *7*(7), 2112–2120.

Ho, K. (2003). Dense phase carbon dioxide processing for juice. In IFT annual meeting book of abstracts. Chicago Ill. Inst. of Food Technologists.

Houška, M., et al. (2006). High pressure and foods—fruit/vegetable juices. *Journal of Food Engineering*, *77*(3), 386–398.

Hsu, K. -C. (2008). Evaluation of processing qualities of tomato juice induced by thermal and pressure processing. *LWT-Food Science and Technology*, *41*(3), 450–459.

Ishikawa, H., et al. (1995). Sterilization of microorganisms by the supercritical carbon dioxide micro-bubble method. *Bioscience, Biotechnology and Biochemistry*, *59*(10), 1949–1950.

Jabbar, S., et al. (2014). Influence of sonication and high hydrostatic pressure on the quality of carrot juice. *International Journal of Food Science & Technology*, *49*(11), 2449–2457.

Jaramillo Sanchez, G. M., et al. (2018). Inactivation kinetics of peroxidase and polyphenol oxidase in peach juice treated with gaseous ozone. *International Journal of Food Science & Technology, 53*(2), 347–355.

Jaramillo-Sánchez, G. M., et al. (2018). Ozone processing of peach juice: Impact on physicochemical parameters, color and viscosity. *Ozone: Science & Engineering, 40*(4), 305–312.

Jeyamkondan, S., Jayas, D., & Holley, R. (1999). Pulsed electric field processing of foods: a review. *Journal of Food Protection, 62*(9), 1088–1096.

Jia, M., Zhang, Q., & Min, D. (1996). Effect of pulsed electric field processing on orange juice flavor analyzed by dynamic headspace gas chromatography. In Annual IFT meeting. Louisiana, New Orleans.

Jordan, S., et al. (2001). Inactivation and injury of pressure-resistant strains of *Escherichia coli* O157 and *Listeria monocytogenes* in fruit juices. *Journal of Applied Microbiology, 91*(3), 463–469.

Kadam, P., et al. (2012). Review on the high pressure technology (HPT) for food preservation. *Journal of Food Processing & Technology, 3*(1), 135.

Kaur, N., & Singh, A. (2016). Ohmic heating: concept and applications—A review. *Critical Reviews in Food Science and Nutrition, 56*(14), 2338–2351.

Kaya, Z., Yıldız, S., & Ünlütürk, S. (2015). Effect of UV-C irradiation and heat treatment on the shelf life stability of a lemon–melon juice blend: multivariate statistical approach. *Innovative Food Science & Emerging Technologies, 29*, 230–239.

Keenan, D. F., et al. (2012). Effect of high hydrostatic pressure and thermal processing on the nutritional quality and enzyme activity of fruit smoothies. *LWT-Food Science and Technology, 45*(1), 50–57.

Keyser, M., et al. (2008). Ultraviolet radiation as a non-thermal treatment for the inactivation of microorganisms in fruit juice. *Innovative Food Science & Emerging Technologies, 9*(3), 348–354.

Kim, J. -G., Yousef, A. E., & Dave, S. (1999). Application of ozone for enhancing the microbiological safety and quality of foods: a review. *Journal of Food Protection, 62*(9), 1071–1087.

Kimball, D. A. (1999). In *Citrus processing. A complete guide* (2nd ed., pp. 257–264). Gaithersburg, Maryland: Aspen Publishers, Inc.

Kincal, D., et al. (2006). A continuous high-pressure carbon dioxide system for cloud and quality retention in orange juice. *Journal of Food Science, 71*(6.).

Knorr, D., et al. (2004). Applications and potential of ultrasonics in food processing. *Trends in Food Science & Technology, 15*(5), 261–266.

Koutchma, T. (2008). UV light for processing foods. *Ozone: Science and Engineering, 30*(1), 93–98.

Koutchma, T., et al. (2016). Effects of ultraviolet light and high-pressure processing on quality and health-related constituents of fresh juice products. *Comprehensive Reviews in Food Science and Food Safety, 15*(5), 844–867.

Kulshrestha, S. A., & Sastry, S. K. (2006). Low-frequency dielectric changes in cellular food material from ohmic heating: effect of end point temperature. *Innovative Food Science & Emerging Technologies, 7*(4), 257–262.

La Cava, E. L., & Sgroppo, S. C. (2015). Evolution during refrigerated storage of bioactive compounds and quality characteristics of grapefruit [*Citrus paradisi* (Macf)] juice treated with UV-C light. *LWT-Food Science and Technology, 63*(2), 1325–1333.

Lacroix, N., Fliss, I., & Makhlouf, J. (2005). Inactivation of pectin methylesterase and stabilization of opalescence in orange juice by dynamic high pressure. *Food Research International, 38*(5), 569–576.

Lee, S. K., & Kader, A. A. (2000). Preharvest and postharvest factors influencing vitamin C content of horticultural crops. *Postharvest Biology and Technology*, *20*(3), 207–220.

Linton, M., McClements, J., & Patterson, M. (1999). Inactivation of *Escherichia coli* O157: H7 in orange juice using a combination of high pressure and mild heat. *Journal of Food Protection*, *62*(3), 277–279.

Liu, Y., et al. (2013). Effect of high hydrostatic pressure on overall quality parameters of watermelon juice. *Food Science and Technology International*, *19*(3), 197–207.

luan Chen, J., et al. (2010). Changes in microorganism, enzyme, aroma of hami melon (*Cucumis melo* L) juice treated with dense phase carbon dioxide and stored at 4°C. *Innovative Food Science & Emerging Technologies*, *11*(4), 623–629.

Majid, I., Nayik, G. A., & Nanda, V. (2015). Ultrasonication and food technology: A review. *Cogent Food & Agriculture*, *1*(1), 1071022.

Mansor, A., et al. (2014). Efficacy of ultraviolet radiation as non-thermal treatment for the inactivation of *Salmonella typhimurium* TISTR 292 in pineapple fruit juice. *Agriculture and Agricultural Science Procedia*, 2, 173–180.

Mastwijk, H., & Bartels, P. (2005). Pulsed electric Field (FEF) processing in the fruit juice and dairy industry. *The International Review of Food Science and Technology*, *2004*, 106–108.

McDonald, C., et al. (2000). Effects of pulsed electric fields on microorganisms in orange juice using electric field strengths of 30 and 50 kV/cm. *Journal of Food Science*, *65*(6), 984–989.

McGlynn, W. G. (2003). In *The importance of food pH in commercial canning operations*. Oklahoma Cooperative Extension Service, Division of Agricultural Sciences and Natural Resources, Oklahoma State University.

Meneses, N., Jaeger, H., & Knorr, D. (2011). pH-changes during pulsed electric field treatments—Numerical simulation and in situ impact on polyphenoloxidase inactivation. *Innovative Food Science & Emerging Technologies*, *12*(4), 499–504.

Min, S., Jin, Z. T., & Zhang, Q. H. (2003). Commercial scale pulsed electric field processing of tomato juice. *Journal of Agricultural and Food Chemistry*, *51*(11), 3338–3344.

Mohd-Hanif, H., Shamsudin, R., & Adzahan, N. M. (2016). UVC dosage effects on the physicochemical properties of lime (*Citrus aurantifolia*) juice. *Food Science and Biotechnology*, *25*(1), 63–67.

Mohideen, F. W., et al. (2015). Effect of continuous ultra-sonication on microbial counts and physico-chemical properties of blueberry (*Vaccinium corymbosum*) juice. *LWT-Food Science and Technology*, *60*(1), 563–570.

Morris, C., Brody, A. L., & Wicker, L. (2007). Non-thermal food processing/preservation technologies: a review with packaging implications. *Packaging Technology and Science: An International Journal*, *20*(4), 275–286.

Ngadi, M., Smith, J. P., & Cayouette, B. (2003). Kinetics of ultraviolet light inactivation of *Escherichia coli* O157: H7 in liquid foods. *Journal of the Science of Food and Agriculture*, *83*(15), 1551–1555.

Nienaber, U., & Shellhammer, T. (2001). High-pressure processing of orange juice: Kinetics of pectinmethylesterase inactivation. *Journal of Food Science*, *66*(2), 328–331.

O'donnell, C., et al. (2010). Effect of ultrasonic processing on food enzymes of industrial importance. *Trends in Food Science & Technology*, *21*(7), 358–367.

Ochoa-Velasco, C. E., & Guerrero-Beltrán, J. Á. (2012). Ultraviolet-C light effect on pitaya (*Stenocereus griseus*) juice. *Journal of Food Research*, *1*(2), 60.

Oms-Oliu, G., et al. (2009). Effects of high-intensity pulsed electric field processing conditions on lycopene, vitamin C and antioxidant capacity of watermelon juice. *Food Chemistry*, *115*(4), 1312–1319.

Pala, Ç. U., & Toklucu, A. K. (2013). Microbial, physicochemical and sensory properties of UV-C processed orange juice and its microbial stability during refrigerated storage. *LWT-Food Science and Technology, 50*(2), 426–431.

Park, S. J., Lee, J. I., & Park, J. (2002). Effects of a combined process of high-pressure carbon dioxide and high hydrostatic pressure on the quality of carrot juice. *Journal of Food Science, 67*(5), 1827–1834.

Patil, S. (2010). Efficacy of ozone and ultrasound for microbial reduction in fruit juice. Doctoral Thesis. Technological University Dublin.

Patil, S., et al. (2009). Inactivation of *Escherichia coli* in orange juice using ozone. *Innovative Food Science & Emerging Technologies, 10*(4), 551–557.

Piyasena, P., Mohareb, E., & McKellar, R. (2003). Inactivation of microbes using ultrasound: a review. *International Journal of Food Microbiology, 87*(3), 207–216.

Polydera, A., et al. (2004). Inactivation kinetics of pectin methylesterase of Greek Navel orange juice as a function of high hydrostatic pressure and temperature process conditions. *Journal of Food Engineering, 62*(3), 291–298.

Quitão-Teixeira, L. J., et al. (2008). Inactivation of oxidative enzymes by high-intensity pulsed electric field for retention of color in carrot juice. *Food and Bioprocess Technology, 1*(4), 364.

Rampersaud, G. C., & Valim, M. F. (2017). 100% citrus juice: Nutritional contribution, dietary benefits and association with anthropometric measures. *Critical Reviews in Food Science and Nutrition, 57*(1), 129–140.

Rao, L., et al. (2014). Enzyme activity and nutritional quality of peach (*Prunus persica*) juice: effect of high hydrostatic pressure. *International Journal of Food Properties, 17*(6), 1406–1417.

Rastogi, N., et al. (2007). Opportunities and challenges in high pressure processing of foods. *Critical Reviews in Food Science and Nutrition, 47*(1), 69–112.

Rawson, A., et al. (2011). Effect of thermal and non thermal processing technologies on the bioactive content of exotic fruits and their products: Review of recent advances. *Food Research International, 44*(7), 1875–1887.

Riahi, E., & Ramaswamy, H. S. (2004). High pressure inactivation kinetics of amylase in apple juice. *Journal of Food Engineering, 64*(2), 151–160.

Rivas, A., et al. (2006). Effect of PEF and heat pasteurization on the physical–chemical characteristics of blended orange and carrot juice. *LWT-Food Science and Technology, 39*(10), 1163–1170.

Rodrigo, D., et al. (2003). Pectin methyl esterase and natural microflora of fresh mixed orange and carrot juice treated with pulsed electric fields. *Journal of Food Protection, 66*(12), 2336–2342.

Roobab, U., et al. (2018). The impact of nonthermal technologies on the microbiological quality of juices: a review. *Comprehensive Reviews in Food Science and Food Safety, 17*(2), 437–457.

Saldana, G., et al. (2011). Defining treatment conditions for pulsed electric field pasteurization of apple juice. *International Journal of Food Microbiology, 151*(1), 29–35.

Salinas-Roca, B., et al. (2017). Quality changes in mango juice treated by high-intensity pulsed electric fields throughout the storage. *Food and Bioprocess Technology, 10*(11), 1970–1983.

Samaranayake, C. P., & Sastry, S. K. (2016). Effects of controlled-frequency moderate electric fields on pectin methylesterase and polygalacturonase activities in tomato homogenate. *Food Chemistry, 199*, 265–272.

Sánchez-Moreno, C., et al. (2005). Impact of high pressure and pulsed electric fields on bioactive compounds and antioxidant activity of orange juice in comparison with traditional thermal processing. *Journal of Agricultural and Food Chemistry, 53*(11), 4403–4409.

Santhirasegaram, V., et al. (2015). Comparison of UV-C treatment and thermal pasteurization on quality of Chokanan mango (*Mangifera indica* L) juice. *Food and Bioproducts Processing, 94*, 313–321.

Shah, N. N. A. K., et al. (2016). Fruit juice production using ultraviolet pasteurization: A review. *Beverages*, *2*(3.).

Shamsudin, R., et al. (2014). Effect of repetitive ultraviolet irradiation on the physico-chemical properties and microbial stability of pineapple juice. *Innovative Food Science & Emerging Technologies*, *23*, 114–120.

Stanley, D. W., & Parkin, K. L. (1991). Biological membrane deterioration and associated quality losses in food tissues. *Critical Reviews in Food Science & Nutrition*, *30*(5), 487–553.

Sukarminah, E., et al. (2017). Ozonization technology and its effects on the characteristics and shelf-life of some fresh foods: A review. *KnE Life Sciences*, *2*(6), 459–470.

Terefe, N. S., et al. (2009). The kinetics of inactivation of pectin methylesterase and polygalacturonase in tomato juice by thermosonication. *Food Chemistry*, *117*(1), 20–27.

Tiwari, B., et al. (2008). Effects of sonication on the kinetics of orange juice quality parameters. *Journal of Agricultural and Food Chemistry*, *56*(7), 2423–2428.

Tiwari, B., et al. (2009). Inactivation kinetics of pectin methylesterase and cloud retention in sonicated orange juice. *Innovative Food Science & Emerging Technologies*, *10*(2), 166–171.

Torlak, E. (2014). Efficacy of ozone against *Alicyclobacillus acidoterrestris* spores in apple juice. *International Journal of Food Microbiology*, *172*, 1–4.

Torres, B., et al. (2011). Effect of ozone processing on the colour, rheological properties and phenolic content of apple juice. *Food Chemistry*, *124*(3), 721–726.

Torres, E., et al. (2015). Effects of high hydrostatic pressure and temperature increase on *Escherichia coli* spp. and pectin methyl esterase inactivation in orange juice. *Food Science and Technology International*, *22*(2), 173–180.

Tran, M. T. T., & Farid, M. (2004). Ultraviolet treatment of orange juice. *Innovative Food Science & Emerging Technologies*, *5*(4), 495–502.

Truong, T., et al. (2002). Effects of carbon dioxide in high-pressure processing on pectinmethylesterase in single-strength orange juice. *Journal of Food Science*, *67*(8), 3058–3062.

Turtoi, M. (2014). Inactivation of *Saccharomyces cerevisiae* using new non-thermal technologies. A review. *Romanian Biotechnological Letters*, *19*(1), 8901.

Unluturk, S., & Atilgan, M. R. (2015). Microbial safety and shelf life of UV-C treated freshly squeezed white grape juice. *Journal of Food Science*, *80*(8), M1831–M1841.

Varela-Santos, E., et al. (2012). Effect of high hydrostatic pressure (HHP) processing on physicochemical properties, bioactive compounds and shelf-life of pomegranate juice. *Innovative Food Science & Emerging Technologies*, *13*, 13–22.

Welti-Chanes, J., Ochoa-Velasco, C., & Guerrero-Beltrán, J. (2009). High-pressure homogenization of orange juice to inactivate pectinmethylesterase. *Innovative Food Science & Emerging Technologies*, *10*(4), 457–462.

Williams, R. C., Sumner, S. S., & Golden, D. A. (2004). Survival of *Escherichia coli* O157: H7 and *Salmonella* in apple cider and orange juice as affected by ozone and treatment temperature. *Journal of Food Protection*, *67*(11), 2381–2386.

Yagiz, Y., Lim, S., & Balaban, M. (2005). Continuous high pressure CO_2 processing of mandarin juice. In *Annual meeting book of abstracts*USA: Institute of Food Technologists Chicago, IL.

Yen, G. C., & Lin, H. T. (1998). Effects of high pressure and heat treatment on pectic substances and related characteristics in guava juice. *Journal of Food Science*, *63*(4), 684–687.

Yeom, H. W., et al. (2000). Effects of pulsed electric fields on the quality of orange juice and comparison with heat pasteurization. *Journal of Agricultural and Food Chemistry*, *48*(10), 4597–4605.

Yeom, H., Zhang, Q., & Chism, G. (2002). Inactivation of pectin methyl esterase in orange juice by pulsed electric fields. *Journal of Food Science*, *67*(6), 2154–2159.

Zhang, Q., et al. (1996). Batch mode food treatment using pulsed electric fields. Google Patents.

Zhang, J., et al. (2006). Sterilization using high-pressure carbon dioxide. *The Journal of Supercritical Fluids*, *38*(3), 354–372.

Zhang, Y., et al. (2010). Pulsed electric field processing effects on physicochemical properties, flavor compounds and microorganisms of longan juice. *Journal of Food Processing and Preservation*, *34*(6), 1121–1138.

Zhang, C., et al. (2011). Comparison of thermal, ultraviolet-c and high pressure treatments on quality parameters of watermelon juice. *Food Chemistry*, *126*(1), 254–260.

Zhou, Z. (2012). *Citrus fruits nutrition*. Beijing, China: Science Press.

Zhou, L., et al. (2009). Alterations in the activity and structure of pectin methylesterase treated by high pressure carbon dioxide. *Journal of Agricultural and Food Chemistry*, *57*(5), 1890–1895.

Zinoviadou, K. G., et al. (2015). Fruit juice sonication: Implications on food safety and physicochemical and nutritional properties. *Food Research International*, *77*, 743–752.

Zou, Z., et al. (2016). Antioxidant activity of citrus fruits. *Food Chemistry*, *196*, 885–896.

Chapter 3

High-Pressure Processing; Principle, Applications, Impact, and Future Prospective

Adnan Khaliq[a], Muhammad Farhan Jahangir Chughtai[a], Tariq Mehmood[a], Samreen Ahsan[a], Atif Liaqat[a], Muhammad Nadeem[b,*], Nimra Sameed[a], Kanza Saeed[a], Junaid Ur Rehman[a] and Ayesha Ali[a]

[a]*Department of Food Science and Technology, Khwaja Freed University of Engineering and Technology, Rahim Yar Khan, Pakistan;* [b]*Department of Environmental Sciences, COMSATS University Islamabad, Vehari, Pakistan*

**Corresponding author*

Section 1

Outline

1 Introduction

High-pressure processing (HPP) is defined as a non-thermal food preservation approach, that utilizes pressure instead of heat to induce preservation effect similar to pasteurization that inactivates harmful pathogens as well as vegetative food spoilage agents. HPP employs immensely high pressure of about 400–600 MPa at temperature range varying from 4°C to <45°C. Thus, preserving food quality parameters, that is, nutritional value, texture, appearance, and taste. This technique is equally effective for the preservation of high moisture liquid and solid foods. Although fatal for microorganisms, this treatment does not harm covalent bonds. As a result of which food chemistry and quality preserved; minimizing the use of chemical preservatives and thermal treatments in food processing (Table 3.1).

Sustainable food processing and engineering challenges. http://dx.doi.org/10.1016/B978-0-12-822714-5.00003-6

TABLE 3.1 Globally available HPP processed food products (Leadley et al., 1997).

Food products	Manufacturer	Product manufacturer country
Jams, fruit sauces, yogurt, and jelly	Meida-Ya Japan	Japan
Beef	Fuji Ciku Mutterham	Japan
Guacamole, salsa dips, ready meals, and fruit juices	Avomex	United States of America
Hummus	Hannah International	United States of America
Fruit and vegetable juices	Odwalla	United States of America
Processed poultry products	Purdue Farms	United States of America
Oysters	Motivatit Seafoods	United States of America
Sliced dry-cured meat and cooked meat	Santa Maria Foods	Canada
Fruit juices	Pampryl	France
Apple juice	Frubaca	Portugal
Sliced ham and tapas	Espuna	Spain
Fruit juices and smoothies	Orchard House	United Kingdom
Meat and cheese snack products	Deli 24	United Kingdom
Fermented sausages	Chris Tim	Romania
Deli meats infused with extra virgin olive oil	Creta Farm	Greece
Carpaccio	Zwaneberg	Netherlands

1.1 HPP equipment

HPP systems are available in both batch and semi-continuous equipment lines. High strength steel alloy is the most common material used in equipment manufacturing because of corrosion resistance and high fracture toughness. Components of batch design equipment comprises of following parts (Ting, 2011):

- pressure vessel which is a thick-wall cylinder
- two end closures to cover the cylindrical pressure vessel
- yoke (structure for restraining end closures while under pressure)
- high-pressure pump and intensifier for generating target pressures
- process control and instrumentation
- a handling system for loading and removing the product

For solid and liquid foods, batch processing technique is practiced while for pumpable foods semi-continuous process is preferred. In operations HPP batch process is analogous to retort systems in thermal processing. For HPP food is vacuum packed and then loaded in a basket which is then placed in pressure vessel carrying pressure-transmitting fluid. An end closure is used in the pressure vessel to contain the fluid. Then the yoke structure glides across the closed vessel to retain pressure on both top and bottom closure. A pump and intensifier together generate a compression effect on the pressure-transmitting fluid to attain desired pressure. During the entire process, pressure is maintained for the desired time, and released after process completion. The product is unloaded from the vessel after treatment. For developing a commercially viable product usually a pressure holding time of less than 10 min is enough.

To attain better results, a modified HP treatment is designed in which product is subjected to alternate compression and decompression cycles at constant pressure, and time is used, this treatment is called pulsed high-pressure processing (PHPP). Various studies have provided clear evidences that PHPP at the same working conditions (time and pressure) is more effective against microorganisms. The only disadvantage of this technique is that the rapid compression and decompression cycles induce an enormous stress on the vessel walls, thus decreasing the life of the equipment.

In contrary to that in a semi-continuous system used for processing of pumpable foods instead of a single pressure vessel, usually 2–3 vessels are used depend on the production capacity. In this design vessels with free-floating compression piston, are inter-connected so that when the product discharges from the first vessel, a pressure is generated in second, while the food sample is loaded in the third vessel to maintain a continuous product output (Ting, 2011).

1.1.1 Pressure-transmitting fluid

The key component of the design is the fluid that transmits the pressure uniformly throughout the product, on industrial-scale water is the most commonly used pressure transmitting fluid. For laboratory scale and pilot plant assemblies, castor oil, glycol, aqueous glycol, sodium benzoate solution, and silicon solution are used as pressure-transmitting fluids, to protect low-cost steel equipment against corrosion and frictional damages. While selecting the pressure-transmitting fluid number of factors are kept in mind like pressure endurance of seals, anti-corrosion properties of material, heat compression, and change in fluid viscosity under pressure. In addition to it, fluid to sample ratio, the composition of the fluid, its thermal characteristics play the key role in determining food's thermal behavior under pressure and compression heating properties of pressure-transmitting fluid inactivate kinetics of microorganisms and their spores which is directly dependent on the heat of compression and compressibility values (Balasubramanian & Balasubramaniam, 2003; de Heij et al., 2003).

1.2 HPP principles

Under high pressure, two major principles control the food behavior. Le Chatelier's Principle according to which phenomena like chemical reactions, interactions that alter the molecular configuration or phase transition that reduce volume are escalated by pressure. On the contrary, the Isostatic Principle explains that phenomena in which boast volume are decelerated by pressure (Karim, 2011). Pressure is promptly but steadily transmitted throughout the food irrespective of the shape and size of the sample; prime edge in comparison to thermal treatment. In the procedure, consistent pressure is applied which uniformly radiates in all directions of the samples; due to even impact distribution in the sample, pressure does not affect the molecular bonding and configuration. As a result, the product retains its original shape once the pressure is released; isostatic pressure, key working principle of HPP (Howson, 2009) (Fig. 3.1).

HPP is a post packaging process, that is, is done after the final packaging of a product, so food commodity does not come in direct human contact after packaging. After the pressure application procedure, package is transferred to finals baskets, which are then loaded in a chamber. Baskets are then pushed into a water-filled vessel. Once the water surrounds the package is pumped up with a pressure 600 MPa (87,000 psi) that is 5–6 times deeper from the deepest ocean (Howson, 2009; Sharma, 2011). Instantaneous and uniform pressure transmission does not depend on product geometry (Ramirez-Suarez & Morrissey, 2006). This practice inactivates the enzymes, destroys microorganisms by altering the temperatures, thus, enhancing product shelf life without affecting organoleptic and nutritive food properties (Daher et al., 2017). Per 100 MPa increase in compression elevates temperature uniformly by 30°C of the food sample, rise in temperature inactivates the microorganisms and enhance the shelf life of food material (Kadam et al., 2012). The schematic explanation of pressure in HPP is shown in the following Fig. 3.2.

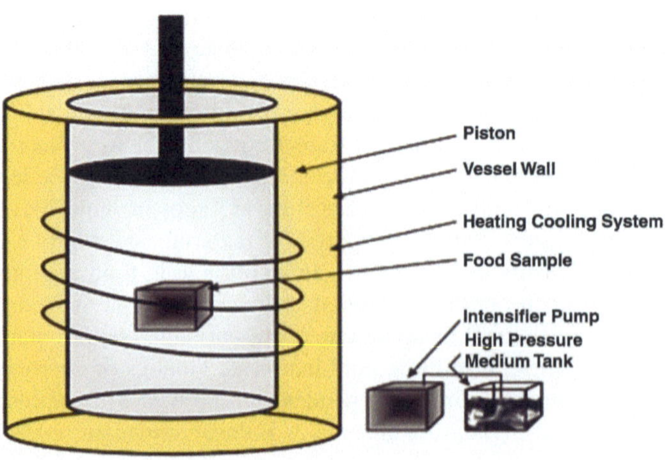

FIGURE 3.1 **High pressure processing equipment.**

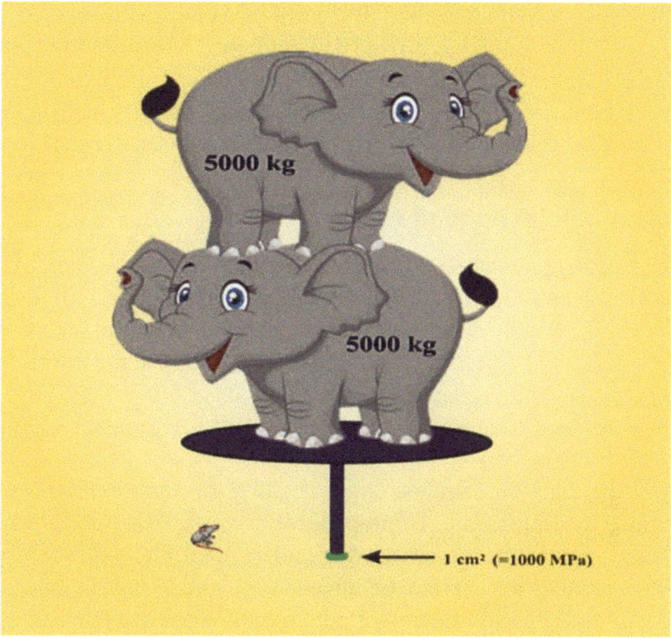

FIGURE 3.2 A 10,000 kg weight on 1 cm^2 equals a pressure of 1000 MPa.

1.3 Impact of HPP on food quality

HPP inactivates micro-organisms, while preserving nutritional and other organoleptic food characteristics like color, appearance, and texture. These are the properties on the bases of which most of the consumers make their food choices. These colors and texture maintain properties of HPP, which is the most beneficial outputs in the fruit and vegetable processing industry (Guerrero-Beltan et al., 2005; Rastogi et al., 2007; Norton and Sun, 2008).

Researchers are focusing their research on the effect of high pressure on fruit enzymes, that is, pectin methylesterase (PME) and polygalacturonase (PG) through instrumental means (Norton and Sun, 2008; Perera et al., 2010). In meat processing procedures elevated pressures alter muscle protein structure, thus affecting its functional properties and texture. Similar research evidence is available for seafood where $\geq$450 MPa pressure result in gummy, chewy, and hard meat (Yagiz, Kristinsson, Balaban, & Marshall, 2007).

According to another research, HPP processing reduce redness in muscle foods (Andres et al., 2006; Cava et al., 2009) in case of fish, elevated color lightness is associated with cooked appearance (Ramirez-Suarez & Morrissey, 2006; Yagiz, Kristinsson, Balaban, & Marshall, 2007). A similar effect of lightness after HP processing has been observed in cow's milk. These color changes are directly linked to structural alterations in muscle protein after HPP. In case of HPP processed fruit, these color changes are negligible, despite the effect of high pressure on plant pigments (Patras et al., 2009a,b; Perera et al., 2010).

Previously, it was thought that HPP have any considerable effect on food flavors. But recent research has proved that combined effect of HPP and enzyme immobilization helps in de-bittering grapefruit juice. A pressure of 160 MPa, while maintaining the temperatures at 37°C for 20 min reduces the naringin content of fruit juice and thus reducing bitterness (Ferreira et al., 2008).

1.4 Impact of HPP on food safety

Commonly effectiveness of pressure treatment on microbial growth relies on the magnitude of the combined thermal-pressure usage. Different microorganisms resist pressure treatment differently. Mild treatment controls (400–600 MPa) are efficacious toward inactivation of spoilage and pathogenic microorganisms (yeasts, viruses, vegetative bacterial cells, and molds). For inactivation of bacterial spores combined treatment controls (400–600 MPa pressure, and >70°C heat) were more effective.

In vegetative bacterial strains, Gram-negative bacteria exhibit lower pressure resistance as compared to Gram-positive bacteria. Among bacterial strains crucial variations can be observed. Vegetative microbial cells exhibit lower pressure resistance in the exponential phase compared to the stationary phase.

In thermal and pressure treatment controls different mechanisms of inactivation are involved, and within them, no significant correlation has appeared. As with ancient thermal applications, the rate of microbial cell reduction in the course of pressure application is also affected by food composition along with its water activity and pH. So, food processors investigating pressure-thermal applications should indulge their work with high pressure controlling experts for conducting experimental trials.

Alone pressure with ambient temperature treatment showed no effective spore destruction. Generally, the pressure effect initiates the spore's germination process through triggering nutrient germinant receptors (Farkas & Hoover, 2000). Mild pressure (100–400 MPa) and temperature (20°C and 50°C) treatment can initiate spore germination (Black et al., 2007). For spore germination, different rates of pressure may require. Further, less-lethal treatment controls may require once the spores get germinate into resultant vegetative cells. So, it can be easily inactivated with combined mild pressure-heat treatment.

Thus, it is problematic to ensure 100% pressure trigging spore germination during the application and difficult to assure the shelf life of the product. So, initial attempts for applying mild pressure (a process Tyndallization, 200–400 MPa with standard temperature) application for spore germination with proceeding thermal pasteurization cannot be commercially favorable for low acid foods sterilization process (Daryaei et al., 2013).

When the magnitude of application is above a certain level (pressure 400–600 MPa and temperature 90°C–120°C), the process is inferred as PATP (pressure-assisted thermal sterilization). Studies until now disclosed that PATP treatment targeted microbes are Clostridium botulinum (Stewart et al., 2015; Reddy et al., 2013; Skinner et al., 2014). Bacillus sterothermophilus (surrogate microbes) are more sensitive toward the combined effect of thermal-pressure processing, in comparison to heat processing. According to another research, bacterial non-pathogenic strains like Bacillus amyloliquefaciens and Clostridium sporogenes spores manifests high heat-pressure resistance. Spores from such strains are prospective surrogate candidates for the evaluating pressure assisted thermal processing (PATP) sterility. For validation of the PATP process, more suitable surrogate microbes need to be identified (Wan, 2014).

Unlike the regular linear profile of spore inactivation on a semi-log behavior plot, during heat processing a unique tailing for inactivation curves of microorganisms has been observed during thermal-pressure combined treatment (Rajan et al., 2006). Therefore, the use of an irregular kinetic model to explain the mutual effect of thermal-pressure microbial killing has been suggested to be more pertinent. According to different studies some empirical non-linear models reported to have the Weibull distribution, biphasic model, the log-logistic, and nth order kinetics (Ahn et al., 2007). Nonlinear survivor curves narrated earlier is thought to be due to the synergistic effect of high pressure and heat on spores and vegetative cells, thus resulting in shoulders and tails. Another study proposed an integrated process lethality model that considered lethal effects of both pressure and temperature on spore killing (Nguyen et al., 2013).

1.5 Applications of HPP

HPP is being practiced as a food preservation technique from centuries. For years this technology is implemented in the processing of plastics, carbon graphite, ceramic, and composite material-based products. The application of thermal processing technology has emerged recently, but it is gaining immediate popularity because this technology gives superior quality products than conventional thermal processing (Ramaswamy et al., 2008). As no external heating is applied in this processing method, it preserves nutritional as well as sensory product attributes, thus improving product quality and yield in comparison to conventional methods (Adzahan and Benchamaporn, 2007).

HPP shortens the processing time, minimizes physical and chemical changes, retain color, flavor, appearance, texture, and freshness, eradicates vitamin C losses, reduce damages to ice crystals, edges off functional alternations in comparison to conventional thermal processing procedures (Avure Technologies Inc, 2007). This technique is also applied for non-thermal preservation of delicate seafood (raw or fresh oysters) by tapering off bacterial loads without causing notable effects on its texture, color, and nutritional profiles (Flick, 2003) (Table 3.2).

TABLE 3.2 Quality attributes of some high-pressure processing (HPP) treated food items.

Food group	Processing conditions MPa/min/°C	Impact on food items as compared to controlled experiment
Orange milk juice beverage	• 400°C • 9 min • 26.6	Ascorbic acid contents and color was retained
Apple puree	• 400 • 5 min • 20°C	Phenolics, vitamin C and ascorbic acid contents was retained
Cheese	• 400 • 20 min • 20°C	No color changes, pH, and yield were increased. Microbiological content was reduced
Chicken	• 400 • 15 min • 5°C	Minor changes in aroma and resistant to lipid oxidation when re-exposed to aerobic conditions
Meat of Alligator	• 200 • 10 min • 20°C	No changes in flavor, low impact on the texture and color, slightly increment in tenderness
Beef	• 200 • 20 min • 20°C	Meat hygienically improved, fatty acids profile retained with minimum effect the quality parameters
Coho salman	• 135 • 0.5 min • 150°C	Color, odor, and textural properties were maintained, quality was improved in terms of gaping, putrid, and cohesivity

1.6 Conclusions and future aspects

HPP offers fresh, mildly processed, preservative-free food commodities and value-added products. Improve product keeping quality by inactivating both food spoiling and pathogenic bacteria. Constant and periodic compression and decompression cycles, reduce the thermal exposure thus maintaining product quality attributes. Only equipment cost is limiting the commercial usage of this advanced processing technique.

This technique has its application for the development of electronic sensors and chemical and biological process indicators used in pressure-thermal process validation. This technique provides an inventive approach for developing nanocomposite based high pressure and thermal resistant packaging materials, establish process uniformity, reduce operation cost while elevating process outputs remarkably. This technique helps in defining clear lines between benefits and limitations of procedures like high-pressure homogenization and pressure-ohmic-thermal sterilization (pressure-based pasteurization) on product quality and safety.

Section 2

Outline

2 High-pressure processing of fruits and vegetables

2.1 Introduction

The key technique for food pasteurization and sterilization is thermal processing. However, heat use diminishes the quality of food. An effective alternative of thermal processing is; high-pressure processing (HPP), which uses high pressures, may or may not encompasses the use of heat, to inactivate microbes or to enhance food desirable attributes. Breakage of covalent bonds does not happen in HPP, so, the freshness and quality of food remain maintained up to the standards, having increased shelf life of food microbiological attributes. This process is recognized as HHP (high hydrostatic pressure) processing and UHP (ultra-high pressure) (Balasubramaniam and Farkas, 2008).

In the current era, consumers show their interest in processed fruit and its products have high quality with some fresh flavor, safe in terms of microbiologically and zero level of additives. (Patterson 2005; Perera et al., 2010). The only conventional and popular method that has been utilized to increase the shelf life of such products is a thermal treatment, and it remarkably reduces microbes in food and enzymatic activity gets vanished. This method has its failure, as it imparts detrimental effects on nutritional characteristics. By taking into account food preservation, many technologies have been introduced to meet consumer demand and to overcome limitations of the use of thermal processing. HHP is a kind of non-thermal technique, which permits the production of food products like fruit juices of standard quality (Oey et al., 2008). Without damaging the nutritional characteristics of food, HPP was successful incapacity to kill or inactive different types of fungal and bacterial microorganisms (Oey et al., 2008; Li et al., 2010).

Due to this reason, food companies and researchers are intended to invest time and capital on this process, HPP, considering it as a very effective alternative of the thermal process (Patterson, 2005; Trejo-Araya et al., 2009). As a result of the particular interest of companies, many products can be seen available in the global markets that are high pressure processed products like jellies and juices (Hogan et al., 2005).

2.2 Impact of HPP on enzymes and microorganisms

Stability of enzymes, like other proteins, relies on a certain range of temperature and pressure. Three-dimensional protein structure disrupts by crossing these limits, resulting in enzyme denaturation and hence, enzyme inactivation (Sila et al., 2008). High-pressure effects are divided into two main classes. In the very first class, 100 MPa low pressure has been proved to activate certain enzymes.

But this simulating effect has only worked for monomeric enzymes. Enzymes inactivation occurs by applying much high pressure (Hendrickx et al., 1998). Following are the enzymes that are responsible for the deterioration in texture and sensory attributes of vegetables and fruits; lipoxygenase (LOX), PG, peroxidase (POD), pectate lyase (PL), polyphenol oxidase (PPO), and PME (Sila et al., 2008).

Demethoxylation of pectin is done by PME that results in demethoxylated pectin, which can make cross-link with calcium for the formation of supermolecular assemblies and behave as substrate for PG depolymerization. PME causes cloud destabilization of juices and products consistency loses. While, during storage, peroxidase induces negative changes in flavor in vegetables (Hendrickx et al., 1998).

Mutual effects of high pressure and thermal processing are observed as effective tools against enzyme inactivation. Enzyme system tagged as model, *Bacillus subtilis* α-amylase, has been studied for kinetic portrayal of temperature and/or pressure for enzyme inactivation by (Ludikhuyze et al., 1997). Effects of enzyme inactivation by using HPP (100–600 MPa) and heat treatment (0°C–60°C) on peroxidase and polyphenol oxidase (PPO) enzymes, for development of red grapes juice through the high-pressure process showing stable shelf life has been experimented by Rastogi et al. (1999).

Results of studies depicted that at 60°C, the lowest activities of POD and PPO were observed at 600 MPa pressure to 100 MPa pressure, respectively. Castellari et al. (1997) described that the use of high-pressure processing along with optimum thermal heat (40°C–50°C) is mandatory for full inactivation of PPO in grapes. In carrots that are thermally processes, HPP plays a crucial role with mild temperature for declining the thermos-softening (Sila et al., 2005).

A very effective and promising non-thermal technique that, for sure, inactivate the microorganisms in vegetables is HPP. Pressure range of 100–1000 MPa is used for HPP either with or without heat, helping out most foods to retain desirable attributes like taste or texture. The main priority of this process over thermal sterilizations and some kind of pasteurization is; retention of nutritive characteristics and sensory attributes of treated food products (Balasubramaniam et al., 2008).

High-pressure technology mainly aims at microbial inactivation. As a result of inactivation, the shelf life is increased and it helps in improving the safety of food products. Some researchers suggest that the elimination of Listeria monocytogenes could be attained by using a high-pressure process in cheese and meat products (Szczawiński et al., 1997; Fonberg-Broczek et al., 2005). This high-pressure technique is also important in the inactivation of other microorganisms like E. coli, Vibrio, and Salmonella, which are hazardous for food quality. In this perspective, shelf life and food quality can be improved by using HPP

A big challenge for high-pressure technology is bacterial spores and information about resistance is not so clear. Bacterial or other microbial spores that are suspended in any food or model systems could effectively be inactivated by using HPP but under very extreme conditions as high temperature, high pressure with long exposure time. When temperature and pressure are combined at 80°C and 690 MPa for 20 min, it showed effective results with a reduction in the number of C. sporogenesspores (Crawford et al., 1996).

When pressure is combined with temperature 70°C, better treatment of Bacillus staerothermophilus is resulted (Hayakawa et al., 1994). There is no or less data on the resistance of spores of C. botulinum and on capability to bear the high pressure at high or low temperatures (Yordanov and Angelova, 2011).

2.3 Impact of HPP on physicochemical parameters of fruits and vegetables

2.3.1 Impact of HPP on color

There is a limited impact of HPP treatment on pigments that are casual agents for fruit and vegetable colors. It is noticed that as compared to thermal treatments, HPP causes much less harm to such food compounds having low molecular weight such as pigments (Tauscher, 1995). The color of a diamond could be changed by HPP, so the change in color of food can be occurred by using this technique (Collins et al., 2000). But, during the storage of HPP processed vegetables and fruits, color change occurs due to partial inactivation of microbes and enzymes. This can cause certain chemical reactions in food that may be undesirable.

Many experiments have been done to check the effects of HPP on many vegetables and fruits products to study their color changes. Palou et al. (1999) observed the effect of HPP on the color of banana puree and concluded that the initial color was maintained after treatment. Rodrigo et al. (2007) found that under the mutual effect of heat and high-pressure process (65°C, 300–700 MPa for 60 min) enhanced 8.8% for La/b fix parameters for strawberry juice. Weemaes et al. (1999) described the deteriorating kinetics of broccoli juice, particularly greenness, by using mutual pressure and temperature (850 MPa and 30°C–90°C).

There was no notable reduction in green color at 800 MPa pressure and 30°C–40°C temperature. There are varying stability limits of chlorophyll a and b for temperature and pressure. Both chlorophylls show high-pressure stability at room temperature and stability vanished at higher temperatures than 50°C, causing a decline of chlorophyll in broccoli juice (Butz et al., 2002).

During storage, the green color of HPP processed vegetables turned light yellow due to oxidation. However, vegetables that are stored by applying pressure at higher temperatures, showed no change in color. Carotenoids, for orange-yellow color, are pressure stable. Their extraction yields could be increased by HPP treatments (Garcia and Barrett, 2002).

Butz et al. (2003) conducted experiments on HPP on different fruit juices individually and in combination as well. He observed that there was no considerable difference in carotenoid content of control samples and HPP treated samples. After HPP treatment, tomato retained its color as its original (700 MPa, 65°C for 1 h (Rodrigo et al., 2007).

2.3.2 *Impact of HPP on texture*

Cell permeability of vegetables and fruits can be disrupted by HPP treatment. It may disrupt cells and helping out the enzymatic and non-enzymatic reactions. Transformation in polymers of the cell wall could be blamed for texture change in vegetables and fruits due to reactions (Oeya et al., 2008). Pectin is a component that helps in binding of cells together, can be degraded by depolymerization and demethoxylation by PG and PME, respectively (Sila et al., 2008).

Parenchyma cells are disorganized due to HPP. There occur disintegration of plant cells and no gas could be found in intercellular spaces. Gap formation is followed by soaked texture after HPP. Basak and Ramaswamy (1998) experimented to check the impact of HPP on the rigidness of different vegetables and fruits such as orange, pear, carrot, and pineapple. They found that during compression, firmness was lost. For 30–60 min, the firmness was decreased or regained at 100 and 200 MPa, for carrot, celery, orange, and pear.

Increased firmness is resulted due to PME activity. Many fruits and vegetables exhibited softening at above 200 MPa pressure at room temperature for 5–60 min. Pear showed most pressure sensitivity followed by pineapple, apple, and orange. Apple was far more sensitive that pear at 200 MPa. Cherry tomatoes showed softening under high pressure (Tangwongchai et al., 2000). Ahmed et al. (2005) suggested that after HPP at 100 or 200 MPa (20°C), the viscosity of mango pulp was increased, while, after HPP at 300 and 400 MPa at the same temperature, viscosity was declined.

Polydera et al. (2005) showed by experiment that higher viscosity was the result of pressure treatment (600 MPa) rather than thermal treatment (80°C/60 s) and during storage, very less reduction in viscosity was observed in HP treated juices even at higher temperatures (30°C). During storage, quality loss of juices is caused by PME activity. By applying different combinations of temperatures and pressures, for activation or inactivation of pectinases during processing could be helpful for better texture creation.

2.3.3 Impact of HPP on flavor

Flavor is an important impression of food that is being sensed by taste and smell. It is thought that HPP does not alter the flavor of fruits and vegetables as flavor compounds have low molecular weight so these are not acted upon by HPP (Oeya et al., 2008). Many experiments were conducted to check the effect of HPP on flavor compounds of vegetables and fruits such as acids, ketones, esters, and hexanal. Navarro et al. (2002) found that HPP at optimum temperature and 400 MPa doubled the contents of hexanal in strawberry puree. Porretta et al. (1995) showed in experiments that HPP at 500, 700, or 900 MPa at room temperature for 3, 6, or 9 min, of tomato juice has resulted in strong rancid taste, n-hexanal was considered to be its responsibility.

In strawberry puree, at cold storage, HPP at 800 MPa for 20°C/20 min, volatile compounds (funeral, linalool, nerolidol, and some esters) concentration was remarkably lower than in unprocessed food, that are responsible for flavor. The relative concentrations of acids such as hexanoic acid, 2-methyl-butanoic acid, and butanoic acid, and ketone compounds were lower in HPP (200, 400, 600, or 800 MPa/18°C–22°C/15 min) than in un-processed strawberries (Winai et al., 2005). Comparative study of effects of HPP (600 MPa, at ambient temperature, 5 min) and pasteurization on volatile components of strawberry, blackcurrant purees, and raspberry were analyzed by using electronic nose at room temperature and it was concluded that HPP purees have better flavor maintenance.

Resulted data showed that heat altered volatile compounds more than HPP. Similar results were obtained in the case of processed raspberry and blackcurrant (Dalmadi et al., 2007). HPP is a very effective processing technique for fruit and vegetable preservation. HPP treated vegetables and fruits to retain the high quality of flavor than any other products preserved by conventional heat methodology.

2.3.4 Impact of HPP on vitamins

In this modern era, the introduction of new techniques involving a combination of non-thermal processes with natural safe additives to get safer food products has been one of the vital goals of the food industry (Barros-Velazquez, 2011; Norton & Sun, 2008). HPP can be useful to attain this goal. Microorganisms can be inactivated and even killed at room temperatures (Saucedo-Reyes et al., 2009). Low molecular weight compounds such as vitamins remain unchanged (Rastogi et al., 2007). In liquid foods, the transmission of pressure is uniform and instant thus following isostatic rule (Toepfl et al., 2006). HPP is out of the limitations of the geometry and size of food products that minimizes the time of processing of large food quantities (Rastogi et al., 2007).

2.4 Conclusions and future aspects

In this fast-moving world, people are very conscious about their food and health. The big threat of foodborne diseases and heavy uses of hazardous chemicals forced consumers to rely on healthy and quit natural foods like vegetables and fruits. To fulfill this demand of people, many non-thermal approaches have been tested for preservative methods, and HPP is proved to be much appreciated.

HPP is applied to inactivate microorganisms and to maintain the quality of food at its maximum standards. It is a very good alternative to heat treatments for vegetables and fruits. It significantly enhances the shelf life of foods, thus, giving a homemade taste. HPP enables the food processors to apply high pressure to pasteurize foods at more or less room temperature. High pressure along with the ambient temperature is a promising strategy to alleviate the shelf life of food.

It helps the food processors to process it with neat and clean ingredients by using minimum additives. HPP can be applied to a broad range of foods including vegetables, fruits, seafood, and sauces. For the last decade, this technique was quite beneficial for the killing of E. coli, Listeria, Vibrio, and Salmonella in different food products.

Section 3

Outline

3 Impact of high-pressure processing on fresh salads and ready to eat foods

3.1 Importance of salads and ready meals

Ready to eat (RTE) foods or ready meals include the various food items that are prepared or processed inform that they do not require any further processing before their consumption. There are many food items available in RTE form depending upon the eating habits and availability in a particular region. Nowadays, the popularity of RTE foods has been growing day by day as they are more convenient to prepare and require very little time for cooking. With the increasing demand for RTE foods, a wide range of RTE products are available in the superstores and markets including fresh salads, fruit salads, leafy green vegetable salads, fresh-cut vegetables, and other meat products. Ready to eat salads or vegetables were prepared by trimming, peeling, slicing or shredding, washing, disinfecting, and packaging of raw vegetable that were kept in refrigerated storage afterward and must be consumed within few days of their harvest. Ready to eat salads and vegetables are also getting popular in developing countries due to convenience in their consumption good taste, and an increase in the awareness about the health benefits of vegetables. Ready to eat salads prepared from leafy green vegetables are also very much popular due to their good nutritional profile as they provide vital phytonutrients, minerals, and water-soluble vitamins. Consistent daily consumption of these salads helps in the prevention of cardiovascular diseases and cancer. Despite the health benefits of ready to eat salads, they are also considered as a source of foodborne illnesses as several outbreaks were associated with the consumption of RTE salads. Reasons for their microbial contamination include: introduction of microflora from the raw vegetables into the salads; they are not even undergone any heat treatment and improper decontamination practices were adopted during their preparations. Microbial contamination can occur at any stage from harvest to its preparation, transportation, storage, and serving at home. RTE foods were most frequently found to be contaminated with Listeria monocytogenes, Salmonella, Campylobacter jejuni, C. botulinum, Staphylococcus aureus, and Yersinia enterocolitica. To preserve the natural attributes and prevent the microbial contamination in RTE salads, many advanced processing technologies were used including the modified atmosphere packaging, use of bio-preservatives or plant extracts, ionizing irradiation, and HPP.

HPP is an innovative non-thermal technology used to increase the shelf life and maintain the freshness of RTE salads and vegetables. It involves the uniform and simultaneous application of elevated pressure up to 900 MPa through the food sample may be in direct contact or packed in flexible pouches. The time required for HPP is not dependent on the size. As a result of high pressure, the compression of RTE foods takes place that increases the temperature of the food product and affect the electrostatic interaction by braking hydrogen bond. Therefore, the small molecules like flavoring compounds and vitamins remain unaffected but the disruption of large molecules in microbial cells like cell membrane takes place as the result of HPP. This technique is used as a substitute for the heat treatment of RTE meals as it preserves the nutritional profile and sensory attributes of RTE salads along with the destruction of foodborne pathogens. However, the cost of this technique is much higher than the heat treatment that is outweighed by the advantages of HPP (Norton and Sun, 2008).

3.2 Impact of HPP on microorganisms

One of the major goals of HPP is the microbial inactivation which is dependent upon number of factors including microbial taxonomy, processing condition (fresh, processed), food structure, and its group (animal or plant). Microbial propagation in the food product is greatly dependent upon the structural properties of food including distribution of water and chemical distribution of preservatives. Due to HPP, the disruption in cell membrane takes place that causes the interruption of the normal cell functioning which is responsible for cell reproduction. Application of pressure of 30 and 50 MPa also interrupts the gene expression as well as protein synthesis ultimately causing microbial death. Yeast and mold are spoilage causing microorganism but they are not foodborne pathogens. They show the least resistance to HPP as compared to bacteria and can be inactivated at a pressure of 200–400 MPa (Daher et al., 2017).

Bacteria are the smallest, single-celled, free-living microorganism. The bacteria responsible for causing the foodborne illnesses are Campylobacter spp, Salmonella, L. monocytogenes, S. aureus, and E. coli. Relatively low pressure ranging from 330 to 700 MPa can inactivate the most vegetative cells, however, the bacterial spores required the very high pressure of 1000 MPa as they are more resilient. As a result of HPP, all specie was not affected in the same way as Gram-positive bacteria are more resistant to HPP as compared to Gram-negative bacteria. Certainly, the bacterial resistance to HPP is highly dependent upon the taxonomy (species, strains) and the growth phase like cells shows more resistance to high pressure in the stationary phase as compared to the exponential phase. Nevertheless, the combination of heat treatment with high pressure and two-exposure treatment to HPP was used to inactivate the resistant bacterial spores (Daher et al., 2017).

High pressure can be employed with both type heat treatments (pasteurization and sterilization). In the case of high-pressure pasteurization, the intermediate pressure from 400–600 MPa was applied at a temperature of 60°C, resulted in the inactivation of vegetative form of bacteria with minimum effect on the nutritional profile of food. After high-pressure pasteurization, the product must be kept at refrigeration temperature during transportation and distribution. Conversely, high-pressure sterilization involves the application of elevated pressure up to 800 MPa at a relatively high temperature of 120°C to inactivate the resistant bacterial spores and produce the shelf-stable foods.

Bacterial spores can also be inactivated by two-exposure treatment with HPP. The first treatment was done at relatively low pressure which resulted in the germination or growth of bacterial spores followed by high-pressure exposure that destroys vegetative and spore-forming cells (Daher et al., 2017). The complete inactivation of mesophilic bacteria, total yeast and mold were reported at a pressure of 400 MPa applied at 45°C for 15 min in vegetable-based infant foods. Similarly, Stratakos et al. reported that the growth of L. monocytogenes can be lowered by employing high pressure of 400 MPa by maintaining low pH during storage as high pH may induce the bacteriostatic effect. The effectiveness of the inhibitory effect caused by HPP can also be enhanced by employing *Weissella viridescens* which is found to be a very effective approach to maintain the safety of RTE salads.

3.3 Impact on enzyme activity

Quality of RTE fruits and vegetables are greatly affected by enzymatic activity as they are not confined with the protective covering and undesirable changes will occur as the result of free mixing of enzymes with substrate. Certain enzymes are of great significance in fruits and vegetables, like PG and PME. They are responsible for the breakdown of cell walls result in a decrease in viscosity and change in the organoleptic properties. While other enzymes like; LOX, peroxidase and polyphenol oxidase result in the color changes and breakdown of lipids. Enzymes are protein in nature and hydrophobic interactions that are responsible for keeping the protein structure intact is the breakdown by HPP and resulted in the structural changes and loss of activity of enzymes. However, the high-pressure processing may result in the inactivation of enzymes in some cases or activation in others. So, for the complete inactivation of enzymes the application of pressure in combination with heat has been recommended.

PG is an enzyme bounded by a cell wall that catalyze or speed up the breakdown of α-1 1,4- galacturanosyl linkages and produces the reducing sugars. In tomatoes, it results in a reduction in the viscosity due to pectin solubilization. Suitability of HPP for tomato products and inactivation of PG was reported by employing the pressure-temperature combination of 300–600 MPa at 5°C–50°C. Stoforos et al. also investigated the inactivation of another enzyme PME in tomato by the application of high pressure up to 800 MPa and intermediate temperature of 60°C–75°C. The temperature and the pressure do not show synergetic effect, as PME inactivation rate increase with the increase in temperature and decreases considerably with the increase in pressure until the pressure surpassed the 700 MPa.

LOX is another important enzyme that catalyzes the oxidation of polyunsaturated fatty acid resulting in the development of off-flavor, whereas PG and pectinesterase affect the textural properties. Shook et al. reported the effectiveness of HPP in diced tomatoes for the inactivation of LOX and PG as they were completely inactivated by pressure of 800 MPa irrespective of processing temperature and time. While pectinesterase was found to be insensitive to HPP.

3.4 Impact on color

Color is an important quality parameter that directly influence the consumer acceptance for the food product. During HPP, there occurs the protein denaturation which further results in the alteration in functionality and color of processed products as compared to the raw product. Color changes appear in food significantly dependent upon the pressure, time, and temperature employed during processing. HPP of raw meat results in the development of color changes, rubbery consistency, and cooked like appearance occur due to denaturation of myoglobin and oxidation of ferrous atom. Similar effects were not visible in semi-cooked or cooked meals as their protein was already denatured during processing so the HPP of raw meat was not recommended. HPP of fish meat also showed the crucial color changes (Hogan et al., 2005).

To minimize the color changes in vegetables and fruits, temperature and pH must be keenly observed. Likewise, the discoloration of broccoli takes place due to the degradation of chlorophyll after exposure to pressure and temperature higher than 50°C. If the temperature is kept below 50°C, no color changes were observed even after the application of 800 MPa pressure. In some products, the HPP negatively impacts the color like in the case of onion as it turns brown upon its exposure to pressure and getting browner with the increase in the intensity of pressure (Norton and Sun, 2008). Color changes in tomato juice was reported when treated at a pressure of 700 MPa for 1 min at 80°C–90°C. While no color changes were shown by the tomato treated at 300–700 MPa for 60 min at 65°C and maximum color changes were reported in strawberry (Rodrigo et al., 2007). So, it is concluded that the color changes as the result of HPP can be reduced by the application of circumspect treatments.

3.5 Impact on texture

The textural properties of food are significant quality parameters that can directly influence product sales and acceptability. The influence of HPP on the textural properties of food depends upon the intensity of applied pressure and type of commodity. No visible textural changes were observed in high moisture food as no shear force was produced during HPP. While the products containing gas showed the color and textural modifications due to the displacement of gas during HPP. For the foods containing air voids, the physical shrinkage and shape distortion takes place due to the damage of air pockets. And for the foods having no air voids, minimum or no textural changes were detected. If required new product development can be done through HPP as it can modify the texture and structure, so desirable textural changes can be achieved in a certain food. The texture of fruits and vegetables depends upon the cell wall structure. So, as the result of HPP, softening of fruits and vegetables occur and they can also lose tissue firmness (Hogan et al., 2005; Norton and Sun, 2008).

For example, the high-pressure treatment of 600 MPa combined with a high temperature of 80°C for 90 min reported to not affect the hardness of carrot. While in the case of lettuces, the pressure treatment at 360 MPa does not affect the textural properties but the application of higher pressure of 600 MPa resulted in its softening.

3.6 Impact on nutrients/nutritional value

Heat treatment of food products results in the detrimental changes in quality due to the destruction of vitamins, off-flavor development, alteration in the texture, and color. So the HPP was adopted to inactive the microorganism with the minimal changes in the nutritional characteristics of produce (Norton and Sun, 2008). In comparison of thermal processing, the HPP exerts a minute effect on the compounds with low molecular weight like vitamins, antioxidants, flavoring compounds, and pigments. So, this processing technique is beneficial in preserving the nutritional profile of vegetable salads as they are a good source of vitamins, minerals, and antioxidants.

However, the increase in the intensity of pressure treatment results in microbial destruction in a short time but the application of higher pressure may affect the quality and nutritional parameters of processed food as compared to unprocessed food. Nonetheless, the least changes were observed in the structural properties of high moisture as no shear forces were applied in HPP (Norton and Sun, 2008). The effect of HPP on the color and antioxidant compounds in vegetables like carrots, onion, olive oil, green papers, lemon, and green celery by employing the four different pressure treatments of 100, 200, 300, 400 MPa was investigated and compared with thermal treatment. The HPP treated sample showed less color modifications and more retention of ascorbic acid as compared to the thermally treated sample. While no significant difference was observed in phenolic content.

Similarly, the effect of high pressure combined with heat treatment on the nutritional value of three vegetables broccoli, tomatoes, and carrot was studied. In broccoli, no loss of chlorophyll occurred by the application of 600 MPa pressure and 75°C temperature for 40 min. While the carotene and lycopene content of tomatoes treated with 600 MPa at 25°C for 60 min was similar to that of samples heat-treated at 95°C for 60 min. And a minute loss in the antioxidant capacity was observed in pressure treated carrots at 800 MPa for 5 min as compared to the untreated sample (Butz et al., 2002). So, it can be concluded that the HPP has the least effect on the nutritional characteristics of fruits and vegetables.

3.7 Conclusions and future aspects

Ready to eat salad and vegetables are the most famous food commodities in developing countries as they are convenient to prepare and liked by the consumers. They can also act as a source of foodborne disease as a large number of food outbreaks were linked with the consumption of RTE salads and meals. HPP is identified as innovative technology to improve the microbial safety, nutritional quality, and organoleptic properties of RTE foods.

Section 4

Outline

4 Impact of high-pressure processing on dairy/egg

4.1 Importance of HPP on dairy/egg

HPP has elucidated the positive effect on the safety and quality of dairy and seafood products as compared to conventional heat treatments. Today, per annum half out of 200,000 tons high pressure processed food products are seafood or fish and meat products (Heinz and Buckow, 2010). The founder of research work done on HPP. Researchers selected milk in their study to determine the impact of this technique on food products (Hite, 1899). In the past few decades, owing to the rising demand for fresh, nutritive, and minimally processed innovative food products, much emphasis has been placed on the development of temperature-independent food preservation methods. Consumer demands minimally processed safe and nutritious food products without added food preservatives. In HPP pressure is transmitted throughout the food uniformly, irrespective of the geometry or size of particles, resultantly product with homogeneous characteristics having increased shelf life is obtained. The shelf life of food is enhanced by low or moderate temperature treatment under pressure hence it is also referred to as cold pasteurization. Several modifications in HPP are employed in food industries including ohmic heating, microwave, intense light pulses, controlled instantaneous decompression (CID), X-rays, and electron beams. HPP has an advantage over other processing techniques as the nutritional, sensory, and organoleptic properties of food being processed remain unaltered. HPP inactivates the food spoilage microorganisms including E. coli O157: H7, L. monocytogenes, and C. jejuni. The combination of chemical and physical food preservation techniques with HPP increases the lethality of microorganisms present in food. HPP can be used along with modified packaging and natural anti-microbial, that is, Nisin, to preserve the natural characteristics and quality of food. This food preservation technique is of special importance for those products having ingredients likely to be modified or damaged by heat treatment. Along uniform transmission of pressure to food ingredients, energy efficiency is also a positive factor linked to HPP technology. Traditional food processing techniques employ the use of high temperatures which affects the nutritional quality as many nutrients are heat sensitive. Pasteurization is the most common technique used for milk preservation, destroys pathogenic microbes thus increasing the shelf life to 7–20 days. Ultra-high temperature (UHT) is also practiced in the dairy industry. Processing at high temperature lowers the nutritional quality of dairy products. High hydrostatic pressure (HHP) food processing technique has been developed to ensure the safety of products without damaging the nutritional profile of milk. In this technique, pressure about 400–600 MPa is applied resulting inactivation of foodborne pathogens. To denature all kinds of enzymes 800 MPa pressure is required (Liepa et al., 2016). In the United States, HPP is employed in poultry, meat, and seafood industry as it can reduce the microbial load, stabilize the ready to eat meat and poultry products thus fulfill the requirement of regulatory authorities (Campus, 2010). In meat and poultry industry, HPP finds various applications including decontamination of high acid foods without thermal operations, inactivating food spoilage microbes, meat texturizing, and thawing resultantly improves the overall quality of the final product. Food processing at ambient temperature can be achieved by employing this technique. Furthermore, functional food ingredients having novel characteristics can be prepared by employing this technique (Rastogi et al., 2007). HPP also have successfully applied in the seafood processing industry as the processing of oysters by employing HPP. However, to obtain high quality product it is needful to study the effect of HPP on the quality of processed food components.

4.2 HPP impact on the texture and water retention of meat and seafood

HPP is employed in the meat and seafood industry as an environmentally friendly and safe preservation technique by completing decontaminating the product from pathogenic microbes. However, it is required to study the impact of HPP on the texture, structure, and overall characteristics of food components in processed food. The impact on protein in the meat is of special significance to get a product of superior quality. It is confirmed that in comparison to commercial thermal processing techniques for extraction of crab meat HPP has several benefits. The results of the study indicated that as compared to thermal processing at 90°C for 20 min, HPP at 100–300 MPa yielded the increased extraction of crab protein in less time. Overall quality and functional properties of crab meat were preserved by employing HPP at 100 MPa (Martínez et al., 2017). HPP also called ultra-high pressure (UHP) and HHP food in a flexible package in vacuum conditions placed in pressure transmitting solution under the pressure of 100–900 MPa. In large molecules such as protein, the primary structure is not affected at all along ion disruption occur resultantly decrease in volume. Whereas secondary, tertiary and quaternary structures of protein get modified which are assumed to be a reason for decreased volume and lipid conformation. Vitamins and flavoring compounds remain preserved. A study indicated that HPP in fresh meat results in unacceptable changes in color while in dry-cured meat the changes in color were inconsiderable. The same study carried out on fish elaborated that when increased pressure is applied to fresh fish changes in color take place from decreased red color to appearance of light-yellow color to the point that fish no longer seems to be fresh (Campus, 2010). After the death of an animal, rigor mortis begins and muscle become tough, myofibril proteins are subjected to breakdown by proteolytic enzymes. An enzyme, Caspase is responsible for meat tenderization but their role is still needing to be investigated (Kemp et al., 2010). While calpains as major meat tenderizing enzyme, in a study conducted by Chéret et al. (2006), in HPP treatment to sea bass at a pressure about 300 MPa, calpains got inactivated due to the destruction of calpain subunits. Resultantly water holding capacity of meat reduced due to the destruction of cytoskeletal proteins such as desmin and drip losses increased. When cutting and slicing of meat is done, moisture retention in meat and meat products is called water holding capacity that is directly linked to the moisture losses during cooking of meat, consumer acceptance, and overall quality of meat. Effect of HPP on water holding capacity of meat is not so much clear and still research work is required. Here is some contradiction in this regard. Some authors reported the positive effect of HPP on water binding capacity of meat, negative or no effect at all. Water holding capacity depends on many factors such as the nature of meat, time and temperature as well as pressure employed in HPP, and ingredients used such as nitriles compounds (Hygreeva and Pandey, 2016).While positive effect of HPP can be achieved by using suitable time, temperature and pressure combination for meat and meat products. Meat protein denaturation and oxidation, as well as drop-in pH, are the major factors affecting the water retention (Huff-Lonergan et al., 1996). From the commercial point of view this factor has much importance in determining the quality of fish. HPP application on sea bream at 300–400 MPa resulted in decreased water-holding capacity due to the destruction of desmin (Campus et al., 2010). However, in another study by Ohmori et al. (1991), the application of HPP on fresh meat increased the proteolytic activity of cathepsin probably owing to the release of protease enzyme from the lysosomal membrane. Fish proteolytic enzymes causing spoilage are more sensitive toward HPP as compare to that of mammalian (Low and Somero, 1974). In thermal processing of meat, the meat releases water-soluble proteins along released water which itself is a loss. In HPP treatment of pork meat at 300–600 MPa for 1 and 15 min greatly increase the water holding capacity to fivefolds as compare to untreated meat (Sazonova et al., 2019). HPP effect on enzyme activity depends on various factors including the type of enzyme, processing conditions, and the medium used resultantly texture of the final product is affected (Campus et al., 2010).In meat by employing 300 MPa pressure for few minutes in HPP meat tenderization (without discoloration or browning) can be achieved (Macfarlane, 1985). In bluefish, HP about 200–300 MPa results in a decrease in muscle "strength value" and elasticity as indicated by puncture test (Ashie & Simpson, 1996). Other uses of HPP in the seafood industry include HPP thawing and pressure-assisted freezing. Improvement in textural qualities by HPP can be employed in improving gel strength by 3 times in uncooked surimi (Ashie & Lanier, 1999), moreover can be used to develop new products. HPP is employed in seafood to recover meat from crabs and lobsters' shells thereby improving yield and overall quality. The principle texture of meat is due to myofibrillar and connective tissue proteins. Connective tissues in meat are little affected by HP application. Pre-rigor application can soften the texture of meat while the post-rigor effect of HPP can be employed by combining specific temperature-pressure combination (Sun and Holley, 2010). In both processes, structural degradation of protein components occurs. Seafood undergoes rapid rigor and substantial softening compared to mammalian meat. These products require firmness in texture as they are fragile. Fish meat tenderizing is linked directly to enzyme protease which results in the softening of meat. HPP can modify the enzyme activity as a tertiary and quaternary structure of protein greatly affected by pressure (Duranton et al., 2014).

4.3 Impact of HPP on dairy/egg quality

Thermal processing results in off-flavor and off coloring in the final product by disrupting covalent bonding in food. HPP has an advantage over conventional thermal processing techniques as it only affects non-covalent bonds by forming intermolecular bonding in proteins without damaging the color, flavor, or nutritional properties of food. Eggs are an important source of minerals and vitamins along with essential amino acids. When employing HPP egg white can coagulate and form a gel at or above the pressure of 600 MPa (Bridgeman, 1914). The texture of the egg is an important attribute in determining the quality of the egg while thermal processing completely changes the texture. As compared to thermal processing, HPP forms superior quality gel when eggs are processed at a pressure of 400–600 MPa. HPP induced egg gels possess no off flavor or destruction of necessary vitamins (Hayashi et al., 1989). Moreover, in HPP functional properties of egg white are enhanced by modifying the structure of the protein (Messens et al., 1997). As compared to the thermal processing gel of superior quality attributes and physicochemical characteristics can be achieved at a much lower temperature by employing pressure of 600 MPa (Singh and Ramaswamy, 2013). HPP can completely inactivate harmful microorganisms that may gain entry to the whole egg in liquid form. In HPP combination of pressure-temperature can reduce the bacterial load in food too much extent. Pressure more than 400 MPa with temperature about 50°C reduces bacterial count of Salmonella, E. coli, and Listeria innocua to 8, 7, or 4log units respectively. Moreover, HPP has also shown to improve the foaming and emulsion properties of egg (Guamis et al., 2005).

In dairy, HPP is employed in combination with pressure-temperature of 1000 MPa at 60°C equivalent to sterilization which destroys spores of pathogenic microbes thus extends the shelf life of milk in association of desirable functional properties of milk being preserved (Heinz & Buckow, 2009). Most of the food spoilage microbes are destroyed at pressure 400–600 MPa. Vitamins and minerals owing to their small size not affected to a large extent by HPP while fat and protein fractions got affected (Chawla et al., 2011). When 400 MPa pressure is applied colloidal calcium, phosphate get solubilized resultantly casein micelle disrupted into smaller diameter results in increased viscosity of milk (Huppertz et al., 2006). By increasing the temperature disruption of casein micelle can be decreased (Orlien et al., 2010). Fat globules membrane disruption occurs when HPP is applied hence lipase easily acts on triglycerides and free fatty acids in milk increased. Lactose upon high-pressure application (100–400 MPa for 10–60 min) does not isomerize to lactulose or acid products which otherwise formed by thermal heat processing (Liepa et al., 2016). HPP improves the bioavailability of milk minerals indigenous to milk by increasing the level of ionized calcium in milk. In HPP of milk by increasing pressure to 400 MPa increase in mineral concentration of calcium, magnesium, and phosphorus in serum is reported (Barba et al., 2015). Vitamins are not lost significantly in HP of milk either vitamin B losses are negligible. Losses of vitamins B6 and B1 are reported when treating at 400 MPa (Liepa et al., 2016). Milk alkaline phosphatase enzyme completely denatured at pressure 800 MPa for 8 min (Liepa et al., 2016). HPP has resulted in positive quality attributes of cheese and yogurt by improving mineral equilibrium and protein configuration as well as for the production of dairy products with novel texture and characteristics. Rennet coagulation of milk can be improved by applying pressure of about 100–500 MPa while at 300 MPa rennet coagulation time of milk is increased. Effect of HPP on cheese has been studied and showed the positive results by improving the microbiological quality of cheese, HPP can also be employed in cheese ripening. Cheese prepared through HPP show better texture and protein retention, resultantly lower hardness, and better sensory and quality parameters (Molina et al., 2000). HPP can destroy the lactic acid bacteria present in cream resultantly extend the shelf life of dairy cream when kept under refrigerated conditions. In low-fat ice-creams, pressure-assisted freezing can improve the overall texture of ice cream by preventing the formation of coarse ice crystals (Guamis et al., 2005).

4.4 Safety aspect of HHP for meat/seafood/dairy/egg

The food industry should adopt the highest standard of safety and quality for food processing as the industry has a direct link with the health of the consumer. HPP is commercially used for the preservation of food products due to its positive effects on the overall quality as compared to conventional processing operations by preserving the national and sensory characteristics of the product. Consumer is becoming more health-conscious and demand clean-label food products that are natural, minimally processed, and additive-free. HPP ensures the microbiological safety of food without the addition of any chemical preservative. Moreover, natural flavor and nutritional aspects of food also preserved (Daryaei & Balasubramanian, 2012).

Among all the novel non-thermal food processing technique, HPP is the only technique completely inactivates the microbes present in food for food preservation (Režek Jambrak et al., 2019). Monoamines, diamines, and polyamines are naturally present toxic compounds in meat and fish that may result in serious food poisoning. These amines are produced by food spoilage microorganisms', that is, lactic acid bacteria from amino acids. HPP can delay the formation of these biogenic compounds by controlling the growth and proliferation of these spoilage microbes. Study conducted on meat and seafood showed that HPP treated products contain a lower level of these amines as compare to non-treated products (Duranton et al., 2014). HPP treatment to meat at 600 MPa completely inactivates microorganisms but it would affect largely to the quality of meat (Hygreeva and Pandey, 2016). HPP is regarded as a powerful tool to destroy completely the Listeria and Salmonella from dry-cured meat products (Campus, 2010). HPP inactivates or destroys the pathogenic microbes present in food thus increase the shelf life of milk up to two to threefolds. Thermal degradation of milk components is reduced moreover flavor and color are retained. Along microorganisms, spores and enzymes are inactivated by high pressure. Functional characteristics of an egg are improved thus there is potential to create new products with novel characteristics. HPP has been regarded as the safe process for the reduction of Vibrio, a pathogenic bacterium found in fish and fish products (Raghubeer, 2007). Moreover, it is reported that HPP has the advantage over techniques it can make the raw shellfish-free from infectious viruses (Kingsley et al., 2002). In lobster, microbiological quality and yield can be increased by novel techniques inactivates microorganisms in food is still not clear. Study is required on the shelf life of these non-thermally processed food products including HPP. HPP inactivates microorganisms, they remain in a viable but non-culture able state where they can rejuvenate and form biofilms in food which are cumbersome to destroy. HP processed foods that come under the category of novel foods. Before introducing novel food to the market, the company needs to get approval that the food complies with the food laws and regulations. Authorization is required before launching these food products to the market moreover safety assessment of the product is done. There are still some gaps in HPP treatment that needs to be covered by research work such as negative impacts of HPP, shelf stability of high pressure treated food, and negative sensory properties of food being processed. Work is required to determine the impact of HPP on allergens and toxins present in food as there is no report available regarding this aspect. Before the marketing of HP treated foods in developing countries, where there are no established regulations validation may be required.

4.5 Conclusions and future aspects

HPP has a great potential to be used in the food industry for improving the overall quality and safety of products as compared to conventional processing operations. It is evident from the literature that by employing high pressure in dairy processing functional as well as nutritional properties of the final product can be enhanced without altering the sensory parameters too much extent. While whey protein denatured, casein micelle decreases in size by breakdown, and the level of free fatty acid also increased after high-pressure treatment to milk. Meat tenderization can be improved by employing HPP without discoloration or browning. Product recovery increased in seafood when employing HPP. HHP allows the use of less time and temperature as compared to conventional thermal processing. HP is much convenient for those foods where ingredients can alter extensively by thermal processing such as whole egg. As compared to thermal processing, the gel obtained by high pressure is of superior quality, improved functional properties, without burnt or off-flavor. HPP can enhance the microbiologically quality of liquid egg, improve foaming, and emulsion properties. Resultantly high pressure improves the microbiological, physicochemical, and sensory and nutrition value of the food being processed as well as the functional properties of raw ingredients as discussed previously. While unacceptable changes in the color of fresh meat and fish can occur after a high-pressure application. For better understanding and practical application in meat processing without unacceptable changes in the end product, research work is required to provide shelf-stable, microbiologically safe, and overall superior quality end products.

The food must contain moisture level as moisture act as a medium for transferring pressure to food components. Pathogenic microbes inactivate only when pressure applied exceed to 400 MPa. Some heat resistant and Gram-positive bacteria as well as spores show resistance to pressure thus need to be destroyed properly. The cost of installation is high. Fragile and delicate foods need special attention while processing through high pressure.

Section 5

Outline

5 Conclusions

5.1 Advantages of HPP

Nowadays, consumers are well aware and increasing demand for food with better safety, quality, and nutritional properties. The production requirement of foods with such quality characteristics has given birth to the innovative food processing technologies such as HPP. HPP has several advantages and limitations over the conventional food processing technologies.

HPP eradicates bacteria and yeast in the processed foods which can be used as an antimicrobial technology as alone or in combination with several other processing technologies. This not only reduces the microbial spoilage but also replaces antimicrobial agents, meeting the demand for wholesome and natural foods. The pre-packaged foods being processed through this technology avoid contact with the processing devices which prevent secondary contamination.

During thermal processing/pasteurization of fruit juices, the loss of functional and sensory quality is a major drawback. HPP is thought to be a fascinating technology for the refrigerated storage of these products, preserving nutritional and sensory parameters as well as extending shelf life.

HPP, being the non-thermal and most effectively commercialized processing technology, destroys foodborne pathogens at room temperature. This reduces the consumption of energy allied with heating and cooling subsequently. The products processed through this technology also maintain nutritional and organoleptic properties and increase marketability in combination with existing trend, that is, organic food, healthy food, etc.

In HPP technology, the pasteurization effect is not influenced by the food volume and packaging form so foods of different volumes can be processed in the same batch. Moreover, HPP is considered as minimal processing technology ensuring the food flavor and safety. The recycling of the pressure transfer medium can be done after food processing and this technology is considered environmentally friendly.

The HPP technology strengthens the enzyme activity, for example, protease, which increases aging and tenderization of meat. This also makes the meat easier to be digested by the consumer. HPP provides diversity in food products as applied in ready-to-eat steamed rice for adjusting starch and protein contents. This also increases the water permeation into the rice and degree of gelatinization, making absorption and digestion easier.

This technology can also be used as a pre-treatment before the extraction process. The use of high pressure physically damages the plant cells and makes nutritional components more permeable to the solvents. For example, the extraction of flavonoids from propolis, collagen from fish skin, ginsenoside from ginseng root, anthocyanidin from grape skin, and triterpenes from Antrodia cinnamomea has been successfully done through this technology.

5.2 Limitations of HPP

HPP ensures the nutritional and organoleptic properties of foods along with safety. This non-thermal and energy-conserving technology is also considered environmentally friendly. However, it also has some limitations such as the products of HPP need refrigeration systems for transportation and storage because HPP alone is insufficient to stop the activity of pathogen spores, for example, C. botulinum. There is more risk associated with the processing of low acid foods. Besides, HPP involves food materials to be pre-packaged and for the processing purpose the packaging material must have flexibility to be compressed minimum by 15%. This is only possible in plastic-type packaging. Moreover, HPP also finds limitations in the processing of foods containing low moisture contents such as powdery flavors and flours. The technology is also not applicable to foods containing air bubbles as in HPP water is used as a pressure transfer medium that deforms air bubbles under high pressure.

5.3 Future perspectives of HPP

The food industrial sector is very strongly connected with the wellbeing of the consumer. The purpose of food processing and technology is to provide wholesome food to the consumer. Among the various applied technologies in food industries, HPP is a non-thermal, alternative to conventional thermal processing, technology applicable to foods and beverages. This technique has minimum damaging results on the nutritional and organoleptic parameters compared to the conventional thermal processing. HPP provides the products with safety, fresh like quality, retention of flavor, bioavailability, and above overall the accelerated marketability in the current trend of organic products. However, there are some limitations in the application of this technology such as inactivation of pathogen spores, the compressibility of packaging, processing of some powdery or air containing products and might be some technical aspects along with some country's regulations. However, by addressing/controlling these issues, HPP finds a broad scope in future to be used in the food industries as the current well aware customer's demand of fresh, nutritional, safe, organic food with overall quality parameters.

References

Adzahan, M. N., & Benchamaporn, P. (2007). Potential of non-thermal processing for food preservation in Southeast Asian countries: Review paper. *ASEAN Food Journal*, *14*(3), 141–152.

Ahmed, J., Ramaswamy, H. S., & Hiremath, N. (2005). The effect of high pressure treatment on rheological characteristics and colour of mango pulp. *International Journal of Food Science & Technology*, *40*, 885–895.

Ahn, J., Balasubramaniam, V. M., & Yousef, A. E. (2007). Inactivation kinetics of selected aerobic and anaerobic bacterial spores by pressure-assisted thermal processing. *International Journal of Food Microbiology*, *113*, 321–329.

Andres, A. I., Adamsen, C. E., Moller, J. K. S., Ruiz, J., & Skibsted, L. H. (2006). High-pressure treatment of dry-cured Iberian ham. Effect on colour and oxidative stability during chill storage packed in modified atmosphere. *European Food Research and Technology*, *222*(5), 486–491.

Ashie, I. N., & Lanier, T. C. (1999). High pressure effects on gelation of surimi and turkey breast muscle enhanced by microbial transglutaminase. *Journal of Food Science*, *64*, 704–708.

Ashie, I. N. A., & Simpson, B. K. (1996). Application of high hydrostatic pressure to control enzyme related seafood texture deterioration. *Food Research International*, *29*, 5–6.

Avure Technologies Inc. (2007). High pressure processing of food. Available from: http://www.avure.com/food/applications/default.asp.

Balasubramaniam, V. M., & Farkas, D. (2008). High pressure food processing. *Food Science and Technology International*, *14*(5), 413–418. https://doi.org/10.1177/1082013208098812. SAGE Publications: on the behalf Spanish Council for Scientific Research (SCSR). ISSN:1082-0132.

Balasubramaniam, V. B., Farkas, D., & Turek, E. J. (2008). Preserving foods through high pressure processing. *Food Technology*, *62*, 32–38.

Balasubramanian, S., & Balasubramaniam, V. M. (2003). Compression heating influence of pressure transmitting fluids on bacteria inactivation during high pressure processing. *Food Research International*, *36*, 661–668.

Barba, F. J., Terefe, N. S., Buckow, R., Knorr, D., & Orlien, V. (2015). New opportunities and perspectives of high pressure treatment to improve health and safety attributes of foods. A review. *Food Research International*, *77*(1), 725–742. https://doi.org/10.1016/j.foodres.2015.05.015.

Barros-Velazquez, J. (2011). Innovations in food technology special issue. *Food Bioprocess Technology*, *4*, 831–832.

Basak, S., & Ramaswamy, H. S. (1998). Effect of high pressure processing on the texture of selected fruits and vegetables. *Journal of Texture Studies*, *29*, 587–601.

Black, E. P., Setlow, P., Hocking, A. D., Stewart, C. M., Kelly, A. L., & Hoover, D. G. (2007). Response of spores to high pressure processing. *Comprehensive Reviews in Food Science and Food Safety*, *6*, 103–119.

Bridgeman, P. (1914). The coagulation of albumen by pressure. *Journal of Biological Chemistry*, *19*(4), 511–512.

Butz, P., Fernandez Garcia, A., Lindauer, R., Dieterich, S., Bognar, A., & Tauscher, B. (2003). Influence of ultra-high pressure processing on fruit and vegetable products. *Journal of Food Engineering*, *56*, 233–236.

Butz, P., Edenharder, R., Fernández Garcia, A., Fister, H., & Merkel, C. (2002). Changes in functional properties of vegetables induced by high pressure treatment. *Food Research International*, *35*, 295–300.

Campus, M., Addis, M. F., Cappuccinelli, R., Porcu, M. C., Pretti, L., Tedde, V., et al. (2010). Stress relaxation behaviour and structural changes of muscle tissues from Gilthead Sea Bream (*Sparus aurata* L.) following high pressure treatment. *Journal of Food Engineering*, *96*, 192–198.

Campus, M. (2010). High pressure processing of meat, meat products and seafood. *Food Engineering Reviews*, *2*(4), 256–273.

Castellari, M., Matricardi, L., Arfelli, G., Rovere, P., & Amati, A. (1997). Effects of high pressure processing on phenoloxidase activity of grape musts. *Food Chemistry*, *60*, 647–649.

Cava, R., Ladero, L., Gonzalez, S., Carrasco, A., & Ramirez, M. R. (2009). Effect of pressure and holding time on colour, protein and lipid oxidation of sliced dry-cured Iberian ham and loin during refrigerated storage. *Innovative Food Science and Emerging Technologies*, *10*, 76–81.

Chawla, R., Patil, G., & Singh, A. (2011). High hydrostatic pressure technology in dairy processing: A review. *Journal of Food Science and Technology*, *48*(3), 260–268. https://doi.org/10.1007/s13197-010-0180-4.

Chéret, R., Aránzazu, H. A., Delbarre-Ladrat, C., de Lambarrie, M., & Verrez-Bagnis, V. (2006). Proteins and proteolytic activity changes during refrigerated storage in sea bass (*Dicentrarchus labrax* L.) muscle after high-pressure treatment. *European Food Research & Technology*, *222*, 527–535.

Collins, A. T., Hisao, K. H., & Kitawaki, H. (2000). Color changes produced in natural brown diamonds by high-pressure, high temperature treatment. *Diamond and Related Materials*, *9*, 113–122.

Crawford, Y. J., Murano, E. A., Olson, D. G., & Shenoy, K. (1996). Use of high hydrostatic pressure and irradiation to eliminate *Clostridium sporogenes* spores in chicken breast. *Journal of Food Protection*, *59*, 711–715.

Daher, D., Le Gourrierec, S., & Pérez-Lamela, C. (2017). Effect of high pressure processing on the microbial inactivation in fruit preparations and other vegetable based beverages: Review. *MDPI-Agriculture*, *7*, 72. https://doi.org/10.3390/agriculture7090072.

Dalmadi, I., Polyak-Feher, K., & Farkas, J. (2007). Effects of pressure and thermal pasteurization on volatiles of some berry fruits. *High Pressure Research*, *27*, 169–172.

Daryaei, H., Balasubramaniam, V. M., & Legan, J. D. (2013). Kinetics of Bacillus cereus spore inactivation in cooked rice by combined pressure-heat treatment. *Journal of Food Protection*, *76*, 616–623.

Daryaei, H., & Balasubramanian, V. M. (2012). Microbial decontamination of food by high pressure processing. In A. Demirci, & M. Ngadi (Eds.), *Microbial decontamination in the food industry*UK: Woodhead Publishing Limited.

de Heij, W. B. C., Van Schepdael, L. J. M. M., Moezelaar, R., Hoogland, Hans, Matser, A. M., & van den Berg, R. W. (2003). High-pressure sterilization: maximizing the benefits of adiabatic heating. *Food Technology - Chicago*, *57*, 37–41.

Duranton, F., Simonin, H., Guyon, C., Jung, S., & de Lamballerie, M. (2014). High-pressure processing of meats and seafood. In *Emerging technologies for food processing* (pp. 35–63). Academic Press.

Farkas, D., & Hoover, D. (2000). High pressure processing. *Journal of Food Science*, *65*, 47–64.

Ferreira, L., Afonso, C., Vila-Real, H., Alfaia, A., & Ribeiro, M. H. L. (2008). Evaluation of the effect of high pressure on naringin hydrolysis in grapefruit juice with naringinase immobilised in calcium alginate beads. *Food Technology and Biotechnology*, *46*, 146–150.

Flick, G. J. (2003). High hydrostatic pressure processing has potential: Commercial fish & shellfish technology (CFAST) journal reprint. *Originally Appearing in Global Aquaculture Advocate*, *6*(1), . Avure Technologies, Inc.

Fonberg-Broczek, M., Windyga, B., Szczawinski, J., Szczawinska, M., Pietrzak, D., & Prestamo, G. (2005). High pressure processing for food safety. *Acta Biochimica Polonica*, *52*, 721–724.

Garcia, E., & Barrett, D. M. (2002). *Preservative treatments for fresh-cut fruits and vegetables*. Department of Food and Science and Technology, University of California.

Guamis, B., Pla, R., Trujillo, A. J., Capellas, M., Gervilla, R., Saldo, J., et al. (2005). In *High pressure processing of milk and dairy and egg products* (pp. 343–360). New York: CRC Press.

Guerrero-Beltan, J. A., Barbosa-Canovas, G. V., & Swanson, B. G. (2005). High hydrostatic pressure processing of fruit and vegetable products. *Food Reviews International*, *21*, 411–425.

Hayakawa, I., Kanno, T., Yoshiyaina, K., & Fujio, Y. (1994). Oscillatory compared with continuous high pressure sterilization on *Bacillus stearothermophilus* spores. *Journal of Food Science*, *59*, 164–167.

Hayashi, R., Kawamura, Y., Nakasa, T., & Okinaka, O. (1989). Application of high pressure to food processing: pressurization of egg white and yolk, and properties of gels formed. *Agricultural and Biological Chemistry*, *53*(11), 2935–2939. http://dx.doi.org/10.1271/bbb1961.53.2935.

Heinz, V., & Buckow, R. (2009). Food preservation by high pressure. *Journal of Consumer Protection and Food Safety*, *5*(1), 73–81. https://doi.org/10.1007/s00003-009-0311-x.

Heinz, V., & Buckow, R. (2010). Food preservation by high pressure. *Journal fur Verbraucherschutz und Lebensmittelsicherheit*, *5*, 73–81.

Hendrickx, M., Ludikhuyze, L., Van den Broeck, I., & Weemaes, C. (1998). Effects of high pressure on enzymes related to food quality. *Trends in Food Science and Technology*, *9*, 197–203.

Hite, B.H. (1899). The effect of pressure in the preservation of milk: a preliminary report (Vol. 58). West Virginia Agricultural Experiment Station.

Hogan, E., Kelly, A. L., & Sun., D. W. (2005). High pressure processing of foods: an overview. In Da Wen Sun (Ed.), *Emerging technologies for food processing* (1st ed., pp. 30). San Diego: Elsevier Academic Press.

Howson, G. (2009). *High pressure processing for food safety, extended shelf life and all-natural.* USA: Avure Technologies.

Huff-Lonergan, E., Mitsuhashi, T., Beekman, D. D., Parrish, F. C., Olson, D. G., & Robson, R. M. (1996). Proteolysis of specific muscle structural proteins by mu-calpain at low pH and temperature is similar to degradation in postmortem bovine muscle. *Journal of Animal Science, 74*, 993–1008.

Huppertz, T., Fox, P. F., de Kruif, K. G., & Kelly, A. L. (2006). High pressure-induced changes in bovine milk proteins: a review. *Biochimica et Biophysica Acta, 1764*(3), 593–598. https://doi.org/10.1016/j.bbapap.2005.11.010.

Hygreeva, D., & Pandey, M. C. (2016). Novel approaches in improving the quality and safety aspects of processed meat products through high pressure processing technology-A review. *Trends in Food Science & Technology, 54*, 175–185.

Kadam, P. S., Jadhav, B. A., Salve, R. V., & Machewad, G. M. (2012). Review on the high pressure technology (HPT) for food preservation. *Journal of Food Processing and Technology, 3*, 135. https://doi.org/10.4172/2157-7110.1000135.

Karim, P. A. (2011). *High pressure processing as an alternative food preservation technology and its application for fruits and vegetables.* Nakhornpathom, 73170, Thailand: Manhattan, KS: Kansas State University. Mahidol University.

Kemp, C. M., Sensky, P. L., Bardsley, R. G., Buttery, P. J., & Parr, T. (2010). Tenderness–An enzymatic view. *Meat Science, 84*(2), 248–256.

Kingsley, D. H., Hoover, D. J., Papafragkou, E., & Richards, G. P. (2002). Inactivation of hepatitis A virus and calicivirus by high hydrostatic pressure. *Journal of Food Protection, 65*, 1605–1609.

Li, L., Feng, L., Yi, J., Hua, C., Chen, F., Liao, X., et al. (2010). High hydrostatic pressure inactivation of total aerobic bacteria, lactic acid bacteria, yeasts in sour Chinese cabbage. *International Journal of Food Microbiology, 142*, 180–184.

Liepa, M., Zagorska, J., & Galoburda, R. (2016). High-Pressure processing as novel technology in dairy industry: A Review. *Research for Rural Development, 1*, 46–83.

Low, P. S., & Somero, G. N. (1974). Temperature adaptation of enzymes. A proposed molecular basis for the different catalytic efficiencies of enzymes from ectotherms and endotherms. *Comparative Biochemistry and Physiology, 49*, 307–312.

Ludikhuyze, L. R., Van den Broeck, I., Weemaes, C. A., & Hen-drickx, M. E. (1997). Kinetic parameters for pressure temperature inactivation of *Bacillus subtilis* alpha-amylase under dynamic conditions. *Biotechnology Progress, 13*, 617–623.

Leadley, C. E., & Williams A. (1997). High pressure processing of food and drink: an overview of recent developments and future potential. Bulletin no. 14. Campden and Chorleywood Food Research Association New Technologies, Chipping Campden, Gloucestershire, UK.

Macfarlane, JJ (1985). High pressure technology and meat quality. In R. A. Lawrie (Ed.), *Developments in meat science* (3rd ed., pp. 155–184). London: Elsevier Applied Science.

Martínez, M. A., Velazquez, G., Cando, D., Núñez-Flores, R., Borderías, A. J., & Moreno, H. M. (2017). Effects of high pressure processing on protein fractions of blue crab (*Callinectes sapidus*) meat. *Innovative Food Science & Emerging Technologies, 41*, 323–329.

Messens, W., Van Camp, J., & Huyghebaert, A. (1997). The use of high pressure to modify the functionality of food proteins. *Trends in Food Science & Technology*, *8*(4), 107–112. http://dx.doi.org/10.1016/S0924-2244(97)01015-7.

Molina, E., Álvarez, M. D., Ramos, M., Olano, A., & López-Fandiño, R. (2000). Use of high-pressure-treated milk for the production of reduced-fat cheese. *International Dairy Journal*, *10*(7), 467–475.

Navarro, M., Verret, C., Pardon, P., & ElMoueffak, A. (2002). Changes in volatile aromatic compounds of strawberry puree treated by high-pressure during storage. *High Pressure Research*, *22*, 693–696.

Nguyen, L. T., Balasubramaniam, V. M., & Ratphitagsanti, W. (2013). Estimation of accumulated lethality under pressure-assisted thermal processing. *Food Bioprocess Technology*, *7*, 633–644.

Norton, T., & Sun, D. -W. (2008). Recent advances in the use of high pressure as an effective processing technique in the food industry. *Food Bioprocess Technology*, *1*, 2–34.

Oey, I., Van der, I., Plancken, V., Loey, A., & Hendrickx, M. (2008). Does high pressure processing influence nutritional aspects of plant based food systems? *Trends in Food Science & Technology*, *19*, 300–308.

Oeya, I., Lille, M., Loeya, V. A., & Hendrickxa, M. (2008). Effect of high pressure processing on colour, texture and flavor of fruit and vegetable-based food products: a review. *Trends in Food Science & Technology*, *19*, 320–328.

Ohmori, T., Shigehisa, T., Taji, S., & Hayashi, R (1991). Effect of high pressure on the protease activities in meat. *Agricultural and Biological Chemistry*, *55*, 357–361.

Orlien, V., Boserup, L., & Olsen, K. (2010). Casein micelle dissociation in skim milk during high-pressure treatment: effects of pressure, pH, and temperature. *Journal of Dairy Science*, *93*(1), 12–18. https://doi.org/10.3168/jds.2009-2244.

Palou, E., Lopez-Malo, A., BarbosaCanovas, G. V., Welti-Chanes, J., & Swanson, B. G. (1999). Polyphenyloxidase activity and color of blanched and high hydrostatic pressure treated banana puree. *Journal of Food Science*, *64*, 42–45.

Patras, A., Brunton, N., Da Pieve, S., Butler, F., & Downey, G. (2009). Effect of thermal and high pressure processing on antioxidant activity and instrumental color of tomato and carrot purees. *Innovative Food Science and Emerging Technologies*, *10*, 16–22. https://doi.org/10.1016/j.ifset.2008.09.008.

Patras, A., Brunton, N. P., Da Pieve, S., & Butler, F. (2009). Impact of high pressure processing on total antioxidant activity, phenolic, ascorbic acid, anthocyanin content and colour of strawberry and blackberry purees. *Innovative Food Science and Emerging Technologies*, *10*, 303–313.

Patterson, M. F. (2005). Microbiology of pressure-treated foods. *Journal of Appled Microbiology*, *98*, 1400–1409.

Perera, N., Gamage, T. V., Wakeling, L., Gamlath, G. G. S., & Versteeg, C. (2010). Colour and texture of apples high pressure processed in pineapple juice. *Innovative Food Science and Emerging Technologies*, *11*, 39–46.

Polydera, A. C., Stoforos, N. G., & Taoukis, P. S. (2005). Quality degradation kinetics of pasteurized and high pressure processed fresh Navel orange juice: nutritional parameters and shelf life. *Innovative Food Science and Emerging Technologies*, *6*, 1–9.

Porretta, S., Birzi, A., Ghizzoni, C., & Vicini, E. (1995). Effects of ultrahigh hydrostatic pressure treatments on the quality of tomato juice. *Food Chemistry*, *52*, 35–41.

Raghubeer, E. V. (2007). High hydrostatic pressure processing of seafood. Technical note "Seafood white paper". Available from: http://www.avure.com/archive/documents/Food-products/seafood-white-paper.pdf.

Rajan, S., Ahn, J., Balasubramaniam, V. M., & Yousef, A. E. (2006). Combined pressure-thermal inactivation kinetics of *Bacillus amyloliquefaciens* spores in egg patty mince. *Journal of Food Protection, 69*, 853–860.

Ramaswamy, H. S., Zaman, S. U., & Smith, J. P. (2008). High pressure destruction kinetics of E. coli (O157: H7) and *Listeria monocytogenes* (Scott A) in a fish slurry. *Journal of Food Engineering, 87*(1), 99–106. https://doi.org/10.1016/j.jfoodeng.2007.11.017. 0260-8774. Elsevier.

Ramirez-Suarez, J. C., & Morrissey, M. T. (2006). Effect of high pressure processing (HPP) on shelf life of albacore tuna (*Thunnus alalunga*) minced muscle. *Innovative Food Science and Emerging Technologies, 7*, 19–27. Elsevier Ltd.

Rastogi, N. K., Raghavarao, K. S. M. S., Balasubramaiam, V. M., Niranjan, K., & Knorr, D. (2007). Opportunities and challenges in high pressure processing of foods. *Critical Reviews in Food Science and Nutrition, 47*(1), 69–112. https://doi.org/10.1080/10408390600626420. Taylor and Francis Group.

Rastogi, N. K., Eshtiaghi, M. N., & Knorr, D. (1999). Effect of combined high pressure and heat treatment on the reduction of peroxidase and polyphenoloxidase activity in red grape. *Food Biotechnology, 13*, 195–208.

Reddy, N. R., Marshall, K. M., Morrissey, T. R., Loeza, V., Patazca, E., et al. (2013). Combined high pressure and thermal processing on inactivation of type A and proteolytic type B spores of *Clostridium botulinum*. *Journal of Food Protection, 76*(8), 1384–1392.

Režek Jambrak, A., Donsì, F., Paniwnyk, L., & Djekic, I. (2019). Impact of novel nonthermal processing on food quality: Sustainability, modelling, and negative aspects. *Journal of Food Quality, 2019*, 1–2.

Rodrigo, D., van Loey, A., & Hendrickx, M. (2007). Combined thermal and high pressure colour degradation of tomato puree and strawberry juice. *Journal of Food Engineering, 79*, 553–560.

Saucedo-Reyes, D., Marco-Celdrán, A., Pina-Pérez, M. C., Rodrigo, D., & Martínez-López, A. (2009). Modeling survival of high hydrostatic pressure treated stationary- and exponential-phase *Listeria innocua* cells. *Innovative Food Science and Emerging Technologies, 10*, 135–141.

Sazonova, S., Gramatina, I., Klava, D., & Galoburda, R. (2019). Effect of high pressure processing on raw pork microstructure and water holding capacity. *Agronomy Research, 17*(S2), 1452–1459.

Sharma, R. (2011). *Market trends in high pressure processing (HPP) food*. Australia: The University of Queensland.

Sila, D. N., Smout, C., Vu, S. T., Van Loey, A., & Hendrickx, M. (2005). Influence of pretreatment conditions on the texture and cell wall components of carrots during thermal processing. *Journal of Food Science, 70*, 85–91.

Sila, D. N., Duvetter, T., De Roeck, A., Verlent, I., Smout, C., Moates, K. G., et al. (2008). Texture changes of processed fruits and vegetables: potential use of high-pressure processing. *Trends Food Science & Technology, 19*, 309–319.

Singh, A., & Ramaswamy, H. (2013). Effect of high-pressure processing on color and textural properties of eggs. *Journal of Food Research, 2*(4), 11.

Skinner, G. E., Marshall, K. M., Morrissey, T. R., Loeza, V., Patazca, E., et al. (2014). Combined high pressure and thermal processing on inactivation of type E and nonproteolytic type B and F spores of *Clostridium botulinum*. *Journal of Food Protection, 77*(12), 2054–2061.

Stewart, C., Dunne, P. C., & Keener, L. (2015). Pressure-assisted thermal sterilization validation. In V. M. Balasubramaniam, G. V. Barbosa-Canovas, & H. L. M. Lelieveld (Eds.), *High pressure processing principles, technology and applications*New York: Springer.

Sun, X. D., & Holley, R. A. (2010). High hydrostatic pressure effects on the texture of meat and meat products. *Journal of Food Science*, *75*(1), R17–R23.

Szczawiński, J., Szczawińska, M., Stańczak, B., Fonberg-broczek, M., & Arabas, J. (1997). Effect of high pressure on survival of Listeria monocytogenes in ripened, sliced cheese at ambient temperature. In K. Heremans (Ed.), *High pressure research in biosciences and biotechnology* (pp. 295–298). Leuven, Belgium: Leuven Univ. Press.

Tangwongchai, R., Ledward, D. A., & Ames, J. M. (2000). Effect of high-pressure treatment on the texture of cherry tomato. *Journal of Agriculture and Food Chemistry*, *48*, 1434–1441.

Tauscher, B. (1995). Pasteurization of food by hydrostatic high pressure: chemical aspects. *Zeitschrift fur Lebensmitteluntersuchung und Forschung*, *200*, 3–13.

Ting, E. (2011). High-pressure processing equipment fundamentals. See Zhang et al. 2011, pp. 20-27.

Toepfl, S., Mathys, A., Heinz, V., & Knorr, D. (2006). Review: Potential of high hydrostatic pressure and pulsed electric fields for energy efficient and environmentally friendly food processing. *Food Review International*, *22*, 405–423.

Trejo-Araya, X. I., Smale, N., Zabaras, D., Winley, E., Forde, C., Stewart, C. M., et al. (2009). Sensory perception and quality attributes of high pressure processed carrots in comparison to raw, sous-vide and cooked carrots. *Innovative Food Science and Emerging Technology*, *10*, 420–433.

Wan, J. (2014). High pressure processing for inactivation of *Clostridium botulinum* spores for low-acid foods: Technical Program and Abstract Book of the 2014 International Nonthermal Processing Workshop and Short Course at The Ohio State University, Columbus, OH. October 21–23, V. M. Balasubramaniam (Ed.). Columbus, OH: Ohio University Press.

Weemaes, C. A., Ooms, V., Indrawati, L., Ludikhuyze, I., Broeck, Van den, Van Loey, A., et al. (1999). Pressure temperature degradation of green color in broccoli juice. *Journal of Food Science*, *64*, 504–508.

Winai, S., Paul, K., & Ioannis, Z. (2005). The effect of high hydrostatic pressure on the anthocyanins of raspberry (*Rubus idaeus*). *Food Chemistry*, *90*, 193–197.

Yagiz, Y., Kristinsson, H. G., Balaban, M. O., & Marshall, M. R. (2007). Effect of high pressure treatment on the quality of rainbow trout (Oncorhynchus mykiss) and mahi mahi (Coryphaena hippurus). *Journal of Food Science*, *72*(9), C509–C515.

Yordanov, D. G., & Angelova, G. V. (2011). High pressure processing for foods preserving. *Biotechnology & Biotechnological Equipment*, *24*, 1940–1945.

Chapter 4

Cold Plasma

Nathália M. Coutinho[a], Marcello R. Silveira[a], Ramon S. Rocha[a,b], Mônica Q. Freitas[a], Maria Carmela K.H. Duarte[a], Reverson F. Quero[c], Tatiana C. Pimentel[d], Erick A. Esmerino[a], Marcia Cristina Silva[b] and Adriano Gomes da Cruz[b,*]

[a]*Faculty of Veterinary Medicine, Federal Fluminense University (UFF), Niterói, Brazil;* [b]*Department of Food, Federal Institute of Education, Science and Technology of Rio de Janeiro (IFRJ), Rio de Janeiro, Brazil;* [c]*Department of Analytical Chemistry, Institute of Chemistry, University of Campinas (UNICAMP), Campinas, Brazil;* [d]*Federal Institute of Paraná, Paranavaí, Brazil*

**Corresponding author*

1 Introduction

In recent years, consumers have become the focus of the food industry, and their demands have dictated market trends. There is an increasing demand for innovation, including healthy foods and functional foods, associated with the concepts of naturalness, organic, and traditional foods (Giordano et al., 2018).

Therefore, researchers have faced a challenge to develop innovative food processing techniques aimed at a high nutritional value and maintenance of fresh-food taste (Troy et al., 2016). In this context, the use of emerging non-thermal processing technologies may be an effective approach for reducing additives while maintaining the natural flavors and quality of the processed products (Shahbaz et al., 2018).

Conventional technologies such as pasteurization involve the application of heat into the product, which can affect the sensory and nutritional properties of food, once the long time required to inactivate microorganisms can degrade the thermolabile compounds, including vitamins. In turn, nonthermal technologies are characterized by processing without heating, and that use alternative methods for preservation, therefore, providing the inactivation of the microorganisms and reaching the standards for pasteurization.

Non-thermal processing can reach the standards of microbiological safety of the food and improve the physicochemical, sensory, and nutritional properties of the products, maintaining the bioactive compounds content and modulating the activity of the enzymes. The non-thermal food processing methods include high hydrostatic pressure, pulsed electric field, pulsed-light technology, ultrasound, supercritical carbon dioxide technology, and irradiation. Studies have

Sustainable food processing and engineering challenges. http://dx.doi.org/10.1016/B978-0-12-822714-5.00004-8

shown a reduction of microbial load and enzymatic activity in food matrices subjected to emerging technologies, thus allowing the development of novel food ingredients. Other technologies, such as cold plasma, are still under improvement stages to attend scaling-up. Cold plasma is an emerging non-thermal technology that, in recent years, has been recognized as a promising methodology for decontamination of foods (Liao et al., 2017).

Plasma is an electrically energized matter in a gaseous state, which is generated by electrical discharges between electrodes. It can be achieved by applying energy in different forms, such as heat, electric or magnetic fields, laser light, microwave, and radiofrequency waves. The kinetic energy of the electrons is increased, generating collisions in the plasma products (ions, electrons, radiation of different wavelengths, and radical) at a higher rate (Mir et al., 2016). When performed at low temperatures, this process becomes practical, inexpensive, and suitable for heat-sensitive materials (Thirumdas, Sarangapani, & Annapure, 2014).

In 1922, Irving Langmuir, an American scientist, suggested that electrons, neutrons, and ions from an ionized gas can be characterized as a corpuscular material entrained in some fluid medium and called entertaining medium plasma, looking like a clear fluid as blood. Although there may be no fluid medium compelling neutrons, electrons, and ions in an ionized gas, the term "Plasma" was employed at the first time in 1928 by Langmuir defining the fourth state of the matter (Raviteja et a; l., 2019).

Cold plasma has several applications, including sterilization, disinfection, enzyme inactivation, and modulation of hydrophilic/hydrophobic properties, surface etching or deposition of thin films (Thirumdas et al. 2017; Sarangapani et al., 2016, 2018; Scholtz et al., 2015; Priyadarshini et al., 2018). This technology was first developed to enhance the surface energy of polymers and sterilization of medical devices (Bahrami et al., 2016). More recently, it emerged as a powerful disinfection tool in the food industry, for the dry disinfection of food surfaces, such as dairy products, beef, pork, eggs, fruits and vegetables, grains, poultry, fish, herbs, sprouted seeds, spices, and fresh products.

By properly varying the cold plasma parameters, the food material can be modified to attend the desired characteristics, with maintenance of the nutritional characteristics and texture properties of the products. Furthermore, it can provide microbial decontamination, as this technology can inactivate several microorganisms, as well as, act on spores and viruses (Barba et al., 2017; Misra and Jo, 2017).

This chapter presents the fundamentals and the process parameters associated to the cold plasma technology. The mechanisms involved in the inactivation of the microorganisms and the impact of the cold plasma on the quality of the products is described, from a microbial, physicochemical, sensory, and toxicological perspective. The advantages, disadvantages, the regulatory context, and the scale-up challenges from the laboratory to the industry are also discussed.

2 General aspects

2.1 Fundamentals

Plasma is defined as an ionized gas composed of neutral molecules, electrons, and charged particles in balanced proportions, which present various types of interactions (Pankaj and Keener, 2017). The efficiency of cold plasma for inactivation of the microorganisms is dependent on the parameters of the process, including the gas type, the treatment time, the intensity of the electric current, the gas flow rate, among others (Liao et al., 2017).

Higher rates of the gas flow and longer times of processing can promote the collisions and the alternatives of action of the reactive species on the microorganisms, increasing the effectiveness of the decontamination (Liao et al., 2017). The activity of non-thermal plasma on the microbial food decontamination has already been accepted (Coutinho et al., 2018). Also known as the fourth state of matter, along with solid, liquid, and gas, the plasma state is reached when energy is applied to a neutral gas, which may be of magnetic, thermal, electrical origin, or microwave or radio frequency which increases the kinetic energy of the electrons.

The parameters of the process and the type of gas can affect the plasma characteristics (Phan et al., 2017), resulting in a set of photons, free electrons, ions, and atoms (natural or excited states). They are denominated light species (electrons and photons) or heavy species (other components) (Misra et al., 2017). The number of positive and negative charges in the plasma is approximately equal, which gives it a global neutral character (Ekezie et al., 2017; Pankaj and Keener, 2017).

Plasma is classified as thermal plasma or cold plasma, according to the temperature of the plasma formed. Thermal plasma exhibits a thermodynamic equilibrium, once the temperature reached by electrons is similar to that reached by neutral heavy particles and the gas, thus reaching a macroscopic temperature of thousands of Kelvin (Scholtz et al., 2015). It can be used in applications that require high temperatures, such as in the metallurgical industry or chemical processes. In cold plasma, the electrons and ions are not in thermal equilibrium, therefore maintaining the temperature at much lower levels. Cold plasma can be classified in two groups, the quasi-equilibrium plasma, which have temperatures in the range of 100°C–150°C, and the non-equilibrium plasma, with temperatures up to 60°C (Mandal et al., 2018).

Cold plasma has played an important role in the treatment of food that are sensitive to heat, once uncharged molecules and the ions have little increases in the energy while remaining at low temperatures (Phan et al., 2017; Pankaj et al., 2018). Diverse methods have been used to generate a high amount of non-equilibrium cold plasmas, under different pressure conditions. Some of these methods are presented later, which were selected mainly for the potential industrial applications and germicidal effects.

2.2 Classification of plasma sources

Several methods are used for the generation of cold plasma, including glow discharge, dielectric barrier discharge (DBD), resistive barrier discharge (RBD), corona discharge, plasma jets, and microwave discharge (Fig. 4.1). These methods differ mainly in the shape, number, and arrangement of the electrodes, type of current used (continuous or alternating), and pressure conditions (vacuum or atmospheric pressure). In general, the cold plasma system comprises a treatment chamber, a discharge device, and gas and pressure control systems.

Glow discharge (Fig. 4.1A) is a plasma formed by the passage of electric current through a gas, by applying a voltage (100 V to several kV) between two electrodes in a glass tube containing a low-pressure gas (1–1000 Pa). When the voltage exceeds a defined value, the gas is ionized, and the energy level of electrons is higher than that of ions and neutral gases. The medium can act as a chemical reactor, and the reactions occur at low temperatures. High-energy electrons and other reactive species are involved in many events including oxidation, cleavage, or polymerization reactions. The advantage of glow discharge is the generation of plasma at a low temperature and in a large volume, thus being highly homogeneous (Momeni et al., 2018).

A dielectric barrier discharge (DBD) (Fig. 4.1B) is generated between two electrodes separated by dielectric layers, which prevents sparks from forming at the point of contact (Moreau et al., 2008). It is a direct or alternating current discharge, which is operated at frequencies in the range of 0.05–500 kHz and gas pressures from 104 to 106 Pa (Zhang et al., 2017). The typical distance in the gap varies in the range of millimeters to centimeters, and the equipment operates under reasonably high-power levels, which is the main disadvantage of the process, once the high ignition voltage of at least 10 kV requires attention and isolation depending on the electrode gap. DBD is an ideal plasma source for large surfaces (Phan et al., 2017). The major advantage of DBD is its adaptability, which includes a wide variety of gases that can be used for plasma generation (air, water vapor, noble gases, and almost all combinations of gases), the homogeneous discharge over a large area, the low gas flow rate ($<$100 sccm) or no gas requirement, and the different geometries of electrodes (Phan et al., 2017).

The resistive-barrier discharge (RBD) (Fig. 4.1C) method uses a similar configuration to the DBD, differing by using high resistivity material to cover at least one of the electrodes. This layer should distribute ballast limiting the discharge current, thus preventing arcing. The advantage of the RBD compared to the DBD is the chance of using direct current or alternating current power of low frequency (60 Hz) to conduct the discharge (Raviteja et al., 2019).

Corona discharge (Fig. 4.1D) is a weakly luminous discharge at atmospheric pressure. In this technique, the luminosity and ionization are predominantly located close to the sharp electrode (points, edges, or thin wires), as opposed to a flat electrode. The electric field is sufficient to accelerate the electrons (generated

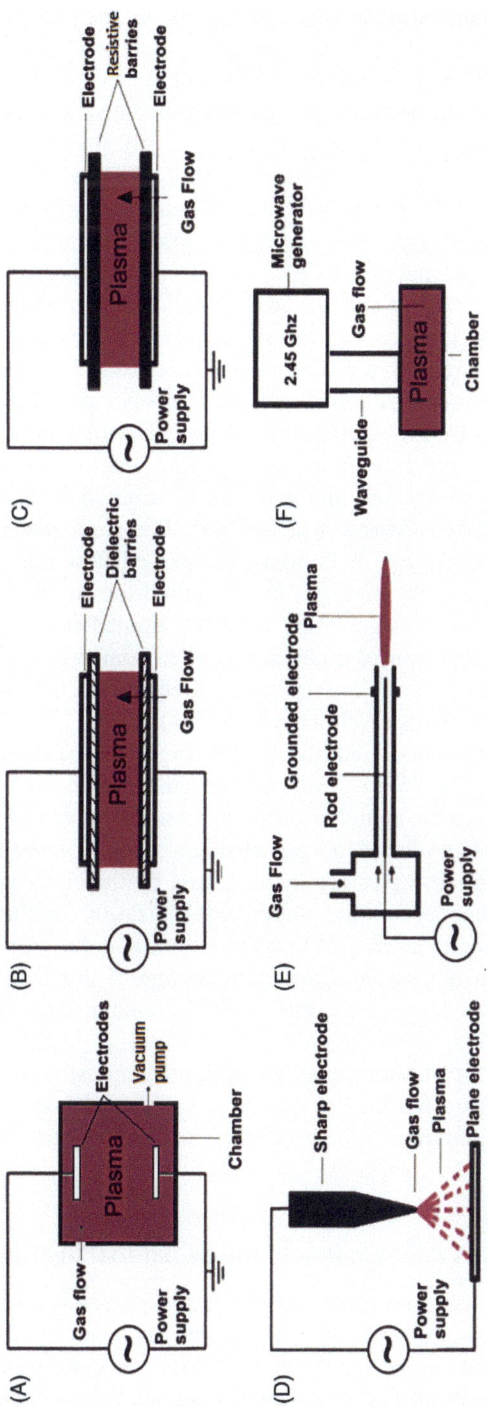

FIGURE 4.1 Cold plasma systems. (A) Glow discharge, (B) dielectric barrier discharge (DBD), (C) resistive barrier discharge, (D) corona discharge, (E) plasma jets, and (F) microwave discharge. *(Adapted from Coutinho et al., 2018.)*

randomly) up to the ionization energy level of the molecules or atoms of the encircle gas (Phan et al., 2017). Corona discharge can work in pulsed voltage mode or direct current, and the electrode may possess negative or positive potential (Pekárek, 2010). Although this system does not need high costs of implementation and operation, and the apparatus is relatively simple, it has restricted treatment capacity, with non-uniform configuration (Scholtz et al., 2010).

Plasma jets (Fig. 4.1E) present several configurations which allow the gas discharge operation in an open (non-sealed) electrode arrangement and the species of the discharge plasma can be projected in an open surrounding, forming a "plasma flame" (Nishime et al., 2017). The non-thermal plasma jets are formed at atmospheric pressure, using two coaxial electrodes, and the gas flows between them at an increased rate. The external electrode is grounded, and the electrode from the center is excited by a radio frequency power of 13.56 MHz. The ionization occurs by the acceleration of free electrons through the radiofrequency field and the electron-molecule collision, generating several reactive species (free radicals, molecules, and atoms in the excited state) that leave the nozzle at increased speed (Misra et al., 2016a). The gases usually employed are noble gases such as argon or helium at increased flow rates (> 10 slm). The main disadvantage is the high gas flow required for the "flame" and, consequently, the high cost. It is generally used for biomedical applications that require more specific and limited applications when compared to food processing. The advantages include the possibility of direct application and the ability to penetrate into narrow gaps (Weltmann et al., 2008).

Microwave-driven (MW-driven) discharges (Fig. 4.1F) can be formed in the absence of electrodes. The MW is resulted from a magnetron working usually at 2.45 GHz; and driven to the chamber of the process by coaxial cable or a wave guide, reaching the electrons of the gas of the process (Tolouie et al., 2017). Thus, the generated microwaves are incorporated by the electrons, leading to an increased kinetic energy and the consequent ionization (Schlüter and Fröhling, 2014). Despite being an easy to handle system, as it does not require electrodes and the spatial limitation is the disadvantage of this technique, which requires an array of discharges for the direct decontamination of large areas (Ehlbeck et al., 2010).

The wide variety of plasma generator designs demonstrates the flexibility and adaptability of the technology to the desired plasma characteristics and the particularities of the materials to be treated, paving the way for the progress of this technology.

2.3 Effectiveness of cold plasma in microbial inactivation

The chemistry of the plasma in the gas and liquid phases is complex and involves several reactions and species. Many agents can be produced during plasma application, as a function of the gas used in the process. Reactive oxygen species (ROS), which are related to the activity against microorganisms include ozone (O_3), hydrogen peroxide (H_2O_2), hydroperoxyl ($HO_2^{\bullet -}$), superoxide anion ($O_2^{\bullet -}$), alkoxyl ($RO^{\bullet -}$), singlet oxygen ($^1O_2$), peroxyl ($ROO^{\bullet -}$), carbonate anion radical ($CO_3^{\bullet -}$), and hydroxyl radical ($^{\bullet}OH$). The reactive nitrogen species (RNS) can include nitrogen dioxide radical ($^{\bullet}NO_2$), nitric oxide ($NO^{\bullet -}$), peroxynitrous acid (OONOH), peroxynitrite ($ONOO^{\bullet -}$), and alkylperoxynitrite (ROONO). The presence of a higher concentration of oxygen in the plasma discharges can result in a higher inhibition of the microorganisms, because a higher level of active species based on oxygen are found, including ozone and atomic oxygen (Lu et al., 2016; Misra and Jo, 2017), and mixtures such as N_2/O_2, N_2/N_2O, He/O_2, Ar/O_2, $He/O_2/H_2O$, and He/N_2 (Guo et al., 2015). The utilization of these agents in combination turns cold plasma an attractive methodology, as it is very difficult to the pathogens to have resistance to all these factors of stress from plasma (Ehlbeck et al., 2010).

Overview of the microbial inactivation mechanisms of cold plasma treatment represents in Fig. 4.2. The inactivation mechanism up to now has not been understood fully, but main effects are due to etching of the surface of the cells promoted by the reactive species that are formed during generation of plasma, intrinsic photodesorption of ultraviolet (UV) photons, diffusion of reactive species, oxidation of macromolecules, volatilization of compounds, and destruction of genetic material (Coutinho et al., 2018; Schlüter & Fröhling, 2014).

After intense exposure of microorganisms to plasma, plasma radicals (OH and NO) form volatile compounds (CO_2 and H_2O), which cause irreversible damage to the bacteria's cell surface, resulting in the death of the cell. Moreover, the membrane of the cell is submitted to electric fields of high intensity, which may cause the rupture of the membrane due to the electrostatic voltage experimented by the increased electrical charge (Misra and Jo, 2017).

The pore formation caused by the action of reactive species increases the permeability of the membrane, altering the transmembrane potential of the cell and the regulation of the pH into the cell, causing an acidification due to the humid air plasma. However, some microorganisms can have a cytoplasm with very dense structure and present cytoplasmic proteins that have buffering capacity, maintaining the pH of the medium. Thus, the pH can be considered a factor non-essential for the inactivation of the microorganisms (Moreau et al., 2008). The pore formation can also contribute to the liberation of the internal fluid, impairing the development of the actions of anti-stress, and restoring the microbial processes. Then, the active species are moved to the internal part of the cell, leading to the destruction of DNA, proteins, and other internal cell components (Phan et al., 2017).

Intrinsic photodesorption can be induced by UV irradiation, depending on the configuration of the plasma source, which can result in the rupture of the chemical bonds in the microorganisms, and the generation of volatile by-products (CHx and CO) from the atoms belonging to the microorganisms (Schlüter and Fröhling, 2014; Misra and Jo, 2017).

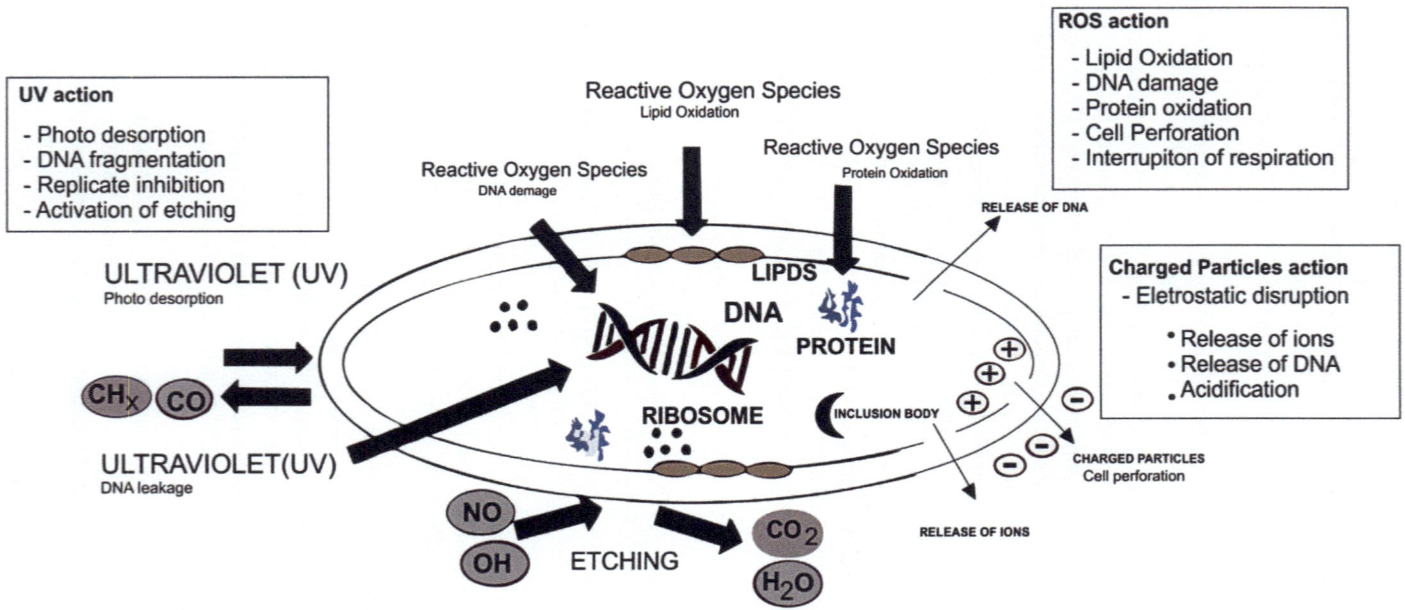

FIGURE 4.2 **Microbial inactivation mechanisms of cold plasma treatment**. *(Photo from Coutinho et al., 2018.)*

Changes in membrane integrity and also the pore formation can directly affect DNA, mainly through the breakdown of interactions between membrane proteins, with consequent DNA liberation from the cell. Furthermore, the DNA can be partially fragmented due to the exposition to the electrical discharges. The impact of plasma on the DNA is associated to the combined effects of UV and free radicals, such as OH and NO, as they may form dimers of thymine and break the strands, decreasing the replication ability of the bacteria (Moreau et al., 2008).

The reactive species react with membrane lipids from the microbial cells, leading to the formation of unsaturated fatty acid peroxides, and oxidation of the amino acids and nucleic acids. Lipids are considered the macromolecules with most vulnerability in the cell membranes, mainly because of the location close to the surface (Liao et al., 2017). Some authors have reported that changes in the lipids of the membrane promote the release of macromolecules (Schlüter and Fröhling, 2014). Therefore, the mechanisms of inactivation are responsible for many reactions in the cells of the microorganisms, such as lipid peroxidation (polyunsaturated fatty acids) and DNA and amino acids oxidation (Misra and Jo, 2017).

2.4 Process parameters

Several factors can affect the lethality of cold plasma technology including power, working gas, gas flow rate, and exposure mode, among others. The plasma generated exhibits specific characteristics depending on the gas used, which can be ambient air, nitrogen, or a mixture of gases, including noble gases, such as helium and argon. The number of reactive species and the cost of the process can also differ (Misra and Jo, 2017). There are many advantages of using noble gases, such as the emission spectra rich in ultraviolet, increased thermal conductivity (for removing heat), and lower voltage in the operating discharge at atmospheric pressure. However, they have higher cost when compared with the ambient air.

Concerning the liquid-gas environment, the reducing agents with higher importance are superoxide radicals (O_2^-) and hydrogen radicals (H), because of the degradation of compounds with increased electron affinity or that contain nitro groups (Misra et al., 2016b).

The generation of reactive species depends on the electrical input (frequency, voltage, and power). Higher electrical input and treatments at longer times can increase the efficiency of the process (Liao et al., 2017). However, the efficacy is limited due to the short lifetime of species, which precludes reaching larger areas, even if longer times of processing are used (Nishime et al., 2017).

The gas flow is significant factor that is related to the speed in which the active species are moved to the aimed location, and can affect the operation of discharge, the retention time, and the mass transfer of reactive species. Some short-lived species may not reach the sample, depending on the gas flow rate

(Nishime et al., 2017), thus increasing this parameter can enhance the collision and reaction possibilities of the reactive species (Zhang et al., 2017).

The sources of plasma can work in direct or indirect modes, being the latter also denominated plasma afterglow or remote plasma. These categories are defined according to the positioning of the sample, which can be located far, near, or within the plasma generation point. The groups are originated from the chemistry of cold plasma and defined according to the half-life and the reactivity of charged active species within the plasma.

The application of direct plasma results in the formation of several reactive species, most of them having noticeably short lifetimes (mainly milliseconds). Surface plasma reactions are also observed, such as deposition and etching (Misra and Jo, 2017). In general, direct exposure is more efficient than indirect exposure. The food is located near to the location where the plasma is generated and the exposure to plasma occurs before the recombination and loss of the active species, which provide higher concentrations of active agents (Laroussi, 2005).

In indirect or remote plasmas, the plasma is generated in a separate chamber and only longer living reactive species such as nitric oxide or ozone contact the sample (plasma exhaust) (Surowsky et al., 2015) because free electrons may recombine with other plasma products during the time of flight, such as heavy ions or atomic species. In addition, many short-lived neutral reactive species do not reach the sample. This operation mode provides a low heat transfer, and there is no effect of the charged particles, as their recombination occurs before the sample is reached. The advantage of this system is that the surface that will be treated is located at a separate point of the point where the plasma is generated, simplifying the device design and its operation and allows the treatment of objects with many sizes and shapes.

Relative humidity (RH) is an important factor related to the microbial inactivation of cold plasma, as it can affect the formation of the reactive species. Suitable water concentrations can result in a higher efficiency of inactivation, while the excess of water can lead to the effect dilution. At higher RH, it is observed higher formation of OH and peroxyl acid groups, because of the decomposition of the excess of molecules of water (Guo et al., 2015; Liao et al., 2017). Therefore, the utilization of water vapor in the system can form chemical species with potential to increase the antimicrobial activity.

Sensitivity to treatment can be different among microorganisms, thus the intrinsic characteristics of the agent to be eliminated must be widely studied. Microorganisms in the exponential phase (vegetative form) are less resistant than those in the stationary phase (sporulated form). Cold plasma is more effective in the inactivation of Gram-negative bacteria when compared to Gram-positive bacteria (Schlüter and Fröhling, 2014; Liao et al., 2017). Gram-positive cells have a thick peptidoglycan structure, which gives resistance to these type of cells to chemical changes. Gram-negative bacteria have a thinner membrane, allowing the electrostatic force to surpass the tensile strength of the membrane of the cell, making this type of microorganism more susceptible to the rupture of the membrane (Nishime et al., 2017). Fungi has a cell wall of chitin, that is more hard than the peptidoglycan of the cell wall of the bacteria, thus the fungal species are more resistant against cold plasma when compared to bacteria strains (Liao et al., 2017).

The concentration of microorganisms present in the food is associated to the efficiency of the cold plasma in the inactivation of microorganisms. Increased concentrations of microorganisms increase the possibility of the formation of cell agglomerates. The penetration ability of reactive plasma species is limited, with a small inactivating effect on cells located within the aggregation, thus it is important to select the process parameters to reduce the microbial load to adequate values (Liao et al., 2017).

The efficiency of the cold plasma treatment depends on the type of material to be processed. In solid samples, in general, cold plasma has limited penetration depth, thus the treatment is restricted to the surface of the product. The capacity of reactive species and plasma to enter in the solid foods depends on many factors, such as porosity, water content, physicochemical composition, among others (Surowsky et al., 2016). Reactive species of plasma with short lifetimes can react with the outer membrane of the cell, enter the cell, and provoke a damage in the where the internal part. Concerning the liquid samples, the penetration is less important because all the food is in contact with the plasma with their products, thus affecting the microorganisms and all other surrounding components (Surowsky et al., 2016).

The food physicochemical characteristics are significant for inactivation of microorganisms associated to the cold plasma. Regarding the moisture, higher moisture levels are associated to a higher decomposition of the molecules of water in hydroxyl radicals, which have a pronounced effect on the microbial inactivation, likely due to. The bacterial resistance is lower in an acidic food matrix. In addition, and the microbial tolerance to plasma is lower in acid products, due to the stress caused by the environment (Liao et al., 2017).

The shape and porosity of the food products can affect the antimicrobial treatment with cold plasma, once samples with irregular shapes and rough surfaces can present cavities that protect microorganisms from the action of plasma (Misra and Jo, 2017). For packaged foods, through-package, open treatments, and in-package plasma treatments can be performed. Concerning the in-package cold plasma treatment, the bactericidal molecules are generated in the package, allowing a large and constant exposure of microorganisms to the reactive species generated by the plasma until the gas returns to its original state. Another advantage of this technique is the prevention of post-processing contamination (Misra et al., 2019).

3 Effects of cold plasma and applicability in food processing

Technologies that operate at room temperature have been studied as alternatives to conventional thermal processing, which inactivates microorganisms and enzymes but negatively impacts the flavor, color, and nutritional value of foods.

Non-thermal technologies have attracted the attention of researchers worldwide to meet consumer demands and increase the shelf-life of products, minimizing the negative impact on the nutritional and quality attributes.

Originally, cold plasma technology was used in non-food industries to increase the printing and adhesion characteristics of electronics and polymers, to improve the energy on the surface of materials, as well as to upgrade the properties of paper and glass, among other applications (Ekezie et al., 2017). Studies have shown that cold plasma technology may be a powerful and profitable technology for the food industry, with effective results in the reduction of the microbial load of some foods, decontaminating food surfaces and packaging, or improving some quality attributes. In addition, cold plasma treatment has been used for toxin degradation, pesticide degradation, and allergen degradation.

3.1 Food decontamination, food surfaces, and packaging

The effect of cold plasma on the cells of the microorganisms is related to the interactions between the plasma ions and the cell. The plasma reactive species have been widely associated to pronounced effects caused by oxidation on the double bonds of the bi-layer of the microbial cell (on lipids), impairing the transit of macromolecules outside and inside the cell (Coutinho et al., 2018; Phan, Phan, Brennan, & Phimolsiripol, 2017; Thirumdas, Sarangapani, & Annapure, 2014).

The internal characteristics of the microorganism play a fundamental role in the efficiency of cold plasma since the sensitivity of the microorganism may differ within species and strains. Gram-positive bacteria were found to have less sensitivity to plasma due to its thick cell wall composed of peptidoglycans when compared with the Gram-negative bacteria, which consist of an outer membrane of lipopolysaccharide and a thin layer of peptidoglycan. Sporulated bacteria are more resistant to plasma treatment when compared with vegetative cells. Food with high bacteria counts forms large colonies of cells, reducing the penetration capacity of the reactive species formed by cold plasma, thus impairing cell destruction. In turn, cell walls of fungal species are composed mainly of chitin, which confers rigidity to the cells, thus presenting stronger resistance to the plasma treatment when compared with bacteria strains (Lee et al., 2017; Ekezie et al., 2017).

In-package cold plasma treatment is a novel alternative for decontamination of foods and prevention against cross-contamination. In a recent review study, Misra et al. (2019) used distinctive designs to study the in-package plasma treatment, including the volumetric DBD, surface dielectric barrier discharge (SDBD) into the package or on the package (less reported). The in-package systems of cold plasma have been used in the inactivation of microorganisms in model systems and foods, such as spores, bacteria, fungi, biofilms, mold, yeast, and viruses. The method was used very successfully for inactivate microorganisms from vegetables and fruits, when compared to dairy products rich in fat and meat products. However, the main limitation of in-package treatments with cold plasma is the oxidation of lipids due to the ROS; in addition, the in-package system should withstand the plasma discharges without causing chemical migration or formation of holes (Misra et al., 2019).

In milk and dairy products, studies on cold plasma have shown the potential of this technology to inactivate harmful microorganisms in several products, such as cheeses in slice (Song et al., 2009; Lee et al., 2012; Yong et al., 2015), and UHT milk (whole, semi-skimmed, and skimmed, Gurol et al., 2012). The microbial characteristics of milk submitted to DBD plasma formed in a container of plastic (15 kHz, 250 W, ambient air) for 5–10 min was evaluated by Kim et al. (2015). The total counts of the aerobic microorganisms of the untreated sample were 0.98 log CFU/mL, and in the milk-treated by plasma no viable cells were observed. The plasma treatment for 10 min resulted in reductions of approximately 2.40 log CFU/mL of *Listeria monocytogenes*, *Escherichia coli*, and *Salmonella typhimurium* in milks inoculated with these microorganisms.

In meat and meat products, the plasma treatment has proven to be effective for disinfection of various products (Frohling et al., 2012; Rod et al., 2012; Choi et al., 2016; Xiang et al., 2018) like ham (Song et al., 2009; Banu et al., 2012), bacon (Kim et al., 2011), and ready-to-eat meat (Rod et al., 2012). The effect of plasma at atmospheric pressure on the counts of *L. monocytogenes* on ham and chicken breast was evaluated by Lee et al. (2011). The authors used four types of carrier gases (N_2, $N_2 + O_2$, He, He + O_2), and stated that the utilization of $N_2 + O_2$ was more effective.

Several authors have reported the effectiveness of cold plasma processing for decontamination of agricultural products, including cucumber, carrot, pear (Wang et al., 2012), nuts, apples, mangoes, and melons (Banu et al., 2012), cabbage, lettuce, and dried fig (Lee et al., 2015), strawberries (Misra et al., 2014a), red pepper (Kim et al., 2014), seeds (Nishioka et al., 2014; Kim et al., 2017), apple juice (Surowsky et al., 2014), as well as fresh tomatoes, lettuce, carrots (Bermúdez-Aguirre et al., 2013).

3.2 Sensory and physicochemical aspects

The sensory and physicochemical properties are parameters associated to the preservation and quality of food products. However, there are few studies about the effect of cold plasma technology on the nutritional characteristics of food.

The cold plasma technology can be used to generate nitrite, which is an essential additive for cured meat products, once nitrite results in the development of a cured color and typical flavor (Parthasarathy and Bryan, 2012). Plasma contains various RNS and nitrogen oxides, including NO_2, NO_3, N_2O, N_2O_3, and N_2O_5, which are relatively stable species. After cold plasma processing, the dissolved nitrogen oxides react with water and form nitric and nitrous acids, then decomposing into nitrate (NO_3) and nitrite (NO_2), and the latter forms nitrous acid, which decomposes into nitrate and nitrogen oxide (Lee et al., 2017).

Emulsion sausages subjected to curing using plasma-treated water containing nitrite maintained the lipid oxidation, color, and sensory characteristics similar to those that used sodium nitrite for curing (Jung et al., 2015). Yong et al. (2017) studied the applicability of atmospheric pressure plasma in pork jerky. The application of cold plasma treatment resulted in the production of nitrite from the brine on the marinated pork, resulting in alterations in the color of the surface (a* value) and the nitroso-heme pigment concentration in the pork jerky. The authors reported that plasma technology can be a potentially safer alternative for pork jerky manufacturing without using sodium nitrite.

The impact of the application of cold plasma on the sensory and/or physicochemical properties of dairy products has also been reported in the literature. The studies included the color of cheese in slices (Lee et al., 2012; Yong et al., 2015), the volatile compounds, protein content, and fatty acid profile of milk (Kim et al., 2015), whey protein isolate (Segat et al., 2015), and milk fat (Saragapani et al., 2017), and pH, bioactive compounds, fatty acid composition, and volatile compounds profile of dairy beverages (Coutinho et al., 2019a; Silveira et al., 2019a). The rheological parameters and particle size, the microstructure, and the thermal properties of dairy beverages were also investigated (Coutinho et al., 2019b; Silveira et al., 2019b). Korachi et al. (2015) reported that the exposure of raw milk to cold plasma increased the concentration of total aldehyde after 20 min of application, with no changes in the levels of ketone and alcohol. Coutinho et al. (2019a) also observed the presence of aldehyde in all treatments; however, the volatile compounds were not identified in the study. The effect of cold plasma on the structure and activity of the alkaline phosphatase (ALP), an intrinsic milk enzyme was evaluated by Segat et al. (2016). The DBD plasma technology application resulted in the enzyme inactivation in few seconds. Enzymes are inactivated by oxidation reactions of peptides, thus changing the protein conformation and reducing the enzymatic activity.

Some recent studies on fruit and vegetable processing have shown that the effect of cold plasma on the food components is dependent on the processing conditions. The effect of cold plasma was investigated in orange juice (Alves Filho et al., 2016, 2019) and cashew apple juice (Rodríguez et al., 2017), which was subjected to nitrogen cold plasma using a lower gas flow rate, and presented increased vitamin C (ascorbic acid), total polyphenol content, and antioxidant activity. Moreover, blueberries submitted to cold plasma (air as ionized gas) presented higher concentration of flavonoids and phenolics after 1 min of process, and higher vitamin C content, which increased from 8.91 mg/100 g (control) to14.01 mg/100 g (80 kV) (Saragapani et al., 2017).

3.3 Mycotoxins degradation

Recently, cold plasma presented potential for food toxin degradation, increasing the number of studies in this area. Mycotoxins are toxic secondary metabolites produced by filamentous fungi, which can contaminate various food products and are classified as carcinogenic, mutagenic, and genotoxic agents, posing serious risks to human health (Pankaj and Keener, 2017).

Previous studies with cold plasma reported that this technology is effective in the mycotoxin degradation, such as aflatoxins (B1, B2, G1, G2) zearalenone, enniatins, deoxynivalenol, fumonisin, ochratoxin, deoxynivalenol, and trichothecene. Furthermore, it can inhibit several microorganisms that produce toxyn, such as *Aspergillus* spp., *Penicillium* spp., *Fusarium* spp., and *Alternaria* spp (Park et al., 2007; Kříž et al., 2015; Ouf et al., 2015; Abbasian et al., 2017; Bosch et al., 2017; Devi et al., 2017; Shi et al., 2017).

Some assumptions have been proposed to understand the plasma-assisted mycotoxin degradation, including the properties of the plasma used (e.g., the oxygen concentration and hydroxyl radicals) and the presence of photons and ultraviolet radiation. Wang et al. (2015) reported similarity between cold plasma and ultraviolet treatment to detect an intermediate product, a common molecule of aflatoxin degradation (AF), after applying UV treatment (C17H15O7) and cold plasma to AFB1. Furthermore, Shi et al. (2017) applied high-voltage atmospheric cold plasma with atmospheric air in gas to pure AFB1 powder on a glass slide and reported about 76% of AFB1 degradation after 5 min of process. The authors proposed that plasma-produced reactive oxygen species (ROS) were the major agents responsible for the degradation of AFB1, and reported that the plasma treatment can modify the ring of lactone, the methoxyl group and cyclopentanone, and eliminate the double bond between the carbons 8 and 9 of the furfural ring, decreasing the bioactivity of the AFB1.

Cold plasma is considered a suitable alternative to thermal processes, however, future studies are necessary to understand the mechanisms of action of the reactive species for the best industrial application of this technology, aimed to produce mycotoxin-free products.

3.4 Pesticide degradation

The frequent consumption of products with pesticides throughout life can bring serious risks to health. Researchers have shown that the cold plasma technology can reduce or degrade various pesticide residues from fresh fruits and vegetables, such as paraoxon, parathion (Kim et al., 2007; Heo et al., 2014), cypermethrin, chlorpyrifos, carbaryl (Phan et al., 2018; Feng et al., 2019), dichlorvos, omethoate (Bai et al., 2009), malathion (Zhu et al., 2010), among others. The cold plasma processing parameters, including the treatment time, the type of gas, the discharge gap, the input voltage, and power, besides the concentration of plasma-generated reactive species (oxygen species-ROS and reactive nitrogen species-RNS), and the energy level of electrons can affect the rate of pesticide decomposition (Gavahian and Khaneghah, 2019).

Zhou et al. (2018) studied the performance of discharge plasma technology in eliminating organophosphorus pesticides from the surface of the goji berry. The plasma treatment provided 96.8% and 99.6% degradation of dichlorvos and omethoate, respectively, after 30 min of plasma exposure at a discharge voltage of 10 kV, without compromising the fruit quality. The authors proposed that the degradation mechanism of organophosphorus during the plasma treatment may be related to various factors, such as the voltage applied, the exposure time, the energy of electrons, and the reactive species produced by the plasma equipment.

Similar results were observed by Misra et al. (2014b) in strawberries containing the pesticides azoxystrobin, cyprodinil, fludioxonil, and pyriproxyfen subjected to in-package dielectric barrier discharge plasma treatment. The increase in the treatment time was related to a decrease in the pesticide dissipation (69%, 45%, 71%, and 46% respectively after 5 min of treatment 80 kV), as observed in the degradation kinetics. Even during the post-discharge period, the excited nitrogen species were favored over the oxygen species, preventing the action of ozone by promoting acidic environments. Therefore, cold plasma has a potential to guarantee the chemical safety of food products.

3.5 Allergen degradation

The non-thermal processing by cold plasma has great potential to affect the allergenicity of food constituents. Some researchers have investigated the use of cold plasma treatment on the elimination of several allergens, included trypsin inhibitor in soymilk and soybeans (Li et al., 2017), β-conglycinin and glycinin in soy protein isolate (Meinlschmidt et al., 2016), and tropomyosin in shrimp (Shriver, 2011).

Nooji (2011) subjected wheat grains to the direct dielectric discharge cold plasma for 5 min, and found a reduction of immunoreactivity up to 37%. Probably, the conformational binding epitope was disrupted or masked by the reactive species generated after the plasma exposure, leading to the removal of IgE-mediated reaction. Venkataratnam et al. (2019) investigated the effects of cold plasma at atmospheric pressure on the antigenicity of the Ara h 1 protein. Whole peanut (WP) and dry defatted peanut flour (DPF) were submitted to plasma treatment (80 kV voltage) and various treatment times (0, 15, 30, 45, and 60 min). The authors reported a reduction of antigenicity up to 43% and 9.3% for DPF and WP, respectively, using competitive ELISA analysis with bovine serum albumin as a standard. The secondary structure modifications were also observed by circular dichroism induced by plasma reactive species.

In contrast, Tammineedi et al. (2013) studied milk α-casein, β-lactoglobulin, and α-lactalbumin subjected to cold atmospheric plasma at 13.56 MHz, using argon gas at a flow rate of 30.7 L/min and maximum treatment time of 15 min. The authors reported that the process did not reduce the allergenicity of milk proteins under the conditions studied, probably due to the low power, short treatment times, and the indirect exposure of the sample to the cold atmospheric plasma. Therefore, the pathway of allergen degradation by cold plasma is different for the various food sources, and the mechanism is uncertain, thus more related studies are needed.

4 Advantages and disadvantages of cold plasma processing

The plasma technology requires less energy and lower costs when compared to thermal processing (Ziuzina et al., 2014; Li et al., 2017). It is a chemical- and solvent-free technology, with a low impact on food quality and no need of water supply. It presents high microbial inactivation efficiency at low temperatures, with low impact on the internal product matrix, which makes its use especially convenient in heat-sensitive materials. The plasma technology can be also used to treat pre-packaged foods, preventing the incidence of post-process recontamination (Misra et al., 2019).

Cold plasma does not generate residues (Thirumdas, Sarangapani, & Annapure, 2014), provided a decrease in the production of effluents and exhaust gases after processing, as it consume reduced concentrations of water and chemical agents, which is very positive for both the economic and environmental aspects. Plasma does not need the application of pressure or heat, resulting in a technology with lower cost to inactivate microorganisms present in areas with large surface. It requires lower processing times when compared to conventional technologies to achieve similar microbial inactivation. Studies have shown that treatment times from few seconds to 2 min can promote higher than 5 log decreases for several microbial agents, such as pathogenic microorganisms (*S. Enteritidis, S. typhimurium, Staphylococcus aureus, E. coli, Campylobacter jejuni, L. monocytogenes, Aeromonas hydrophila, Campylobacter coli*) or sporulated microorganisms (*Clostridium botulinum* and *Bacillus cereus*, López et al., 2019).

Other advantages of plasma technology include the production on-demand of the agent, and the generation of plasma in the quantity and form necessary for the intention of use, besides being a technology efficient when considering the resources. Cold plasma is safe for the environment and fulfills the ecological standards, mainly because the reactive species can disappear after finishing the processing (Coutinho et al., 2018). The plasma generating equipment is scalable and can be used for wide-scale applications. Due to its many advantages, cold plasma has proven to be a very promising alternative technology in food preservation.

The treatment of irregularly and bulky shaped products is the main disadvantage of the cold plasma. The reactive plasma species can only penetrate to a limited extent, thus microbial inactivation occurs on the surface of the material, which demands attention concerning the volume and size of material to be subjected to the plasma treatment (Song et al., 2009). The coarse surface of many food products can provide sites for the microorganisms to attach, allowing them to scape from the cold plasma treatment (Yong et al., 2015). The effects of the reactive species at long time and the mechanisms of action are not totally elucidated. Reactive species can alter the morphology of the cells and impair the regular function (Jayasena et al., 2015).

Other limitations of plasma processing include the higher lipid oxidation (rancidity), the decrease in firmness of fruits, and the decreased peroxide values of chestnuts subjected to higher energy supply and treatment times

(Priyadarshini, Rajauria, O'donnell, & Tiwari, 2018; Thirumdas, Sarangapani, & Annapure, 2014). Therefore, aspects related to the color, nutritional content, overall quality of the food, and chemical changes have to be evaluated (Mason et al., 2015; Korachi et al., 2015; Sarangapani et al., 2018; Misra and Jo, 2017).

5 Current status of cold plasma processing

The development of novel food products can lead to a competitive advantage in the food industries. A survey on the Scopus database from 2000 to 2018 showed an increasing interest in these technologies, once the publications increased in the United States of America, and other countries in Europe and Latin America (Hernández-Hernández et al., 2019). There is a growing demand for studies on cold plasma worldwide, with emphasis for the industry, academia, and government research laboratories (Joshi et al., 2018; Sarangapani et al., 2018), with more than 100 papers published annually (Keener and Misra, 2016).

To supplant the thermal processing, the emerging technologies should offer clear advantages in terms of costs and product quality, or perform additional functions that cannot be performed by heat alone. Thermal processing is well established and has served the industry very well, in most cases, thus the costs of applying cold plasma in the food industry are one of the most important issues to be addressed, especially in relation to equipment, energy used, and general costs (Barba et al., 2017). For a novel processing to be viable, the equipment should have low cost, provide a fast and continuous response, require little maintenance, and work with different gases, mainly ambient air. Moreover, the maintenance of the quality and sensory characteristics of the products should compensate the higher cost (Li and Farid, 2016). Researchers have focused on several factors concerning the cold plasma processing unit, including the industrial use, technology transfer, reproducibility of results, cost, and maintenance of equipment, aimed at economic benefits (Thirumdas et al., 2017).

In relation to the health aspects, several studies have been carried out to characterize the antimicrobial efficacy and to understand the inactivation mechanisms of plasma technology (López et al., 2019). Thus, a scaling up strategy should be considered, to ensure a uniform treatment and to establish a dose for plasma applications (Cullen et al., 2018), which represents a big challenge due to the diversity of food products. Studies on the process safety must be performed, guaranteeing the safety of the consumer against toxicological, chemical, and physical risks.

The interest of the food industry in exploring a novel technology such as cold atmospheric plasma is also guided by the consumers' opinion. To date, the reaction of consumers to plasma technology is unknown. Consumers may perceive new food technologies as more dangerous than the traditional technologies. The positive effects of cold plasma treatments on food, and the economic and environmental aspects should be disseminated to raise the public awareness of the benefits (Keener and Misra, 2016). Efforts are required for a successful implementation of emerging technologies. The variety of literature data using different equipment and operational conditions impair the interpretation of results. The lack of uniformity of plasma in terms of properties and the different antimicrobial efficacy reported by the researchers makes it difficult to compare different studies. Therefore, little findings related to plasma efficacy are obtained, thus limiting to general conclusions.

The industrial implementation of cold plasma should be especially focused on the impact of the technology on the nutritional, sensory, and the stability of the products during storage, and the lack of toxicity of the chemical species generated. The design must consider compact and economical equipment, suitable for use in a processing line, with energy efficiency (López et al., 2019). Furthermore, efficacy tests equivalent to the alkaline phosphatase test for thermal processing of milk must be carried out to allow the use of the technology for "pasteurization" purpose, as established by the regulatory agencies (Coutinho et al., 2018).

Important aspects of this technology are still immature, particularly concerning its use in food. Therefore, plasma processing should be extended to industrial applications after food safety issues are clarified.

6 Regulatory context of cold plasma processing

The adoption of novel technologies is impacted by the regulatory aspect. For plasma-processed food, several factors, such as the inactivation of the pathogenic microorganisms, chemicals induced by plasma, and interactions between process, packaging, and product should be evaluated. The adoption can be discouraged due to the high cost of obtaining precise and consistent data to confirm safety and regulatory compliance. Such procedures are necessary to ensure that the requirements established by the regulatory agencies are met (Keener et al., 2013). All aspects must be supported by scientifically valid data, including actual and desired processing performance measures, advantages and limitations, among others (Niemira et al., 2018; Pankaj et al., 2018; Sarangapani et al., 2018).

Governments should encourage the application of new technologies through public policies with tax reduction on equipment imports and products, in addition to providing infrastructure and partnerships between governmental science and development agencies (Boothby et al., 2010). Better policies can accelerate the adoption of novel technologies. Joint ventures between industry and academy can contribute with cost reduction through the participation of public research laboratories (Milliou and Petrakis, 2011; Barba et al., 2015). Each country has a regulatory approval process for novel technologies to standardize the implementation and facilitate decision-making by suppliers, processors, retailers, and consumers. Despite the potential advantages, the introduction of a novel technology in the food industry sector is a challenge, due to the competitive pressure in a mature sector with low profit margins (Keener and Misra, 2016).

The Expert Group for Technical Advice on Organic Production (EGTOP) of the European Commission (EC) reported no restrictions on using cold plasma to preserve organic foods (Sakudo et al., 2020); however, the regulatory approval remains undetermined. In the United States, the US Department of Food Safety and Inspection Service (USDA-FSIS) is responsible for approval of the use of cold plasma in meat, poultry, or eggs, and the application and dosage of food additives. Although the Food and Drug Administration (FDA) and the Department of Agriculture (USDA) have not issued regulations about the use of cold plasma in food or surfaces which enter in contact with food, the ability of cold plasma for food inactivation of microorganisms has been drawn to attention in 2012 (Niemira, 2012). The atmospheric cold plasma technology used to treat food or food packaging is regulated by US EPA (US Environmental Protection Agency) as a pesticide and by the FDA as a FCS (food contact substance) (Sakudo et al., 2020).

For an effective implementation of plasma technology in the food industry, manufacturers, scientists, engineers, and consumers involved in efforts for regulatory approval of the use of plasma should provide documents addressing an overview of the technology, methodologies for measurement of plasma, collection and analysis of data, and regulatory steps for acceptance of food processing by plasma. These documents can help manufacturers of atmospheric plasma technology to obtain the regulatory approval (Keener and Misra, 2016).

7 Perspectives

Cold plasma is a non-thermal technology which several applications at the food industry. However, its successful implementation requires multidimensional aspects with special attention to the optimization of operational parameters to be used to the different food matrices as well microbiological and sensory studies as well as the evaluation of toxicological aspects and appropriate legislation. Efforts should be performed by the Health Agencies and Food Scientists to proportionate data to improve the knowledge about this technology.

References

Abbasian, E. G., Bayat, M., & Nosrati, A. C. (2017). Study of the effect of plasma jet on Fusarium isolates with ability to produce DON toxins. *World Family Medicine Journal/Middle East Journal of Family Medicine, 15*(9), 204–207.

Alves Filho, E. G., Cullen, P. J., Frias, J. M., Bourke, P., Tiwari, B. K., Brito, E. S., et al. (2016). Evaluation of plasma, high-pressure and ultrasound processing on the stability of fructooligosaccharides. *International Journal of Food Science & Technology, 51*, 2034–2040.

Alves Filho, E. G., Rodrigues, T. H. S., Fernandes, F. A. N., de Brito, E. S., Cullen, P. J., Frias, J. M., et al. (2019). An untargeted chemometric evaluation of plasma and ozone processing effect on volatile compounds in orange juice. *Innovative Food Science & Emerging Technologies, 53*, 63–69.

Bahrami, N., Bayliss, D., Chope, G., Penson, S., Perehinec, T., & Fisk, I. D. (2016). Cold plasma: A new technology to modify wheat flour functionality. *Food Chemistry, 202*, 247–253.

Bai, Y., Chen, J., Mu, H., Zhang, C., & Li., B. (2009). Reduction of dichlorvos and omethoate residues by O_2 plasma treatment. *Journal of Agricultural and Food Chemistry*, *57*(14), 6238–6245.

Banu, S. M., Sasikala, P., Dhanapal, A., Kavitha, V., Yazhini, G., & Rajamani, L. (2012). Cold plasma as a novel food processing technology. *International Journal of Emerging Trends in Engineering and Development*, *4*, 803–818.

Barba, F. J., Parniakov, O., Pereira, S. A., Wiktor, A., Grimi, N., Boussetta, N., et al. (2015). Current applications and new opportunities for the use of pulsed electric fields in food science and industry. *Food Research International*, *77*, 773–798.

Barba, F. J., Koubaa, M., Prado-Silva, L. D., Orlien, V., & Sant'Ana, A. D. S. (2017). Mild processing applied to the inactivation of the main foodborne bacterial pathogens: A review. *Trends in Food Science & Technology*, *66*, 20–35.

Bermúdez-Aguirre, D., Wemlinger, E., Pedrow, P., Barbosa-Canovas, G., & Garcia-Perez, M. (2013). Effect of atmospheric pressure cold plasma (APCP) on the inactivation of *Escherichia coli* in fresh produce. *Food Control*, *34*, 149–157.

Bosch, L., Pfohl, K., Avramidis, G., Wieneke, S., Viöl, W., & Karlovsky, P. (2017). Plasma-based degradation of mycotoxins produced by *Fusarium*, *Aspergillus*, and *Alternaria* species. *Toxins*, *9*(3), 97.

Boothby, D., Dufour, A., & Tang, J. (2010). Technology adoption, training and productivity performance. *Research Policy*, *39*(5), 650–661.

Choi, S., Puligundla, P., & Mok, C. (2016). Corona discharge plasma jet for inactivation of *Escherichia coli* O157:H7 and *Listeria monocytogenes* on inoculated pork and its impact on meat quality attributes. *Annals of Microbiology*, *66*, 685–694.

Coutinho, N. M., Silveira, M. R., Rocha, R. S., Moraes, J., Ferreira, M. V. S., Pimentel, T. C., et al. (2018). Cold plasma processing of milk and dairy products. *Trends in Food Science & Technology*, *74*, 56–68.

Coutinho, N. M., Silveira, M. R., Fernandes, L. M., Moraes, J., Pimentel, T. C., Freitas, M. Q., et al. (2019). Processing chocolate milk drink by low-pressure cold plasma technology. *Food Chemistry*, *278*, 276–283.

Coutinho, N. M., Silveira, M. R., Pimentel, T. C., Freitas, M. Q., Moraes, J., Fernandes, L. M., et al. (2019). Chocolate milk drink processed by cold plasma technology: Physical characteristics, thermal behavior and microstructure. *LWT-Food Science and Technology*, *102*, 324–329.

Cullen, P. J., Lalor, J., Scally, L., Boehm, D., Milosavljević, V., Bourke, P., et al. (2018). Translation of plasma technology from the lab to the food industry. *Plasma Processes and Polymers*, *15*, 1700085.

Devi, Y., Thirumdas, R., Sarangapani, C., Deshmukh, R. R., & Annapure, U. S. (2017). Influence of cold plasma on fungal growth and aflatoxins production on groundnuts. *Food Control*, *77*, 187–191.

Ehlbeck, J., Schnabel, U., Polak, M., Winter, J., Woedtke, T. V., Brandenburg, R., et al. (2010). Low temperature atmospheric pressure plasma sources for microbial decontamination. *Journal of Physics D: Applied Physics*, *44*(1), 013002.

Ekezie, F. -G. C., Sun, D. -W., & Cheng, J. -H. (2017). A review on recent advances in cold plasma technology for the food industry: Current applications and future trends. *Trends in Food Science & Technology*, *69*, 46–58.

Feng, X., Ma, X., Liu, H., Xie, J., He, C., & Fan, R. (2019). Argon plasma effects on maize: pesticide degradation and quality changes. *Journal of the Science of Food and Agriculture*, *99*, 5491–5498.

Frohling, A., Durek, J., Schnabel, U., Ehlbeck, J., Bolling, J., & Schluter, O. (2012). Indirect plasma treatment of fresh pork: decontamination efficiency and effects on quality attributes. *Innovative Food Science & Emerging Technologies*, *16*, 381–390.

Gavahian, M., & Khaneghah, A. M. (2019). Cold plasma as a tool for the elimination of food contaminants: Recent advances and future trends. *Critical Reviews in Food Science and Nutrition, 60*, 1581–1592.

Giordano, S., Clodoveo, M. L., Gennaro, B. D., & Corbo, F. (2018). Factors determining neophobia and neophilia with regard to new technologies applied to the food sector: A systematic review. *International Journal of Gastronomy and Food Science, 11*, 1–19.

Guo, J., Huang, K., & Wang, J. (2015). Bactericidal effect of various non-thermal plasma agents and the influence of experimental conditions in microbial inactivation: A review. *Food Control, 50*, 482–490.

Gurol, C., Ekinci, F. Y., Aslan, N., & Korachi, M. (2012). Low-temperature plasma for decontamination of *E. coli* in milk. *International Journal of Food Microbiology, 157*, 1–5.

Heo, N. S., Lee, M. K., Kim, G. W., Lee, S. J., Park, J. Y., & Park, T. J. (2014). Microbial inactivation and pesticide removal by remote exposure of atmospheric air plasma in confined environments. *Journal of Bioscience and Bioengineering, 117*, 81–85.

Hernández-Hernández, H., Moreno-Vilet, L., & Villanueva-Rodríguez, S. (2019). Current status of emerging food processing technologies in Latin America: Novel non-thermal processing. *Innovative Food Science & Emerging Technologies, 58*, 102233.

Jayasena, D. D., Kim, H. J., Yong, H. I., Park, S., Kim, K., Choe, W., et al. (2015). Flexible thin-layer dielectric barrier discharge plasma treatment of pork butt and beef loin: Effects on pathogen inactivation and meat-quality attributes. *Food Microbiology, 46*, 51–57.

Joshi, I., Salvi, D., Schaffner, D. W., & Karwe, M. V. (2018). Characterization of microbial inactivation using plasma-activated water and plasma-activated acidified buffer. *Journal of Food Protection, 81*, 1472–1480.

Jung, S., Kim, H. J., Park, S., Yong, H. I., Choe, J. H., Jeon, H. J., et al. (2015). The use of atmospheric pressure plasma-treated water as a source of nitrite for emulsion type sausage. *Meat Science, 108*, 132–137.

Keener, L., Nicholson-Keener, S. M., & Koutchma, T. (2013). Harmonization of legislation and regulations to achieve food safety: US and Canada perspective. *Journal of the Science of Food and Agriculture, 94*, 1947–1953.

Keener, K., & Misra, N. (2016). Future of cold plasma in food processing. In *Cold plasma in food and agriculture* (pp. 343–360). Academic Press.

Kim, S. H., Kim, J. H., & Kang, B. K. (2007). Decomposition reaction of organophosphorus nerve agents on solid surfaces with atmospheric radio frequency plasma generated gaseous species. *Langmuir, 23*(15), 8074–8078.

Kim, B., Yun, H., Jung, S., Jung, Y., Jung, H., Choe, W., et al. (2011). Effect of atmospheric pressure plasma on inactivation of pathogens inoculated onto bacon using two different gas compositions. *Food Microbiology, 28*, 9–13.

Kim, J. E., Lee, D. U., & Min, S. C. (2014). Microbial decontamination of red pepper powder by cold plasma. *Food Microbiology, 38*, 128–136.

Kim, H. J., Yong, H. I., Park, S., Kim, K., Choe, W., & Jo, C. (2015). Microbial safety and quality attributes of milk following treatment with atmospheric pressure encapsulated dielectric barrier discharge plasma. *Food Control, 47*, 451–456.

Kim, J. W., Puligundla, P., & Mok, C. (2017). Effect of corona discharge plasma jet on surface-borne microorganisms and sprouting of broccoli seeds. *Journal of the Science of Food and Agriculture, 97*, 128–134.

Korachi, M., Ozen, F., Aslan, N., Vannini, L., Guerzoni, M. E., Gottardi, D., et al. (2015). Biochemical changes to milk following treatment by a novel, cold atmospheric plasma system. *International Dairy Journal, 42*, 64–69.

Kříž, P., Bartoš, P., Havelka, Z., Kadlec, J., Olšan, O., Špatenka, P., et al. (2015). Influence of plasma treatment in open air on mycotoxin content and grain nutriments. *Plasma Medical, 5*, 145–158.

Laroussi, M. (2005). Low temperature plasma-based sterilization: Overview and state-of-the-art. *Plasma Processes and Polymers, 2*(5), 391–400.

Lee, H. J., Jung, H., Choe, W., Ham, J. S., Lee, J. H., & Jo, C. (2011). Inactivation of *Listeria monocytogenes* on agar and processed meat surfaces by atmospheric pressure plasma jets. *Food Microbiology, 28*, 1468–1471.

Lee, H. J., Jung, S., Jung, H., Park, S., Choe, W., Ham, J. S., et al. (2012). Evaluation of a dielectric barrier discharge plasma system for inactivating pathogens on cheese slices. *Journal of Animal Science and Technology, 54*, 191–198.

Lee, H., Kim, J. E., Chung, M. S., & Min, S. C. (2015). Cold plasma treatment for the microbiological safety of cabbage, lettuce, and dried figs. *Food Microbiology, 51*, 74–80.

Lee, J., Lee, C. W., Yong, H. I., Lee, H. J., Jo, C., & Jung, S. (2017). Use of atmospheric pressure cold plasma for meat industry. *Korean Journal for Food Science of Animal Resources, 37*, 477–485.

Li, X., & Farid, M. (2016). A review on recent development in non-conventional food sterilization technologies. *Journal of Food Engineering, 182*, 33–45.

Li, J., Xiang, Q., Liu, X., Ding, T., Zhang, X., Zhai, Y., et al. (2017). Inactivation of soybean trypsin inhibitor by dielectric-barrier discharge (DBD) plasma. *Food Chemistry, 232*, 515–522.

Liao, X., Liu, D., Xiang, Q., Ahn, J., Chen, S., Ye, X., et al. (2017). Inactivation mechanisms of non-thermal plasma on microbes: A review. *Food Control, 75*, 83–91.

López, M., Calvo, T., Prieto, M., Múgica-Vidal, R., Muro-Fraguas, I., Alba-Elías, F., et al. (2019). A review on non-thermal atmospheric plasma for food preservation: Mode of action, determinants of effectiveness, and applications. *Frontiers in Microbiology, 10*, 622.

Lu, X., Naidis, G. V., Laroussi, M., Reuter, S., Graves, D. B., & Ostrikov, K. (2016). Reactive species in non-equilibrium atmospheric-pressure plasmas: Generation, transport, and biological effects. *Physics Reports, 630*, 1–84.

Mandal, R., Singh, A., & Singh, A. P. (2018). Recent developments in cold plasma decontamination technology in the food industry. *Trends in Food Science & Technology, 80*, 93–103.

Mason, T., Chemat, F., & Ashokkumar, M. (2015). Power ultrasonics for food processing. In *Power ultrasonics* (pp. 815–843). Woodhead Publishing.

Meinlschmidt, P., Ueberham, E., Lehmann, J., Reineke, K., Schlüter, O., & Schweiggert-Weisz, U. (2016). The effects of pulsed ultraviolet light, cold atmospheric pressure plasma, and gammairradiation on the immunoreactivity of soy protein isolate. *Innovative Food Science and Emerging Technologies, 38*, 374–383.

Milliou, C., & Petrakis, E. (2011). Timing of technology adoption and product market competition. *International Journal of Industrial Organization, 29*(5), 513–523.

Mir, S. A., Shah, M. A., & Mir, M. M. (2016). Understanding the role of plasma technology in food industry. *Food and Bioprocess Technology, 9*(5), 734–750.

Misra, N. N., Patil, S., Moiseev, T., Bourke, P., Mosnier, J. P., Keener, K. M., et al. (2014). In-package atmospheric pressure cold plasma treatment of strawberries. *Journal of Food Engineering, 125*, 131–138.

Misra, N. N., Pankaj, S. K., Walsh, T., O'Regan, F., Bourke, P., & Cullen, P. J. (2014). In-package nonthermal plasma degradation of pesticides on fresh produce. *Journal of Hazardous Materials, 271*, 33–40.

Misra, N., Pankaj, S., Segat, A., & Ishikawa, K. (2016). Cold plasma interactions with enzymes in foods and model systems. *Trends in Food Science & Technology, 55*, 39–47.

Misra, N. N., Schlüter, O., & Cullen, P. J. (Eds.). (2016). *Cold plasma in food and agriculture: fundamentals and applications*. Academic Press.
Misra, N., & Jo, C. (2017). Applications of cold plasma technology for microbiological safety in meat industry. *Trends in Food Science & Technology*, *64*, 74–86.
Misra, N., Koubaa, M., Roohinejad, S., Juliano, P., Alpas, H., Inácio, R. S., et al. (2017). Landmarks in the historical development of twenty first century food processing technologies. *Food Research International*, *97*, 318–339.
Misra, N., Yepez, X., Xu, L., & Keener, K. (2019). In-package cold plasma technologies. *Journal of Food Engineering*, *244*, 21–31.
Momeni, M., Tabibiazar, M., Khorram, S., Zakerhamidi, M., Mohammadifar, M., Valizadeh, H., et al. (2018). Pectin modification assisted by nitrogen glow discharge plasma. *International Journal of Biological Macromolecules*, *120*, 2572–2578.
Moreau, M., Orange, N., & Feuilloley, M. (2008). Non-thermal plasma technologies: New tools for bio-decontamination. *Biotechnology Advances*, *26*(6), 610–617.
Niemira, B. A. (2012). Cold plasma decontamination of foods. *Annual Review of Food Science and Technology*, *3*(1), 125–142.
Niemira, B. A., Boyd, G., & Sites, J. (2018). Cold plasma inactivation of *Escherichia coli* O157:H7 biofilms. *Frontiers in Sustainable Food Systems*, *2*.
Nishime, T., Borges, A., Koga-Ito, C., Machida, M., Hein, L., & Kostov, K. (2017). Non-thermal atmospheric pressure plasma jet applied to inactivation of different microorganisms. *Surface and Coatings Technology*, *312*, 19–24.
Nishioka, T., Takai, Y., Kawaradani, M., Okada, K., Tanimoto, H., Misawa, T., et al. (2014). Seed disinfection effect of atmospheric pressure plasma and low pressure plasma on *Rhizoctonia solani*. *Biocontrol Science*, *19*, 99–102.
Nooji, J. K. (2011). *Reduction of wheat allergen potency by pulsed ultraviolet light, high hydrostatic pressure, and non-thermal plasma*. Gainesville, FL: University of Florida.
Ouf, S. A., Basher, A. H., & Mohamed, A. A. H. (2015). Inhibitory effect of double atmospheric pressure argon cold plasma on spores and mycotoxin production of *Aspergillus niger* contaminating date palm fruits. *Journal of the Science of Food and Agriculture*, *95*(15), 3204–3210.
Pankaj, S. K., & Keener, K. M. (2017). Cold plasma: background, applications and current trends. *Current Opinion in Food Science*, *16*, 49–52.
Pankaj, S., Shi, H., & Keener, K. M. (2018). A review of novel physical and chemical decontamination technologies for aflatoxin in food. *Trends in Food Science & Technology*, *71*, 73–83.
Park, B. J., Takatori, K., Sugita-Konishi, Y., Kim, I. H., Lee, M. H., Han, D. W., et al. (2007). Degradation of mycotoxins using microwave-induced argon plasma at atmospheric pressure. *Surface and Coatings Technology*, *201*, 9–11.
Parthasarathy, D. K., & Bryan, N. S. (2012). Sodium nitrite: the "cure" for nitric oxide insufficiency. *Meat Science*, *92*, 274–279.
Pekárek, S. (2010). DC corona discharge ozone production enhanced by magnetic field. *The European Physical Journal D*, *56*(1), 91–98.
Phan, K. T. K., Phan, H. T., Brennan, C. S., & Phimolsiripol, Y. (2017). Nonthermal plasma for pesticide and microbial elimination on fruits and vegetables: an overview. *International Journal of Food Science & Technology*, *52*(10), 2127–2137.
Phan, K. T. K., Phan, H. T., Boonyawan, D., Intipunya, P., Brennan, C. S., Regenstein, J. M., et al. (2018). Non-thermal plasma for elimination of pesticide residues in mango. *Innovative Food Science and Emerging Technologies*, *48*, 164–171.
Priyadarshini, A., Rajauria, G., O'donnell, C. P., & Tiwari, B. K. (2018). Emerging food processing technologies and factors impacting their industrial adoption. *Critical Reviews in Food Science and Nutrition*, *59*(19), 3082–3101.

Raviteja, T., Dayam, S. K., & Yashwanth, J. (2019). A Study on cold plasma for food preservation. *Journal of Scientific Research and Reports*, *23*(4), 1–14.

Rod, S. K., Hansen, F., Leipold, F., & Knochel, S. (2012). Cold atmospheric pressure plasma treatment of ready-to-eat meat: inactivation of *Listeria innocua* and changes in product quality. *Food Microbiology*, *30*, 233–238.

Rodríguez, O., Gomes, W. F., Rodrigues, S., & Fernandes, F. A. (2017). Effect of indirect cold plasma treatment on cashew apple juice (*Anacardium occidentale* L). *LWT—Food Science and Technology*, *84*, 457–463.

Sakudo, A., Misawa, T., & Yagyu, Y. (2020). Equipment design for cold plasma disinfection of food products. In *Advances in cold plasma applications for food safety and preservation* (pp. 289–307). Academic Press.

Sarangapani, C., Thirumdas, R., Devi, Y., Trimukhe, A., Deshmukh, R. R., & Annapure, U. S. (2016). Effect of low-pressure plasma on physico-chemical and functional properties of parboiled rice flour. *LWT—Food Science and Technology*, *69*, 482–489.

Saragapani, C., Keogh, D. R., Dunne, J., Bourke, P., & Cullen, P. J. (2017). Characterisation of cold plasma treated beef and dairy lipids using spectroscopic and chromatographic methods. *Food Chemistry*, *235*, 324–333.

Sarangapani, C., Patange, A., Bourke, P., Keener, K., & Cullen, P. (2018). Recent advances in the application of cold plasma technology in foods. *Annual Review of Food Science and Technology*, *9*(1), 609–629.

Schlüter, O., & Fröhling, A. (2014). Cold plasma for bioefficient food processing. In C. A. Batt, & M. -L. Tortorello (Eds.), *Encyclopedia of food microbiology* (pp. 948–953). Academic Press.

Scholtz, V., Julák, J., & Kříha, Vítězslav (2010). The microbicidal effect of low-temperature plasma generated by corona discharge: Comparison of various microorganisms on an agar surface or in aqueous suspension. *Plasma Processes and Polymers*, *7*(3–4), 237–243.

Scholtz, V., Pazlarova, J., Souskova, H., Khun, J., & Julak, J. (2015). Nonthermal plasma—A tool for decontamination and disinfection. *Biotechnology Advances*, *33*(6), 1108–1119.

Segat, A., Misra, N. N., Cullen, P. J., & Innocente, N. (2015). Atmospheric pressure cold plasma (ACP) treatment of whey protein isolate model solution. *Innovative Food Science & Emerging Technologies*, *29*, 247–254.

Segat, A., Misra, N. N., Cullen, P. J., & Innocente, N. (2016). Effect of atmospheric pressure cold plasma (ACP) on activity and structure of alkaline phosphatase. *Food and Bioproducts Processing*, *98*, 181–188.

Shahbaz, H. M., Kim, J. U., Kim, S. -H., & Park, J. (2018). Advances in nonthermal processing technologies for enhanced microbiological safety and quality of fresh fruit and juice products. In *Food processing for increased quality and consumption* (pp. 179–217). Academic Press.

Shi, H., Ileleji, K., Stroshine, R. L., Keener, K., & Jensen, J. L. (2017). Reduction of aflatoxin in corn by high voltage atmospheric cold plasma. *Food Bioprocess Technology*, *10*(6), 1042–1052.

Shriver, S. K. (2011). Effect of selected nonthermal processing methods on the allergen reactivity of Atlantic white shrimp (Litopenaeus setiferus), Doctoral dissertation, University of Florida.

Silveira, M. R., Coutinho, N. M., Esmerino, E. A., Moraes, J., Fernandes, L. M., Pimentel, T. C., et al. (2019). Guava-flavored whey beverage processed by cold plasma technology: Bioactive compounds, fatty acid profile and volatile compounds. *Food Chemistry*, *279*, 120–127.

Silveira, M. R., Coutinho, N. M., Rocha, R. S., Moraes, J., Esmerino, E. A., Pimentel, T. C., et al. (2019). Guava flavored whey-beverage processed by cold plasma: Physical characteristics, thermal behavior and microstructure. *Food Research International*, *119*, 564–570.

Song, H. P., Kim, B., Choe, J. H., Jung, S., Moon, S. Y., Choe, W., et al. (2009). Evaluation of atmospheric pressure plasma to improve the safety of sliced cheese and ham inoculated by 3-strain cocktail *Listeria monocytogenes*. *Food Microbiology*, *26*, 432–436.

Surowsky, B., Frohling, A., Gottschalk, N., Schluter, O., & Knorr, D. (2014). Impact of cold plasma on *Citrobacter freundii* in apple juice: inactivation kinetics and mechanisms. *International Journal of Food Microbiology*, *174*, 63–71.

Surowsky, B., Schlüter, O., & Knorr, D. (2015). Interactions of non-thermal atmospheric pressure plasma with solid and liquid food systems: A review. *Food Engineering Reviews*, *7*, 82–108.

Surowsky, B., Bußler, S., & Schlüter, O. (2016). Cold plasma interactions with food constituents in liquid and solid food matrices. In *Cold Plasma in Food and Agriculture* (pp. 179–203). Academic Press.

Tammineedi, C. V. R. K., Choudhary, R., Perez-Alvarado, G. C., & Watson, D. G. (2013). Determining the effect of UV-C, high intensity ultrasound and nonthermal atmospheric plasma treatments on reducing the allergenicity of a-casein and whey proteins. *LWT—Food Science and Technology*, *54*(1), 35–41.

Thirumdas, R., Kadam, D., & Annapure, U. S. (2017). Cold plasma: An alternative technology for the starch modification. *Food Biophysics*, *12*(1), 129–139.

Thirumdas, R., Sarangapani, C., & Annapure, U. S. (2014). Cold plasma: A novel non-thermal technology for food processing. *Food Biophysics*, *10*, 1–11.

Tolouie, H., Mohammadifar, M. A., Ghomi, H., & Hashemi, M. (2017). Cold atmospheric plasma manipulation of proteins in food systems. *Critical Reviews in Food Science and Nutrition*, *58*(15), 2583–2597.

Troy, D. J., Ojha, K. S., Kerry, J. P., & Tiwari, B. K. (2016). Sustainable and consumer-friendly emerging technologies for application within the meat industry: An overview. *Meat Science*, *120*, 2–9.

Venkataratnam, H., Sarangapani, C., Cahill, O., & Ryan, C. B. (2019). Effect of cold plasma treatment on the antigenicity of peanut allergen Ara h 1. *Innovative Food Science & Emerging Technologies*, *52*, 368–375.

Wang, R. X., Nian, W. F., Wu, H. Y., Feng, H. Q., Zhang, K., Zhang, J., et al. (2012). Atmospheric-pressure cold plasma treatment of contaminated fresh fruit and vegetable slices: inactivation and physiochemical properties evaluation. *European Physical Journal D*, *66*, 276.

Wang, S. Q., Huang, G. Q., Li, Y. P., Xiao, J. X., Zhang, Y., & Jiang, W. L. (2015). Degradation of aflatoxin B1 by low-temperature radio frequency plasma and degradation product elucidation. *European Food Research and Technology*, *241*(1), 103–113.

Weltmann, K. -D., Brandenburg, R., Woedtke, T. V., Ehlbeck, J., Foest, R., Stieber, M., et al. (2008). Antimicrobial treatment of heat sensitive products by miniaturized atmospheric pressure plasma jets (APPJs). *Journal of Physics D: Applied Physics*, *41*(19), 194008.

Xiang, Q., Liu, X., Li, J., Ding, T., Zhang, H., Zhang, X., et al. (2018). Influences of cold atmospheric plasma on microbial safety, physicochemical and sensorial qualities of meat products. *Journal of Food Science & Technology*, *55*, 846–857.

Yong, H. I., Kim, H. -J., Park, S., Alahakoon, A. U., Kim, K., Choe, W., et al. (2015). Evaluation of pathogen inactivation on sliced cheese induced by encapsulated atmospheric pressure dielectric barrier discharge plasma. *Food Microbiology*, *46*, 46–50.

Yong, H. I., Lee, S. H., Kim, S. Y., Park, S., Park, J., Choe, W., et al. (2017). Color development, physiochemical properties, and microbiological safety of pork jerky processed with atmospheric pressure plasma. *Innovative Food Science & Emerging Technologies*, *53*, 78–84.

Zhang, H., Ma, D., Qiu, R., Tang, Y., & Du, C. (2017). Non-thermal plasma technology for organic contaminated soil remediation: A review. *Chemical Engineering Journal*, *313*, 157–170.

Zhou, R., Zhou, R., Yu, F., Xi, D., Wang, P., Li, J., et al. (2018). Removal of organophosphorus pesticide residues from *Lycium barbarum* by gas phase surface discharge plasma. *Chemical Engineering Journal, 342*, 401–409.

Zhu, W. C., Wang, B. R., Xi, H. L., & Pu, Y. K. (2010). Decontamination of VX surrogate malathion by atmospheric pressure radio-frequency plasma jet. *Plasma Chemistry and Plasma Processing, 30*(3), 381–389.

Ziuzina, D., Patil, S., Cullen, P., Keener, K., & Bourke, P. (2014). Atmospheric cold plasma inactivation of *Escherichia coli, Salmonella enterica serovar Typhimurium* and *Listeria monocytogenes* inoculated on fresh produce. *Food Microbiology, 42*, 109–116.

Chapter 5

Pulsed Electric Field

Yogesh Kumar*, A.A. Bashir, Navnath Indore, R.K. Vishwakarma and R.K. Singh
ICAR—Central Institute of Post-Harvest Engineering and Technology (CIPHET), Ludhiana, India
**Corresponding author*

1 Introduction

Thermal treatment is a conventional food processing technology that is used for increasing the shelf-life of various food products. However, it alters the structural and physicochemical properties of raw food considerably (Olsen et al., 2010). Heat-sensitive bioactive compounds are also inactivated by the thermal processing of raw foods (Dadi et al., 2019). The changes in the nutritional and flavor profile affect the consumer acceptability of heat-treated food products (Schindler et al., 2010). Certain foods are consumed as raw because of their superior quality attributes or better nutritional profile, for example, colostrum, fruit juices, vegetable soups, etc. Hence, the thermal treatment is not suitable to enhance the shelf-life of these types of food products. Thus, there is a need for some other technology that improves the shelf-life without affecting other quality attributes, especially the nutritional and sensory profile of food products. In recent times, the concept of non-thermal food processing is given priority over thermal treatment due to various advantages. It produces fresh-like, safe, and stable foods without changing the structural properties. These non-thermal technologies are also environmental friendly, economical, and process food in a shorter duration than conventional thermal processing techniques (Yogesh, 2016).

Pulsed electric field (PEF) food processing is a non-thermal technique which uses the electric field for processing of foods for a variety of purpose (Cholet et al., 2014). It uses very short, high-voltage electric field pulses for processing of foods (Maza et al., 2020; Toepfl et al., 2007). The application of these pulses increases the permeability of cells, formation of hydrophilic pores, and opening of protein channels (Kotnik et al., 2001; Tsong, 1991). This effect of PEF on a biological cell is called as "electroporation" (Saulis, 2010). The electroporation effect disrupts the structural integrity and functionality of the cell membrane (Soliva-Fortuny et al., 2009). PEF technique is useful for inactivation of microorganisms (Monfort et al., 2012; Novickij et al., 2020), extraction of juices

Sustainable food processing and engineering challenges. http://dx.doi.org/10.1016/B978-0-12-822714-5.00005-X

(El Kantar et al., 2018), pressing (Yu et al., 2016), dehydration (Semenoglou et al., 2020), enhancing mass transfer (Janositz et al., 2011), drying (Wiktor et al., 2019), enhancing the concentration and activity of bioactive ingredients (Wang et al., 2020), and inactivation of endogenous enzymes (Meza-Jiménez et al., 2019). PEF technique has also been combined with other hurdle techniques for improving its efficiency (Novickij et al., 2020; Pyatkovskyy et al., 2018). This chapter discusses the principle and applications of PEF technology in various aspects of food processing.

2 PEF: mechanism of electroporation or electropermeabilization

The cell membrane is essential for the normal functioning of cells, synthesis of RNA/DNA, protein, and other cell components. The PEF treatment affects the structure of the cell membrane by increasing in the permeabilization across the membrane (Kotnik et al., 2019). The breakdown of the membrane affects the normal biological functioning of the cell. The changes in the cell membrane by the application of PEF are termed as electroporation or electropermeabilization (Kotnik et al., 2001). Electroporation is initiated when a very short and high-voltage electric field is applied across the cell (Soliva-Fortuny et al., 2009). It eventually increases the transmembrane potential (induced TMP), and if induced TMV is higher than resting TMV (-40 to -70 mV), then there is the formation of pores in the cell membrane (Kotnik et al., 2019). It is a dynamic process, and the number and size of pores increase during the application of PEF. However, the pores heal themselves, and the membrane regains its functionality during short-term exposure to the PEF, which is called as a reversible effect of PEF (Soliva-Fortuny et al., 2009). However, disclosure of PEF for longer duration results in the formation of numerous bigger size pores, which results in intense damage to the membrane. The cell membrane loses its functionality permanently, which results in cell death, which is termed as irreversible PEF damage.

The accumulation of charged particles on one side of the cell membrane produces an electric force that pulls oppositely charged particles; thus, cell membrane functions as a capacitor (Zimmermann et al., 1974). This is called capacitance or dielectric constant of the membrane. The capacitance of a cell is proportional to the area of membrane patch and is more or less equal to 2 $\mu F/cm^2$. The opposite-charged particles accumulate on both sides of the cell membrane, which results in the production of perpendicular TMP. If TMP is greater than some threshold level or critical limit for a cell, then damage to the membrane occurs. The induced-TMP by the application of external electric field on various forms of a cell can be calculated by the equations of various workers as:

$$\Delta\phi_g = f_s E_0 a \cos\theta - \Delta\phi_0 \tag{5.1}$$

(Saulis, 2010)
Where
$\Delta\phi_g$, transmembrane potential; $\Delta\phi_0$, resting potential; E_0, external electric field;

a, radius of spherical cell; θ, the angle between the membrane surface and the direction of field; and f_s, a factor that depends on the form of cell, the electric conductivity of plasma membrane, intracellular and extracellular medium.

$$V_m = 1.5aE_c \tag{5.2}$$

(for spherical cell) (Hülsheger et al., 1983)
Where
a, cell radius (m) of spherical cell surrounded by a non-conducting membrane; E_c, applied external electric field (V).

$$V_m = L(1-0.66\text{a})aE_c \tag{5.3}$$

(for nonspherical cell) (Zimmermann et al., 1974)
Where
L, cell length.
The maximum TMP for spheroid cell (a and b axis) was given as:

$$V_m = \text{f}(\upsilon, \theta_o)\text{RE}_o \tag{5.4}$$

(Kinosita & Tsong, 1977)
Where
R, largest semiaxis (R = a for an oblate and R = b for a prolate cell); f, geometric factor which depends on the axial ratio υ = b/a; and θ_o, the angle between the symmetry axis b and the direction of the field (E_o).

The critical transmembrane potential (TMP_c) was given as:

$$\Delta V_c \approx 1.1\text{aE}_{1/2} \tag{5.5}$$

(Kinosita & Tsong, 1977)

The electromechanical model is the most accepted theory for the PEF inactivation of cells (Zimmermann et al., 1974). The attraction of opposite charges inside the cell and across the cell membrane produces an electro compression force. When electro compression exceeds the elastic resistance of a cell membrane, the formation of pores occurs (Riemann et al., 1975). A higher induced-TMP produces a higher electrocompression force, which increases the permeability of the cell membrane and the formation of pores. TMP is dependent on the electric field strength and the properties of a cell. It is easy to destabilize or damage larger cells in comparison to smaller cells by the application of the external electric field. Plant and animal cells are damaged by low field strength, but a higher field is required to inactivate microbial cells. The critical field strength which initiates membrane permeabilization is 1–2 kV/cm for plant cells having the size in the range of 40–200 μm; 12–20 kV/cm for microorganisms having the size in the range of 1–10 μm (Heinz et al., 2001).

The equation gives the electric field strength on the membrane:

$$E_m = u_m/d \approx ER/d \tag{5.6}$$

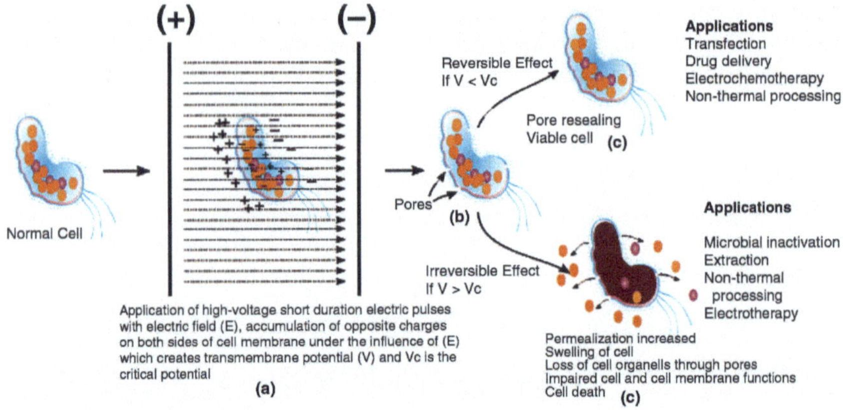

FIGURE 5.1 Reversible or irreversible effects induces by applied PEF (in this figure *Salmonella* cell inactivation is shown), increase in the transmembrane potential, pore formation and permealization of cell membrane and pore resealing or inactivation of cell occur. *(Data taken from Yogesh, (2016).)*

(Barba et al., 2015)
Where d, membrane width (≈5 nm).
(The value of R is approximately 50 μm for plant cells. Hence the value of $E_m \sim 10^4$ E, i.e., effective electric field strength on plant tissue cell is about 10^6–10^7 V/cm).

Apart from electric field strength and size of cells, some other critical factors also affect the mechanism and efficiency of PEF-mediated cell damage. As discussed earlier, there are two types of electric breakdown: reversible and irreversible. There are applications of both effects in food science as well as in medical science (Fig. 5.1). While irreversible electrical collapse is used in microbial inactivation, extraction, drying, pressing, the reversible breakdown is used in drug delivery, vaccine development, and studies related to the host-pathogen interactions (Yogesh, 2016).

3 PEF: factors and their effects on biological tissues

The application of PEF in food processing is growing in recent times (Table 5.1). The PEF technique has been applied for food's non-thermal processing and other applications such as drying and extraction. The effects of PEF on inactivation of microorganisms, extraction of intracellular compounds, improvement in the drying capacity, modifications in the biological molecules, and production of secondary metabolites have been investigated (Dziadek et al., 2019; Martín-García et al., 2020; Simonis et al., 2019). It has been observed during these studies that the electroporation effect of the PEF on various tissues, even of the same shape and size, is not similar (Liu et al., 2017a, 2019). Moreover, the evaluation of electroporation efficiency on various biological tissues is a complicated process. The factors which affect the efficiency are: distribution of

TABLE 5.1 Applications of permeabilization effects of electric fields on cell membranes.

Applications	Materials	Targets to be treated	Treatments	Effects	References
Microbial inactivation	Apple juice	*E. coli*, G^-	PEF: 30.76 kV/cm, 0.2 kHz, 40°C	About 5 log reduction	(Moody et al., 2014)
	Melon juice, watermelon juice	*E. coli* O157:H7, G^- *S. Enteritidis*, G^- *L. monocytogenes*, G^+	PEF: 35 kV/cm, 193 Hz, 2.0% citric acid or cinnamon bark oil	About 5.0 $\log_{10}$ CFU/mL reduction	(Mosqueda-Melgar et al., 2008)
	Milk	*E. coli*, G^-, *S. aureus*, G^+, *P. fluorescens*, G^-	PEF: 40 kV/cm, 89 μs for *E. coli* and *S. Aureus*, and 42.5 kV/cm, 106 μs for *P. fluorescens*, 32.5°C	About 5.0 $\log_{10}$ CFU/mL reduction	(Cregenzán-Alberti et al., 2015)
	Cell suspensions	*Lactobacillus plantarum* 564, G^+	PEF: 31.6 kV/cm, 5 μs × 100 pulses	2.65 Log_{10} units reduction	(Seratlić et al., 2013)
	Emblica officinalis juice	*Z. bailii*	PEF: 26 kV/cm, 1 μs × 500 pulses	5.1 $\log_{10}$ cycles reduction	(Bansal et al., 2015)
Extraction	Potato peels	Steroidal alkaloids	PEF: 0.75 kV/cm, 3 μs × 200 pulses	The maximum yield (1856.2 μg) was 99.9% higher than the control.	(Hossain et al., 2015)
	Rapeseed stems and leaves	Polyphenols, proteins	PEF: 5 kV/cm, 10 μs × 200 pulses	The purity increased from 57.0% to 83.6%	(Yu et al., 2015)
	Rapeseed	Rapeseed oil	PEF: 7.0 kV/cm, 30 μs × 120 pulses	The oil yield increased from 23% to 32% for hulled rapeseed and from 43% to 45% for non-hulled rapeseed.	(Guderjan et al., 2007)
	Plum and grape peels	Phytochemicals	PEF: 25 kV, 26 mm gap, 6 μs pulse width, 10 Hz	The yield of anthocyanins and flavonoids increased several folds, while that of ascorbic acid decreased.	(Medina-Meza & Barbosa-Cánovas, 2015)
	Chlorella vulgaris	Lutein	PEF: 25 kV/cm, for 100 μs, 25°C ~ 30°C	The yield increased around 3.5 ~ 4.2 folds.	(Luengo et al., 2015)

(*Continued*)

TABLE 5.1 Applications of permeabilization effects of electric fields on cell membranes. (*Cont.*)

Applications	Materials	Targets to be treated	Treatments	Effects	References
Acceleration of winemaking	Brandy	—	ACEF: 1 kV/cm, 50 Hz	Aging for 14 months, higher tannin concentration (+54.4% and +43.9%), and chromatic aberration (+19.2% and +24.9%) in 5-L and 2-L barrels were obtained than the control, respectively.	(Zhang et al., 2013)
	Cabernet Sauvignon	—	PEF: 5 kV/cm, 122 Hz, 3 μs pulse duration, below 2°C	The maceration time reduced for PEF wine was 48 h. Higher total anthocyanin content (+11%), color intensity (+38%), and total polyphenol index (+22%) were obtained than the control after aging for 4 months.	(Puértolas et al., 2010a,b)
	Aglianico grapes	—	PEF: 1.5,3.0 kV/cm, 1 kHz, 10 MHz, 25°C ± 3°C	Higher polyphenols (+100%), anthocyanins (+30%), color intensity (+20%), and the antioxidant activity (+40%) were obtained than the control.	(Donsì et al., 2011)
	Cabernet Sauvignon	—	PEF: 5 kV/cm, 1 Hz, 50 monopolar exponential decay pulses	The maceration time reduced for PEF wine was 196 h. The anthocyanin profiles of PEF wine were similar to the control, indicating that the permeabilization of the cell membranes had no selective effect on any anthocyanin.	(López et al., 2009)

Source: Reprinted from Wang et al., 2018 with permission from Taylor & Francis.

local electric field (homogeneous or inhomogeneous), the spatial distribution of cells, difference in the shape and size of the cells within a tissue, local solute concentrations and conductivity (Heinz et al., 2001; Martín-García et al., 2020; Pavlin et al., 2008; Teissie et al., 2005). The increase in the density of cells leads to a decrease in the electroporation effect (Kandušer and Miklavčič, 2008). The electric conductivity also has an impact on electroporation efficiency. The increase in the electric conductivity constant results in a lower electric field threshold for plant tissues (Ben Ammar et al., 2011). In general, the increase in the conductivity results in a lower electroporation effect. The PEF efficiency on plant tissue is also dependent on osmotic flow, water redistribution inside the tissue, and resealing of cells (Barba et al., 2015).

As discussed earlier, the inactivation or damage of smaller cells requires higher electric field strength, while larger cells are inactivated at comparative lower field strength. Thus, the vegetative cells are inactivated at around 150 kV/cm, while some microbes are killed at 35 kV/cm. The smaller bacterium *Listeria innocua* is inactivated at a minimum of 15 kV/cm electric field strength while the larger *Saccharomyces cerevisiae* is inactivated at a field strength around 2–4 kV/cm (Toepfl et al., 2006, 2007).

The electric field strength, form, or shape of the electric field pulse wave, and treatment time are the major processing factors which affect the efficiency of PEF food processing (Martin-Belloso et al., 1997; Pothakamury et al., 1997; Qin et al., 1995). The strength of the electric field is directly proportional to the efficiency of tissue damage and microbial inactivation (Grahl & Markl, 1996).

There are various shapes of electric field pulse waves that also affect the efficiency of PEF treatment. The bipolar or monopolar, exponentially decaying, and square wave pulses are the most common in PEF food processing. The voltage in exponentially decaying wave pulses is unidirectional, which reaches a maximum value rapidly and then decreases to zero at a slower rate. Hence, a log tail of the lower field strength is observed in the exponential decaying wave pulse. This part of the wave with a lower field strength does not inactivate cells; rather, it increases the temperature of food by generating excess heat.

Contrary to this, the voltage is maintained at peak for a longer duration in square wave pulse, and there is no tailing of lower field strength. Thus, square wave electric field pulses are more effective and energy-efficient than exponential decaying wave pulse (Ricci et al., 2018). However, the production of the pulse, as well as the change in field direction, are easier in exponential decaying pulse than square wave pulse. Bipolar pulses are more effective in comparison to the monopolar pulses of similar field strength (Barba et al., 2015). The reversal in the electric field direction results in the bidirectional damage or additional stress to the cell. The application of bipolar pulses also results in a lesser deposition of solids on the electrodes. However, recently it is noticed that bipolar nanosecond electric pulses are far less efficient than monopolar pulses of the similar total duration and amplitude (Ibey et al., 2014; Pakhomov et al., 2018). The bipolar cancellation suggests that the application of the second pulse of the opposite polarity cancels the effect of the first pulse. However, this cancellation effect is not observed for conventional multi-microsecond and millisecond bipolar pulses (Kotnik et al., 2001). The increase in the pulse number and pulse width (treatment time) also increases the PEF efficiency; however, an increase in the treatment duration results in ohmic heating of food products (Gómez et al., 2019).

Apart from PEF related factors, some other treatment conditions or hurdles may affect the electroporation efficiency. The inactivation of cells is higher at an elevated temperature at similar PEF factors (Käferböck et al., 2020). A first-order kinetic model proposed to describe cell inactivation, where the treatment temperature correlated to the first-order constant (Sensoy et al., 1997). The inactivation efficiency is higher in the continuous PEF operation in comparison to the batch operation. A higher reduction in the *Escherichia coli* population was achieved in milk samples by continuous mild PEF operation (25 pulses at 25 kV/cm) than batch operation at higher PEF strength (64 pulses at 35 kV/cm) (Martin-Belloso et al., 1997).

The properties of food also affect the electroporation efficiency of PEF (Saldaña et al., 2009; Toepfl et al., 2007). However, the exact correlation between food properties and electroporation efficiency is not well established yet. The media, such as water and buffer systems, affect the PEF efficiency on similar or different types of cells (Table 5.2). The electric conductivity (σ, Siems/m) of a medium affects the efficiency of PEF as a higher conductivity of medium results in a lesser PEF effect because it generates smaller peaks of the electric field inside the treatment chamber. These smaller peaks do not result

TABLE 5.2 Comparison of the PEF resistance of different pathogenic microorganisms.

Pathogen	Treatment media	Treatment conditions	Log reduction	References
E. coli O157:H7	McIlvaine buffer pH 4.0	30 kV/cm; 150 µs; 24°C	0.5	(Saldaña et al., 2009)
	McIlvaine buffer pH 7.0	30 kV/cm; 150 µs; 24°C	1.7	
	Apple juice	34 kV/cm, 166 µs, 30°C	4.0	(Evrendilek et al., 1999)
	Apple juice	31 kV/cm, 202 µs, 30°C	2.6	(Evrendilek & Zhang, 2005)
	Apple juice	32.2 kV/cm, 35°C	4.96	(Ukuku et al., 2010)
	Liquid egg yolk	30 kV/cm, 310 µs, 40°C	4.9	(Amiali et al., 2007)

TABLE 5.2 Comparison of the PEF resistance of different pathogenic microorganisms. (*Cont.*)

Pathogen	Treatment media	Treatment conditions	Log reduction	References
L. monocytogenes STCC 5672	McIlvaine buffer pH 4.0	30 kV/cm; 150 μs; 24°C	1.5	(Saldaña et al., 2009)
	McIlvaine buffer pH 7.0	30 kV/cm; 150 μs; 24°C	0.6	
L. monocytogenes	Melon juice	35 kV/cm, 2000 μs, 39°C	4.3	(Mosqueda-Melgar et al., 2007)
	Milk	30 kV/cm, 300 μs	3.5	(Reina et al., 1998)
S. aureus STCC 4459	McIlvaine buffer pH 4.0	25 kV/cm; 150 μs; 24°C	2.6	(Saldaña et al., 2009)
	McIlvaine buffer pH 7.0	25 kV/cm; 150 μs; 24°C	2.0	
	LWE	25 kV/cm, 60 μs, 24°C	1.5	(Monfort et al., 2010)
S. aureus	Skim milk	25 kV/cm, 450 μs	3.3	(Evrendilek et al., 2004)
	Skim milk	31 kV/cm, 6 μs	3.0	(Sobrino-López & Martín-Belloso, 2006)
	Orange juice	20 kV/cm, 150 μs	2.0	(Walkling-Ribeiro et al., 2009)
Salmonella Typhimurium STCC 878	McIlvaine buffer pH 4.0	30 kV/cm; 150 μs; 24°C	4.2	(Saldaña et al., 2009)
	McIlvaine buffer pH 7.0	30 kV/cm; 150 μs; 24°C	3.6	
	LWE	30 kV/cm; 50 μs; 24°C	2.0	(Monfort et al., 2010)
Salmonella Typhimurium	Juice-milk based	35 kV/cm, 350 μs	2.0	(Sampedro et al., 2011)
	McIlvaine buffer pH 7.0	28 kV/cm, 150 μs; 24°C	3.2	(Álvarez et al., 2003)

Source: Reprinted from Saldaña et al. (2014), with permission from Taylor & Francis.

in permanent electroporation effect (Barbosa-Canovas et al., 1999). Since the ionic strength of a medium is directly proportional to the electric conductivity (thus at a higher ionic strength of the medium), a lesser PEF efficiency is observed. It has also been reported that acidic pH results in higher efficiency of PEF treatment. The presence of ions also affects PEF efficiency.

Bivalent cations (Mg^{2+} or Ca^{2+}) show a protective effect on the cell membrane by reducing the sensitivity of PEF-treated cells. Hence, a lesser PEF effect is observed in the presence of bivalent cations (Hülsheger et al., 1983). However, this protective effect is not shown by the monovalent cations (Na^+ and K^+). Gram-positive bacteria are covered with a thick peptidoglycan layer, which provides a protective effect. Hence, the Gram-positive bacteria are more resistant in comparison to the Gram-negative bacteria against the PEF treatment. The size of cells affects the efficiency of PEF, and the PEF treatment easily destroys larger cells in comparison to the smaller cells. Hence, yeasts are more susceptible to the electric field in comparison to bacteria. The growth stage of microorganisms also affects the efficiency of microbial inactivation by PEF treatment. The resistance of bacterial cells to PEF in stationary stage is directly related to the membrane fatty acid composition (Liu et al., 2017b).

Moreover, the reduction in the growth temperature (37°C–15°C) results in a higher antimicrobial effect of PEF. However, this effect of fatty acid composition and growth temperature was not observed for the cells in exponential growth stage. Generally, the cells are more susceptible in logarithmic growth phase due to a higher sensitivity of cell membrane to the PEF during this phase.

4 PEF: applications in food processing

4.1 PEF: effects on microbial inactivation

The effects of PEF for inactivation of microorganisms in food have been evaluated. PEF enhance the storage life fruit juices, fruits, vegetables, and other food products without affecting the sensory and nutritional profile. In recent times, the effects of PEF on microbial inactivation have also been evaluated. The combination of PEF with other techniques has also been tested in a variety of raw and processed food products. The effect of continuous-flow system PEF was studied on the inactivation of *E. coli*, *Listeria monocytogenes*, *Lactobacillus plantarum*, *Salmonella* Senftenberg, and *S. cerevisiae* in orange juice (Timmermans et al., 2019). A higher microbial inactivation was observed at moderate intensity electric field (bipolar square wave pulse; 2.7 kV/cm; 15–1000 μs pulse width) in comparison to high intensity electric field (monopolar square wave pulse; 10 kV/cm; 2 μs pulse width). The antimicrobial effect of moderate intensity PEF against *E. coli* and *L. monocytogenes* in watermelon juice and coconut water was also studied at different pH and conductivity of medium. It was found that the PEF effect was independent of pH (3.8–6.0) when moderate intensity PEF was applied. However, the pH of the product was a critical factor for microbial inactivation when high intensity PEF was applied.

Mendes-Oliveira et al. (2020) studied the inactivation rate of *E. coli* O157:H7 and *Salmonella typhimurium* in a continuous PEF system (bipolar square wave pulse; 2 μs pulse width). The authors developed a model which successfully described the survival kinetics of these microorganisms against PEF. The model was also validated in real food matrices using different juices to predict the experimental and literature data. A 5-log reduction of *E. coli* O157:H7 and *S. typhimurium* was found in apple cider at repetition rates of 1000 pps at 35 kV/cm, 1250 pps at 30 kV/cm, and 1500 pps at 30 kV/cm.

A higher PEF efficiency (inactivation of *E. coli*) was observed when treatment solution was allowed to flow perpendicular to the surface of the electrode in continuous-flow PEF reactors (Tanino et al., 2020). Moreover, a lower flow rate of treatment solution led to an increase in the PEF inactivation; however, it also increased the temperature of the solution.

Pyatkovskyy et al. (2018) studied the antimicrobial effects of different combinations of high pressure processing (HPP), PEF, and sonication treatments against non-pathogenic *L. innocua* in water. Individual PEF treatment (30 kV; 1 ms) resulted in about 2-log reduction in the *Listeria* population. However, the HPP treatment (400 MPa for 100 s) showed the highest reduction (8-log) of *Listeria* population. A sequential application of HPP-PEF, and PEF-HPP treatments showed additive antimicrobial effect while simultaneous application demonstrated a synergistic effect. The combined bactericidal effects of essential oils (EOs), organic acids, and PEF were assessed against a range of *Campylobacter jejuni* strains (Clemente et al., 2020). It was found that the application of PEF in combination with oregano EO in liquid showed no significant antimicrobial effect in comparison to individual PEF treatments. However, a significant increase in the microbial inactivation was achieved when a sequential combination approach was applied with PEF treatment followed by re-suspension in oregano EO. Around 1.5-log reduction was achieved in chicken at 1 kV/cm electric field followed by re-suspension in EO. This inactivation was in the same range as reported for conventional thermal treatment. Similarly, the combined antimicrobial effect of PEF (bipolar square wave pulse; 0, 15, 20, and 25 kV/cm; 40 μs pulse width; 1 kHz) and eugenol (2-Methoxy-4-(2-Propenyl) Phenol) was studied against *E. coli* (Debao et al., 2019). The authors noted a strong synergistic effect of PEF and eugenol for inactivation of *E. coli*. The treatment with eugenol increased the membrane permeability and morphological alterations in the *E. coli* cells.

PEF (exponential wave pulses; 20, 25, and 30 kV/cm; 3 μs pulse width; 1 Hz) was observed to produce a synergistic effect (up to 2.5- additional log reduction) when used in combination with nanoemulsions of *Mentha piperita* L. essential oil against *E. coli* O157: H7 in guava and mango juices (de Carvalho et al., 2018). Novickij et al. (2020) evaluated the antimicrobial effect of PEF (30 kV/cm; 200 ns–500 μs) in combination with nisin-loaded nanoparticles and mild thermal treatment (40°C, 2 min) against *S. typhimurium* and *L. innocua*. The PEF showed a synergistic antimicrobial effect in combination with nisin-loaded pectin nanoparticles against Gram-negative bacteria.

Jin et al. (2017) evaluated antimicrobial effect of PEF (bipolar square wave pulse; 2 kV/cm; 1 μs pulse width) in combination with 60 ppm peracetic acid solution on whole fresh blueberries. The application of combined treatment resulted in a 3-log reduction of *E. coli* and *Listeria* as well as 2-log reduction of native microbiota without affecting the color and appearance of blueberries. The PEF treatment also increased the anthocyanins and phenolic compounds by 10% and 25%, respectively. However, a softer texture was observed in PEF-treated blueberries.

Schottroff et al. (2019) studied the effect of PEF (rectangular pulses; 32 kV/cm; 3–8 μs) on whey protein isolate (WPI) formulations at two pH levels (4 and 7). It was shown that the application of PEF resulted in a 6.51-log reduction of *L. innocua* in WPI without affecting the concentration of heat sensitive bioactive compounds (up to 22.2% higher retention of vitamins A and C as well as immunoglobulin M and G). It was also noticed that the increase in the protein concentration and pH values resulted in lower inactivation of microorganisms. A 5.53-log reduction of acid-adapted *E. coli* was achieved in model solution composed of citric acid (8 g/L), fructose (35 g/L), glucose (35 g/L), Na_2HPO_4 (0.2 M) and NaCl (5%) when treated by PEF (monopolar; 35 kV/cm; 2 μs pulse width) (Yildiz et al., 2019). At similar PEF processing condition, a 5.16-log reduction was recorded in *E. coli* population in strawberry juice.

The antimicrobial effect of one train of PEF pulses (square wave pulse; 15, 20, and 30 kV/cm; 3 μs pulse width; 1 Hz) in comparison to pulses applied in two trains was studied (Delso et al., 2020). It was found that the tailing in the survival curves of *S. typhimurium* was due to the transient increment of the microbial resistance to the PEF. However, the tailing disappeared after sometime because microbial cells became sensitive to the second train again. This effect was also validated on whole milk and orange juice.

4.2 PEF: effects on extraction of bioactive compounds

PEF has been evaluated for enhancing the extraction of bioactive ingredients from different food and agricultural products and by-products. PEF enhances the extraction of bioactive compounds without inactivating them, unlike thermal processing. PEF increases diffusion, mass transfer, pressing ability etc., which improve the extraction yield and concentration of bioactive compounds. The PEF treatment (bipolar square wave pulse; 25 kV/cm), when applied in combination with freeze-thawing for extraction of valuable components (phycocyanin, proteins, oils, and carbohydrates) from a cyanobacteria *Arthrospira maxima* resulted in 90% increase of C-phycocyanin content in comparison to bead milling (Käferböck et al., 2020). A 90% increase was noticed in C-phycocyanin content during PEF-assisted extraction in comparison to the bead milling. Moreover, the purity of the extracts obtained after PEF treatment was also higher in comparison to non-PEF-treated samples. The application of PEF treatment resulted in higher extraction yield of proteins and phycocyanin from

Arthrospira platensis (*Spirulina platensis*). A higher extraction of C-phycocyanin into aqueous media was observed after application of PEF (monopolar square wave pulse; 15, 20, 25 kV/cm; 3 μs pulse width; 0.5 Hz) to the fresh biomass of *A. platensis* (Martínez et al., 2017). The extraction was higher at higher field strength and a longer PEF treatment duration. The PEF treatment (square wave pulses; 40 kV/cm; 1 μs; 2–6 Hz) also enhanced the antioxidant activity of extract from *A. platensis* in comparison to the extract obtained after bead milling (Jaeschke et al., 2019). Another study revealed that the homogeneity of the suspension before PEF treatment (square wave pulses; 40 kV/cm; 1 μs pulse width; 4 or 2 Hz) on *A. platensis* affected the extraction efficacy for proteins and C-phycocyanin (Akaberi et al., 2020). C-phycocyanin was also extracted from cyanobacteria *Nostoc commune* TUBT05 and *Oscillatoria okeni* TISTR8549 using the application of freezing/thawing and PEF (Chittapun et al., 2020). However, the PEF treatment was found to be effective for the extraction of C-phycocyanin from *N. commune* only.

The extraction of flavan-3-ols, flavonoids, phenolic acid derivatives and total free phenolic compounds from Brewers' spent grain was facilitated with the application of PEF as a pre-treatment strategy (Martínez et al., 2020). The highest extraction of total free phenolic compounds was achieved at 2.5 kV/cm, 50 Hz, and 14.5 s PEF processing conditions. Around 2.7 and 1.7 times higher concentration of total free and bound phenolic compounds from Brewers' spent grain were observed in PEF-treated samples in comparison to control samples. Kumari et al. (2019) observed that the application of PEF pre-treatment (bipolar square wave pulse; 2.8 kV/cm; 20 μs pulse width; 10 Hz) on Brewer's spent grains followed by aqueous extraction at 55°C, 220 rpm for 16 h led to an increase in the yields of carbohydrate, protein, starch, and reducing sugar in comparison to untreated samples. A higher antimicrobial activity was observed for PEF-treated light Brewers' spent grain extracts against *S. typhimurium* and *L. monocytogenes* in comparison to the untreated extracts.

PEF-assisted extraction from apple peels was also carried out (Wang et al., 2020). The PEF treatment (0.48–1.2 kV/cm) was given to aqueous suspensions of apple peels. At higher PEF strength, the values of °Brix and total polyphenol content were higher, while at lower PEF strength, the extraction of soluble matter was more effective. The comparative efficacy of conventional extraction (CE) and extraction assisted by infrared irradiation (IR), ultrasound (US), pulsed electric fields (PEF), and high-voltage electrical discharges (HVED) for aqueous extraction of bioactive compounds from pomegranate peels was evaluated (Rajha et al., 2019). It was noticed that the saturation in extraction required similar specific energy input ($W \approx$ 90–100 kJ/kg) for US, PEF, and HVED-assisted extraction. However, the application of PEF (exponential decay pulses; 10 kV/cm) resulted in selective extraction and enhanced recovery of ellagic acid. The application of PEF (monopolar square wave pulse; 5 kV/cm) before solvent extraction (acetone and ethyl lactate) led to an increase in the extraction rate (27%–37%), lycopene yields (12%–18%), and the antioxidant activity (18.0%–18.2%) from industrial tomato peel residues (Pataro et al., 2020).

The PEF treatment (exponential decaying wave pulses; 3 kV/cm and 10 kV/cm) prior to pressing led to an increase in the yield of juice by 25% for orange, 37% for pomelo, and 59% for lemon (El Kantar et al., 2018). A higher extraction of polyphenols up to 22 mg GAE/g dry matter was also recorded after application of higher electric field on orange peels. The application of high-intensity electric field strengths (exponential decaying wave pulses; >30 kV/cm) on Merlot grapes (musts) using a continuous-flow system and a 4-day cold maceration resulted in instantaneous extraction of considerable amount of anthocyanins with different chemical features (Leong et al., 2020). The amount of anthocyanins was higher in PEF-assisted extraction in comparison to the non PEF-assisted extraction.

The application of PEF treatment (0.5–1.5 kV/cm; 15 μs pulse width) on whole tomatoes led to an increase in the peeling efficiency and up to 72.3% reduction in the work required for peel detachment (Andreou et al., 2020). The application of PEF (0.5–2.5 kV/cm; 15 μs pulse width) on chopped tomatoes led to an increase in the tomato juice yield up to 20.5%. Moreover, the PEF treatment to the residues of the first juicing step further improved the juice yield and 90.2% overall yield was observed. A 56.4% increase in the carotenoid extraction yield was noticed after PEF treatment (1.0–5.0 kV/cm; 15 μs pulse width) of juicing residues. A higher extraction efficiency of lycopene (9.84 vs. 14.31 mg/100 g tomato residue) and total phenolic compounds (doubled) was also recorded after PEF treatment. The application of PEF improved the extraction of bioactive compounds and individual phenols (chlorogenic acid, coumaric acid, and neochlorogenic acid) from thinned peach fruits when water was used as a solvent (Redondo et al., 2018). However, a negative effect on the extraction yield was observed when methanol was used as a solvent with PEF.

The application of moderate intensity PEF (monopolar exponential decaying wave pulse; 2 kV/cm; 4 μs pulse width; 0.1 Hz) to whole pumpkin led to an increase in the carotenoid content of puree made from PEF-treated pumpkins (García-Parra et al., 2018). PEF treatment (monopolar exponentially decaying wave pulse; 4 μs pulse width; 0.1 Hz) for a longer duration (2 kV/cm, 140 μs total treatment time; 3 kV/cm; 260 μs total treatment time) led to discoloration, loss of firmness, and decrease in the total phenolic content (except flavonol content) of apples. However the treatment for a shorter duration (0.4 kV/cm; 20 μs total treatment time) enhanced the total phenolic content and total flavan-3-ols after 24 h without affecting the physicochemical properties (Ribas-Agustí et al., 2019).

The PEF-assisted sunflower oil extraction carried out in continuous-flow mode (Shorstkii et al., 2020). The application of PEF (bipolar pulse; 7 kV/cm; 30 μs pulse width) on the sunflower seed led to an increase in the extraction yield by 2.3% in comparison to control samples. A higher diffusion coefficient (55.5%), micropore volume (32.2%), and disintegration index (43%) were recorded in PEF-treated samples. PEF-assisted (exponentially decaying wave pulse; 2, 4 and 6 kV/cm; 1 Hz) extraction of total phenolics from cinnamon was carried out (Pashazadeh et al., 2020). After optimization of the processing conditions, the yield, total phenolic contents, DPPH activity, and EC_{50} of the cinnamon extract were 5.06%, 505.9 mg k/g, 91.7%, and 0.51 mg/mL, respectively.

4.3 PEF: effects on drying of food and food products

Conventionally, the drying of food products is carried out by mechanical or thermal processes. However, these processes change the properties of foods considerably, as thermal treatment inactivates bioactive constituents and mechanical process results in higher energy consumption. PEF as a non-thermal processing technique not only preserves the functionality of active compounds present in the food but is also energy efficient in comparison to thermal treatment. The effects of PEF, separately or in conjunction with other technologies, have also been studied on the drying and dehydration of food. Recently, the effect of PEF (bipolar rectangular wave pulses; 1.6 kV/cm; 15 μs pulse width; 20 Hz) and osmotic solution (40%–60% glycerol and 5% NaCl) concentration was studied on the mass transfer during osmotic dehydration of sea bass fillets (Semenoglou et al., 2020). The increase in glycerol concentration increased the mass transfer expressed by the effective diffusivities of water (D_{ew} values) and solubles (D_{es} values). The application of PEF further increased the mass transfer by increasing D_{ew} and D_{es} values up to 50% and 66%, respectively. Similarly, the PEF treatment (0.9–2.8 kV/cm) improved the mass transfer during subsequent osmotic dehydration and air drying (increased effective diffusivity coefficients and drying rate) of goji berry (Dermesonlouoglou et al., 2018). The combined application of osmotic dehydration (55°C, 60 min), PEF (2.8 kV/cm), and air drying (60°C) led to a decrease in the total processing time of drying by 33% with improvement in the color retention, antioxidant capacity, and total phenolic content.

The PEF pre-treatment (bipolar square wave pulse; 3 kV/cm; 1 μs pulse width) before osmotic dehydration of fresh blueberries resulted in the reduction of degradation loss. It also led to an increase in phenolic acid content, flavonols, total phenolics, and increased migration loss of anthocyanins, in comparison to thermally pre-treated and control samples (Yu et al., 2018). PEF pre-treatment also resulted in a higher water loss and solid gain during osmotic dehydration of fresh blueberries which reduced the dehydration time from 130 to 48 h, without affecting the quality and appearance of blueberries. Similar observations were recorded during osmotic dehydration of kiwifruit (*Actinidia deliciosa cv* Hayward) pre-treated by PEF (100, 250, and 400 V/cm) (Traffano-Schiffo et al., 2017). The authors observed the process of plasmolysis which was proportional to the electric field applied. The plasmolysis process eventually increased the rate of dehydration. A lesser dehydration time was also noticed during osmotic dehydration of strawberries pre-treated by PEF (near-rectangular shaped pulse; 100, 200, and 400 V/cm; 100 μs pulse width) (Tylewicz et al., 2017). The application of PEF before osmotic dehydration resulted in higher mass transfer even at a lower electric field strength without affecting the characteristics of strawberries.

PEF and ultrasound pre-treatments before osmotic dehydration and convective drying were found to retain a higher amount of bioactive compounds (polyphenols, anthocyanins, and flavonoids) in cranberries in comparison to reference samples. Color and taste were also found to be better with the PEF pre-treatment (Nowacka et al., 2019). Similarly, the application of PEF (monopolar exponential decaying pulse; 5 kV/cm; 7 μs pulse width) and ultrasound before convective drying of carrot at 70°C led to a decrease in the drying time (Wiktor et al., 2019). The application of PEF before ultrasound treatment resulted in higher (63%) water diffusion coefficient in comparison to reference samples. A higher carotenoid retention (60.5%–80.6%) and redness values were also found in PEF pre-treated dried samples in comparison to untreated dried carrots. Similarly, the PEF was used to enhance the internal diffusion during convective drying process of fresh onions (Ostermeier et al., 2018). At 1.07 kV/cm PEF strength (monopolar exponential decaying pulse; 40 μs pulse width; 2 Hz) and a drying temperature of 60°C, shortest drying time was observed. In another study, the application of PEF (monopolar near-rectangular pulse; 600 V/cm) reduced the vacuum drying time (by 22%–27% at 40°C–70°C) of potato tissue (Liu et al., 2018b).

The improvement in the quality of freeze-dried fruits and vegetables has been carried out using PEF treatment. The application of PEF (exponential decaying wave pulse; 1.0 kV/cm; 100 μs pulse width (strawberries); 143 μs pulse width (bell peppers); 1 Hz) prior to freeze-drying of strawberries and red bell peppers resulted in a reduction of shrinkage phenomena for both bell peppers and strawberries in comparison to untreated samples with 30% and 50% lower volume losses, respectively (Fauster et al., 2020). A 50% increase in the rehydration capacity and a 60% decrease in the firmness of PEF-treated freeze-dried samples were also noticed for strawberries and red bell peppers. The PEF treatment (1.07 kV/cm; 10 ms pulse width) of apple tissue before vacuum freezing increased freeze-drying kinetics and a 57% reduction in the drying time was recorded in PEF-treated apple slices in comparison to untreated samples (Lammerskitten et al., 2019). A higher water absorption was also recorded in the PEF-treated freeze-dried samples during rehydration process. The application of PEF (monopolar pulses; 600 V/cm) prior to vacuum-drying of potato resulted in a lesser absorption of oil during frying (Liu et al., 2020). The PEF treatment also reduced the time duration for vacuum-drying of potato samples and a lesser starch gelatinizing was found during drying of potato samples. The PEF pre-treatment (0.65–1.25 kV/cm; 20 μs pulse width; 100 Hz) led to increased drying rate of fresh apricots without affecting the polyphenol oxidase activity (Huang et al., 2019). The PEF treatment for a longer duration (60 s) resulted in higher β-carotene content and antioxidant activity while a decrease in the browning was observed during processing.

4.4 PEF: effects on quality parameters

PEF treatments have been reported to alter the properties of food considerably. PEF results in electroporation of cells and the damage of cells results in the modification of food properties. PEF also activates or deactivates certain enzymes or their precursors that results in altered properties of PEF-treated food products. The application of PEF (bipolar square wave pulse; 2 and 4 kV/cm; 20 μs pulse width; 100 Hz) led to an increase in the β-glucan content in supernatant fraction of thermally processed oat flour (Duque et al., 2020a). Moreover, the chemical composition (in terms of total starch, protein, and fat contents) was not much affected by the PEF treatment. The middle fraction exhibited an increase in the particle size; reduction in the gelatinisation enthalpy and percent relative crystallinity in comparison to untreated counterpart. The bottom fraction showed the highest gelatinisation enthalpy and the lowest changes in the particle size. PEF treatment (bipolar square wave pulse; 2 and 4.4 kV/cm; 20 μs pulse width; 100 Hz) affected the structural properties of raw oat flour (Duque et al., 2020b). An increase in the particle size, damage of starch granule morphology, modifications in starch short-range molecular order, and protein secondary structures was observed in PEF-treated samples. The PEF treatment also resulted in reduction of gelatinization enthalpy, overall viscosity profile, and reduction in pasting temperature while an increase in thermal transition temperatures was recorded. Hence, PEF treatment improved the pasting stability of raw and thermally processed oat flours.

PEF treatment modifies the structure and other properties of various food starch. In a recent study, the PEF (2.86–8.57 kV/cm; 6 μs pulse width; 600 Hz) effects on three types of starch (wheat starch for type A, potato starch for type B, and pea starch for type C) were investigated (Li et al., 2019). The changes in the structure of all three types of starch, especially potato starch was observed while no change in the crystalline type of starch granules was noticed. The PEF treatment influenced the digestibility of wheat starch and potato starch more pronouncedly than pea starch. PEF (monopolar pulse; 2.86, 5.71, and 8.57 kV/cm; 6 μs pulse width; 600 Hz) also decreased the gelatinization enthalpy of rice starch, and the retrogradation value of rice starch paste. An increase in the rapid digestible starch and slight changes in the resistant starch was also observed (Wu et al., 2019).

The application of PEF (exponentially decaying wave pulse; 0.8, 2, and 3.5 kV/cm; 4 μs pulse width) on carrots resulted in a higher phenolic contents (López-Gámez et al., 2020). The PEF at higher field strength increased the softening of carrot tissue without affecting the color attributes. Higher total carotenoid (50%) and lycopene (53%) contents, were recorded in tomatoes after PEF treatment (monopolar exponential decaying wave pulse; 200 kV/m; 0.1 Hz) in addition to a higher lipophilic antioxidant capacity (González-Casado et al., 2018). At 200 kV/m electric field strength, a higher total soluble solids and pH were recorded with a marked reddening and softening of the tomato fruit. The PEF treatment also caused a stress injury to the tomato tissue leading to a decrease in the RO_2 and RCO_2, a drop in the ethylene production, and the induction of acetaldehyde synthesis. Similarly, a higher total phenolic content and a higher antioxidant activity found in PEF-treated (0.5, 1.5, and 2.5 kV/cm; 25 μs pulse width; 50 Hz) raw *Panax ginseng* roots (Kim et al., 2019). At 2.5 kV/cm field strength, a lower hardness and chewiness values were observed in *ginseng* roots.

PEF has also been evaluated for improving the tenderisation process of muscles (Table 5.3). Recently it has been reported that the application of PEF (0.36 and 0.60 kV/cm; 20 μs) led to an increase in the calpain activity of beef *Semimembranosus* samples (Bhat et al., 2019b). The PEF treatment enhanced the proteolysis of troponin-T and desmin, however, the shear force and myofibrillar fragmentation index were not affected. This suggested that PEF affects only enzymatic process and does not improve the tenderness attributes

TABLE 5.3 Findings of different studies elucidating the effect of pulsed electric field on the tenderness of different muscles.

Muscle studied	Muscle status	PEF treatment	Findings	Authors
Beef Longissimus lumborum	Pre-rigor	5 kV, 10 kV × 20, 50, 90 Hz	Toughness increased with treatment intensity	(Suwandy et al., 2015b)
Beef Semimembranosus	Pre-rigor	5 kV, 10 kV × 20, 50, 90 Hz	Tenderness increased regardless of intensity, Shear force reduced by 21.6%	(Suwandy et al., 2015b)
Beef Longissimus lumborum	Pre-rigor	Repeated (1 ×, 2 ×, 3 ×) PEF treatment (10 kV, 90 Hz, 20 μs)	Tenderness reduced with 3 × PEF treatment, 1 × and 2 × treatment had no effect	(Bekhit et al., 2016)
Beef Semimembranosus	Pre-rigor	Repeated (1 ×, 2 ×, 3 ×) PEF treatment (10 kV, 90 Hz, 20 μs)	Tenderness increased with the lowest shear force in 3 × PEF treatment	(Bekhit et al., 2016)
Beef Longissimus lumborum	Post-rigor	5 kV, 10 kV × 20, 50, 90 Hz	Tenderness increased regardless of intensity, shear force reduced by 19.5%	(Bekhit et al., 2014)
Beef Semimembranosus	Post-rigor	5 kV, 10 kV × 20, 50, 90 Hz	Tenderness increased with treatment intensity, shear force reduced by 4.1%, 10.4%, 19.1% for 20, 50, 90 Hz, respectively	(Bekhit et al., 2014)

TABLE 5.3 Findings of different studies elucidating the effect of pulsed electric field on the tenderness of different muscles. (*Cont.*)

Muscle studied	Muscle status	PEF treatment	Findings	Authors
Beef Longissimus lumborum	Post-rigor	Repeated (1 ×, 2 ×, 3 ×) PEF treatment (10 kV, 90 Hz, 20 μs)	Tenderness increased with repeats, shear force decreased by 2.5 N for each additional treatment	(Suwandy et al., 2015d)
Beef Semimembranosus	Post-rigor	Repeated (1 ×, 2 ×, 3 ×) PEF treatment (10 kV, 90 Hz, 20 μs)	Tenderness was not affected by PEF, shear force not affected by PEF	(Suwandy et al., 2015d)
Beef Longissimus lumborum	Post-rigor	(10 kV, 90 Hz, 20 μs)	No effect on shear force due to fibre direction or initial muscle pH	(Suwandy et al., 2015c)
Beef Longissimus lumborum	Post-rigor	5 kV, 10 kV × 20, 50, 90 Hz	Tenderness increased regardless of intensity, shear force reduced by 19.0%	(Suwandy et al., 2015a)
Beef Semimembranosus	Post-rigor	5 kV, 10 kV × 20, 50, 90 Hz	Tenderness increased with treatment intensity, shear force reduced by 19.0%	(Suwandy et al., 2015a)
Beef Semitendinosus	Post-rigor	1.9 kV/cm, 65 Hz, 20 μs	Tenderness was not affected by PEF, shear force not affected by PEF	(O-Dowd et al., 2013)
Beef Longissimus thoracis et lumborum	Post-rigor	1.4 kV/cm, 10 Hz, 20 μs, 300 and 600 pulses	Samples showed tendency toward reducing toughness, shear force not affected by PEF	(Arroyo et al., 2015a)
Beef Longissimus thoracis	Post-rigor	0.2–0.6 kV/cm, 1–50 Hz, 20 μs	Tenderness was not affected by PEF, shear force was not affected by PEF	(Faridnia et al., 2014)
Beef Semitendinosus	Post-rigor	1.4 kV/cm, pulse width 20 μs, 50 Hz	Freezing-thawing + PEF improved tenderness, Tenderness was not improved by PEF alone	(Faridnia et al., 2015)
Biceps femoris	Post-rigor	1.7kV/cm, 50 Hz, 20 μs	Tenderness increased in treated samples, shear force reduced in PEF samples	(Faridnia et al., 2016)

(*Continued*)

TABLE 5.3 Findings of different studies elucidating the effect of pulsed electric field on the tenderness of different muscles. (*Cont.*)

Muscle studied	Muscle status	PEF treatment	Findings	Authors
Beef Longissimus et lumborum	Post-rigor	2.5 kV, 200 Hz, 20 μs-LPEF 10 kV, 200 Hz, 20 μs-HPEF	Tenderness was not affected by PEF, shear force of HPEF samples was higher than LPEF and untreated samples	(Khan et al., 2017)
Pork Longissimus thoracis et lumborum	Post-rigor	1.2 or 2.3 kV/cm × 100 or 200 Hz × 150 or 300 pulses	No significant effect of PEF was observed on the texture profile analysis	(McDonnell et al., 2014)
Turkey breast meat	Post-rigor	7.5, 10, 12.5 kV (fresh meat), 14, 20, 25 kV (frozen meat) × 10, 55, 110 Hz	Tenderness was not affected by PEF in both fresh and frozen samples, shear force not affected by PEF	(Arroyo et al., 2015b)
Pork haunches and shoulder, beef meat	Post-rigor	0.5–5 kV/cm, 50–1000 pulses, 1–25 kJ/kg	Tenderness increased in treated samples	(Töpfl, 2006)

Source: Reprinted from Bhat et al. (2019a), with permission from Taylor & Francis.

of excessively tough muscles. Similarly, the application of PEF (bipolar square wave pulse; 2.5 kV, 50 Hz and 10 kV, 90 Hz; 20 μs pulse width) did not enhance the tenderization process of *Longissimus dorsi* obtained from red deer (*Cervus elaphus*) (Bhat et al., 2019). Although a slight increase in the calpain activity and proteolysis of troponin-T was noticed but the shear force and myofibrillar fragmentation index did not change in PEF-treated muscles. Contrary to this, it was noticed that the application of PEF (bipolar square wave pulse; 0.8–1.1 kV/cm; 20 μs pulse width, 50 Hz) improved the tenderness as well as color properties of beef *biceps femoris* and *semitendinosus* muscles; however, a higher fat oxidation and saturated fatty acid contents were also recorded in PEF-treated frozen muscle samples (Kantono et al., 2019). In this study, the PEF was applied perpendicular to the orientation of muscle fibers.

The weakening of the Z-disk and I-band junctions as well as 25%–38% longer sarcomeres were observed in PEF-treated (1.00–1.25 kV/cm; 20 μs pulse width; 50 Hz) beef *longissimus thoracis* muscle in comparison to control samples (Chian et al., 2019). Moreover, a higher in vitro protein digestibility was recorded in PEF-treated muscle which suggested the role of PEF in improvement of digestive properties of foods of animal origin. The solubility, surface hydrophobicity, and sulfhydryl group content of the myofibrillar proteins extracted from pale, soft, exudative-like chicken breast meat were significantly improved after application of PEF (monopolar square wave pulse; 8–28 kV/cm) (Dong et al., 2020). At higher PEF intensity (> 18 kV/cm), these properties were found lower which might be due to the protein aggregation. The PEF-treated samples showed higher α-helix content and a lower content of β-turn and random coil.

PEF (bipolar square wave pulse; 0.3–1.2 kV/cm; 20 μs pulse width; 50 Hz) application was tested for reduction of oxalate levels in *Oxalis tuberosa* tubers (Liu et al., 2018). It was found that the PEF treatment resulted in reduction of oxalate contents by almost 50% in some tissue. In addition, the softening of the tissue was also noticed; however, the effect of PEF was uneven on the tuber tissue. The *Pru p 3* is the major allergenic protein of peach and the allergenicity of this protein was reduced by the application of 25 kV/cm PEF at 50°C (50 square waveform pulses of 3 μs) (Tobajas et al., 2020). The application of PEF (bipolar square wave pulse; 0.3, 0.7, and 1.2 kV/cm; 20 μs pulse width; 50 Hz) on bunching onion bulb tissues led to an increase in the fructan contents (Liu et al., 2019). This study also revealed that the effect of PEF was not same on similar types of tissue and affected by the spatial location of the tissue.

PEF treatment (bipolar square wave pulse; 0.52 kV/cm, 10 kV; 20 μs; 20 Hz) of beef muscles also improved the salt diffusion and sodium delivery in the beef jerky without affecting color, yield (%), oxidative and microbial stability (Bhat et al., 2020). The sensory scores of PEF-treated low-sodium meat product were comparable with control samples. The application of PEF (monopolar exponential decaying wave pulses; 24 kV/cm; 0.1 ms pulse width; 1 Hz) also increased the removal of lipid from Asian sea bass skin (Chotphruethipong et al., 2019).

The PEF (15 kV/cm) and soaking in 1% chamuang leaf extract treatment provided a higher oxidative stability and sensory scores to pacific white shrimp (Ahmad Shiekh & Benjakul, 2020). This combined PEF and extract treatment also resulted in lower microbial counts as a lower increase in the mesophile, psychrophile, *Pseudomonas,* Enterobacteriaceae and H_2S producing bacteria were recorded. Similarly, PEF (0.66 kV/cm, 1.38 kV/cm, and 2.00 kV/cm; 20 μs pulse width; 50 Hz) treatment also resulted in lower lipid oxidation, and a lower value of free amino acids and saturated fatty acids in abalone muscle (Luo et al., 2019). Contrary to this, the high intensity PEF (10 kV; 20 μs; 200 Hz) negatively affected the shear force values as well as color and oxidative properties of cold-boned beef loins at 1 and 14 days after treatment (Khan et al., 2017).

4.5 PEF: miscellaneous effects on food and food products

Various properties of fruits and vegetables are affected by the PEF treatment (Table 5.4). PEF influences the rheological properties of fruits and vegetables (Wu et al., 2020). During stress relaxation measurement of the PEF-treated (410 V/cm; 50 μs pulse width) fruits and vegetables, it was found that the

TABLE 5.4 Effect of PEF treatment on different characteristics of fruit and vegetable product (Salehi, 2020).

Juice	Processing conditions	Obtained results	References
Apple	4 μs, 35 kV/cm and 1200 pps	Minimal variability was observed in pH, and no significant changes were detected in acidity, phenolics content, and volatile compounds concentration showed statistically significant differences between treatments.	(Aguilar-Rosas et al., 2007)
Apple	12.5 kV/cm, 76.4 kJ/L, 62 Hz; flow 27.6 L/h	Polyphenol oxidase and peroxidase activities were largely reduced by HIPEF.	(Wibowo et al., 2019)
Apple	15, 25, and 35 kV/cm, 3 μs, 8.5–65.5 kJ/kg	Maximum polyphenoloxidase deactivation of 48% was achieved when the juices were preheated to 40°C and PEF-treated at 30 kV/cm (100 kJ/kg).	(Schilling et al., 2008)
Apple	28–40 kV/cm, 25–100 μs	An analysis of the normalized energy densities for all UV/PEF processing conditions indicated a maximized energy efficiency at higher electric field strengths and shorter treatment times.	(Walkling-Ribeiro et al., 2008)
Apple	1–5 kV/cm, n = 30 pulses	Polyphenol contents and antioxidant capacities of the apple juices extracted from PEF-treated apple mash did not significantly differ from the controls.	(Schilling et al., 2007)
Apple	35 kV/cm, frequency of 1200 PPS	PEF treatment caused a 14.49% reduction in phenolics content, and a significant decrease in volatile compounds in apple juice.	(Aguilar-Rosas et al., 2007)
Apple	20, 30, and 40 kV/cm; 25, 50, and 100 μs	Highest level of decrease in the enzymatic activity of 71% and 68%, for polyphenoloxidase and peroxidase, respectively by using a combination of preheating to 50°C, and a PEF treatment time of 100 μs at 40 kV/cm, and the kinetic data for the inactivation of both enzymes could be well described using a 1st-order model.	(Riener et al., 2008)
Apple	0, 25, 30, and 35 kV/cm for 75 μs with pulse rise time of 2 μs or 0.2 μs	PEF treatment can inactivate the activity of enzymes and keep quality attributes such as vitamin C, total phenols content, antioxidant capacity, and color of apple juice.	(Bi et al., 2013)

TABLE 5.4 **Effect of PEF treatment on different characteristics of fruit and vegetable product (Salehi, 2020) (*Cont.*)**

Juice	Processing conditions	Obtained results	References
Apple	100 square-wave pulses (1 μs, 15 Hz) at 40 kV/cm	Application of PEF or a combination of UV irradiation and PEF to freshly squeezed apple juice resulted in a similar total microbial reduction compared to the severe heat treatment.	(Noci et al., 2008)
Apple	100 μs and 28 kV/cm	Authors did not report any significant color difference (expressed as ΔE) in apple juice treated by PEF.	(Cserhalmi et al., 2004)
Apple	27–33 kV/cm, 3, 6, or 10 L/h, 200 pulses/s	A significant decrease (33%) in the activity of PPO in freshly squeezed apple juice when heated at 50°C and exposed to PEF at 27 kV/cm for 58.7 μs.	(Liang et al., 2006)
Blueberry	0–400 V	During storage of 30 days at 4°C, vitamin C decreased by 11.5% and 26.8% after PEF and HTST (high temperature short time) treatments, all treated juices showed to be microbiologically safe, whereas PEF can protect nutritionally related properties and color of blueberry juice samples more efficiently.	(Zhu et al., 2019)
Carrot	0.25 to 1 kV/cm; 2–40 ms; 1.5–576 J/kg	PEF treatment at 1 kV/cm for 2 ms (29 J/kg) increased concentrations of sucrose and β-glucose in 43% and 48%, respectively.	(Aguiló-Aguayo et al., 2014)
Carrot–orange	25, 30, 35, and 40 kV/cm; 30–340 μs	PEF treatment at 25 kV/cm for 280–330 μs extended the half-life of the carrot–orange juice stored at 2°C for 50 days.	(Torregrosa et al., 2006)
Carrot and orange (blended)	25 kV/cm and 280 μs, P1; and 25 kV/cm and 330 μs, P2	Reduction in PME activity in addition to the preservation of the sensory characteristics, hydroxymethylfurfural and physicochemical quality similar to the fresh juice.	(Rivas et al., 2006)
Grape	3 kV/cm; 10 kJ/kg	Extraction of individual anthocyanins was studied showing a selective extraction based on the glucose moieties linked to the anthocyanidins; anthocyanin monoglucosides were better extracted by PEF.	(Corrales et al., 2008)
Grapefruit	1 kHz, 20 kV/cm for 600 μs	°Brix, pH and acidity showed non-significant effects irrespective of the treatment. But, color values changed significantly after the treatment.	(Aadil et al., 2015a)

(*Continued*)

TABLE 5.4 Effect of PEF treatment on different characteristics of fruit and vegetable product (Salehi, 2020) (*Cont.*)

Juice	Processing conditions	Obtained results	References
Grape-fruit	1 kHz; 0, 5, 10, 15, 20, and 25 kV/cm for 600 μs; 80 mL/min	PEF at 25 kV/cm could improve the quality of grapefruit juice. Significant decrease in non-enzymatic browning and viscosity while an increase in cloud value, DPPH, total antioxidant capacity, total phenolics, and total carotenoids, while decreases in the activity of micro rganisms, was also observed with the increase in pulsed electric strength.	(Aadil et al., 2015b)
Grape-fruit	20, 30, and 40 kV/cm for 25, 50, 75, and 100 μs	Highest level of inactivation of PME (96.8%) was reported using a combination of preheating to 50°C, and a PEF treatment time of 100 μs at 40 kV/cm.	(Riener et al., 2009)
Melon	35 kV/cm for 1709 μs at 193 Hz and 4 μs pulse duration	Melon juice processed by HIPEF and citric acid or cinnamon bark oil exhibited the lower scores in odor, taste, sourness, and total acceptance.	(Mosqueda-Melgar et al., 2008)
Opuntia dillenii cactus fruit	56 exponential decay pulses; maximum electric field strength of 3 kV/cm; total specific energy input of 5 kJ/kg fruit mash	Treating fruit mash with a moderate electric field strength increased juice yield and improved juice characteristics. Promisingly, the highest release of isorhamnetin 3-O-rutinoside from fruit's peel into juice was maximally achieved by PEF.	(Moussa-Ayoub et al., 2016)
Orange	0–35 kV/cm, 2.2 ms pulse duration time	A combination of PEF treatment at 25 kV/cm and a water bath temperature of 50°C caused 90% inactivation of PME.	(Yeom et al., 2002)
Orange	40 kV/cm for 97 ms	PEF processed juice retained more ascorbic acid, flavor, and color than thermally processed juice.	(Min et al., 2003)
Orange	30 kV/cm for 240 or 480 μs	The average losses of flavor compounds in orange juice processed by 240, 480 μs PEF and heat process were 3.0%, 9.0%, and 22.0%, respectively.	(Jia et al., 1999)
Orange	35 kV/cm for 59 μs	Packaging material had a considerable effect on the retention of orange juice aroma compounds, color, and vitamin C. PEF-treated orange juice had a shelf life of >16 weeks in glass and polyethylene terephthalate at 4°C.	(Ayhan et al., 2001)

TABLE 5.4 Effect of PEF treatment on different characteristics of fruit and vegetable product (Salehi, 2020) (*Cont.*)

Juice	Processing conditions	Obtained results	References
Orange	15, 25, and 35 kV/cm; 100, 400, and 1000 μs; using pulses of 4 μs width at 200 Hz	There were no differences in antioxidant capacity between HIPEF treated and untreated products, whereas heat-treated foods showed lower values of antioxidant capacity.	(Elez-Martínez & Martín-Belloso, 2007)
Orange	13.82–25.26 kV/cm, 10.89–51.32 J	Unlike heat treatments, PEF treatments did not cause hydroxyl methyl furfural formation during storage. PEF treatments accomplished ascorbic acid preservation with half-life of 1112 days.	(Agcam et al., 2016)
Orange	13.82–25.26 kV/cm; 10.89–51.32 J; 1033.9 and 1206.2 μs	The samples treated with PEF had more stable flavonoids and phenolic acids than those treated with the thermal pasteurization. The PEF treated samples had higher sensory scores than the heat treated samples.	(Agcam et al., 2014a)
Orange	35 kV/cm for 1,000 μs; bipolar 4-μs pulses at 200 Hz	Vitamin C retention was outstandingly higher in juice processed by HIPEF fitting recommended daily intake standards throughout 56 days storage at 4°C.	(Elez-Martínez et al., 2005)
Orange, pomelo, and lemon	3 kV/cm and 10 kV/cm	The PEF treatment increased the yield of juice obtained after pressing by 25% for orange, 37% for pomelo, and 59% for lemon. Also, significant increase of polyphenols extraction from PEF treated peels was reported.	(El Kantar et al., 2018)
Peach	0, 17, 20, 23, 27, and 30 kV/cm; 50 mL/min flow rate; 3 ms pulse duration, 500 pps; 131 ms	Suggested that the PEF process can successfully be applied to peach nectar with significant amount of microbial inactivation without a significant adverse effect on physical and chemical properties.	(Altuntaş et al., 2011)
Peach	0–5 kV/cm; 10–50 pulses of 3μs (30–150 μs)	PEF-assisted extraction of bioactive compounds from thinned peach fruits using water as a solvent is an alternative to conventional extraction methods which require dried products, large amounts of organic solvents, and long extraction times.	(Redondo et al., 2018)

(*Continued*)

TABLE 5.4 Effect of PEF treatment on different characteristics of fruit and vegetable product (Salehi, 2020) (*Cont.*)

Juice	Processing conditions	Obtained results	References
Peach, apricot, and sour cherry	125, 250, and 400 Hz; 24 kV/cm	Initial mean values of the physical properties did not change significantly with the increasing treatment times for sour cherry juice (64%), apricot nectar (60%), and peach nectar (30%). The increasing treatment time did not significantly exert an adverse impact on any of the sensory scores.	(Evrendilek, 2016)
Peach, apricot, and sour cherry	0, 13, 17, 20, 23, 27, 30, and 34 kV/cm; 0, 62, 94, 123, 163, 198, and 218 μs	Light and scanning electron microscopy observations revealed considerable morphological alterations in fungal conidia such as cytoplasmic coagulation, vacuolations, shrinkage, and protoplast leakage.	(Evrendilek et al., 2008)
Pomegranate	35 and 38 kV/cm, 281 μs at 55°C, flow rate of 100 L/h	PEF processing didn't significantly impact total phenolic, anthocyanin, and antioxidant compounds in comparison to those in untreated pomegranate juice. PEF processing had significantly less impact on the color of pomegranate juice than thermal processing.	(Guo et al., 2014)
Pomegranate	Bench top (7.2 L/h, 35 kV/cm, 72 μs) and pilot scale (100 L/h, 35 kV/cm, 281 μs)	Physicochemical analysis showed that there were no significant differences in total phenolic and antioxidant compounds, pH, and total soluble solid among untreated and PEF treated pomegranate juice.	(Jin et al., 2014)
Pomegranate	0, 17, 23, and 30 kV/cm	Total monomeric anthocyanins, total ascorbic acid content and total antioxidant capacity remained constant during PEF treatment. Sensorial properties (flavor, taste, aftertaste, and total acceptance) were not affected by the PEF treatment.	(Evrendilek, 2017)
Tomato	100 kV/cm; 2 μs width, 0.5 Hz frequency	Enzyme polygalacturonase was unaffected by PEF, but the activity of PME was reduced by 55%.	(Nguyen & Mittal, 2007)
Tomato	0.5–2.5 kV/cm, 0–4000, pulses, 15 μs pulse width	Carotenoid extraction yield increased up to 56.4%. Lycopene extraction increased from 9.84 mg lycopene/100 g to 14.31 mg/100 g tomato residue for a PEF treatment at 1.0 kV/cm for 7.5 ms. The concentration of extracted total phenolic compounds doubled (56.16 mg gallic acid/kg) with a 2 kV/cm, 700 pulses treatment.	(Andreou et al., 2020)

TABLE 5.4 Effect of PEF treatment on different characteristics of fruit and vegetable product (Salehi, 2020) (*Cont.*)

Juice	Processing conditions	Obtained results	References
Tomato	1 kV/cm using 16 monopolar pulses of 4 μs at a frequency of 0.1 Hz	Combination of moderate-intensity PEF and high-intensity PEF treatments could be used not only to produce tomato juices with high carotenoid content but also, to maintain higher the carotenoid content during storage time.	(Vallverdú-Queralt et al., 2013)
Tomato	0.4 to 2.0 kV/cm and number of pulses from 5 to 30	A 44% increase in total polyphenol content was achieved under 30 pulses at 1.2 kV/cm. The hydrophilic antioxidant capacity was also enhanced by 44% applying 18 pulses at 1.2 kV/cm, and the lipophilic antioxidant capacity was increased by 37% under 5 pulses at 1.2 kV/cm.	(Vallverdú-Queralt et al., 2012)
Tomato	35 kV/cm for 1500 μs, using 4 μs bipolar pulses at 100 Hz	High intensity PEF treated juice exhibited higher levels of compounds contributing to tomato aroma than untreated and heat treated juices throughout storage.	(Aguiló-Aguayo et al., 2010b)
Tomato	35 kV/cm for 1500 μs in bipolar 4-μs pulses at 100 Hz, with an energy density of 8269 kJ/L	HIPEF treated tomato juice maintained higher lycopene and vitamin C content than the thermally treated juices during the storage time.	(Odriozola-Serrano et al., 2008)
Tomato	40 kV/cm for 57 μs	PEF processed tomato juice had significantly lower nonenzymatic browning and higher redness than thermally processed or control juice. Sensory evaluations indicated that the flavor of PEF processed juices was preferred to that of thermally processed tomato juices.	(Min & Zhang, 2003)
Water-melon	35 kV/cm for 1727 μs applying 4-μs pulses at 188 Hz in bipolar mode	Use of HIPEF technology could be an alternative to thermal treatments and could contribute to maintain valuable attributes of watermelon juice better.	(Aguiló-Aguayo et al., 2010a)
Water-melon	35 kV/cm for 1682 μs at 193 Hz and 4μs pulse duration	Watermelon juice processed by HIPEF and citric acid or cinnamon bark oil displayed the lower rating in taste, sourness, and overall by the panelists.	(Mosqueda-Melgar et al., 2008)

(Continued)

TABLE 5.4 Effect of PEF treatment on different characteristics of fruit and vegetable product (Salehi, 2020) (*Cont.*)

Juice	Processing conditions	Obtained results	References
White grape	0.25–1 kV/cm	PEF treatment results in increase of the juice yield from 49%–54% to 76%–78% at 45 min of pressing.	(Praporscic et al., 2007)
White grape	0.4 kV/cm	PEF treatment accelerated development of grape deformation. Statistical analysis showed no significant effect of PEF treatment on turbidity and content of polyphenols for the constant pressure regime.	(Grimi et al., 2009)

values of the equilibrium elastic modulus and relaxation time values decreased by more than 92% and 17%, respectively. The application of PEF led to a decrease (70%) in the instantaneous elastic modulus and retarded elastic modulus (92%) during creep parameter estimation. Moreover, a decrease (94%) in the viscosity coefficient was also noticed. The application of PEF (rectangular pulses; 1.0 kV/cm; 10 μs pulse width) on potato led to an increase in the tissue softening which improved the smoothness of cutting surface and up to 80% less feathering was noticed due to an improvement in the cutting behavior (Fauster et al., 2018). The improvement in the elastic properties of potato sticks after PEF treatment resulted in lower breaking loss (11.0%–6.0%). However, peeling behavior of potato and rheological properties of puree were not affected after PEF treatment.

The pre-treatment of potato disk with PEF (monopolar pulse; 600 V/cm; 100 μs pulse width) prior to convective air-drying was done to see the effects during frying of potato disks (Liu et al., 2018a). A decrease in the oil uptake was noticed in the PEF pre-treated potatoes (≈0.30 g/g dry matter) during frying in comparison to untreated potatoes (≈0.47 g/g dry matter). The pre-treatment of potatoes with PEF and subsequent convective air-drying resulted in a denser and lesser oil penetrable skin of potato which prevented the oil uptake during frying. The effects of PEF (bipolar square wave pulse; 0.3–1.2 kV/cm; 20 μs pulse width; 50 Hz) treatment on whole sweet potato tubers were investigated in terms of microstructural properties of whole tuber as well as its quality after frying (Liu et al., 2017a). The application of PEF (1.2 kV/cm) prior to frying (190°C) resulted in 18% lower oil content than non-PEF treated sweet potato samples. PEF treatment led to an increase in the reaction of polyphenol oxidase and phenols in the potato tissue. It also increased the activation energy of the browning rate of potato chips during frying. Thus, PEF treatment resulted in similar brown color intensity at a lower frying temperature in comparison to non-PEF treated samples. The PEF (rectangular pulses; 1.5 kV/cm; 10 μs pulse width; total treatment time 10 ms; 100 Hz frequency) was also evaluated in combination with blanching (85°C for 3.5 min) for reduction of acrylamide precursors (reducing sugars and free asparagine) in potato to reduce the final acrylamide content in fried potato crisps (Genovese et al., 2019). It was noticed that the application of PEF enhanced the leaching of acrylamide precursors and resulted in 30% lesser acrylamide content after frying of potato slices without significantly affecting the textural and color properties.

The PEF processing (bipolar square wave pulse; 20–40 kV/cm; 2.5 μs pulse width) of fruit juice-stevia beverage (0%–2.5% stevia) resulted in higher retention of ascorbic acid, total anthocyanins, and carotenoids (Carbonell-Capella et al., 2017). The ratio of rebaudioside A/stevioside was found higher at 30 kV/cm field strength applied for 230 μs. However, at higher electric field strength (40 kV/cm), the hydroxymethylfurfural content and total color differences were higher. PEF processing at 21 kV/cm for 360 μs with 2.5% stevia resulted in highest content of bioactive compounds and sweetening properties without affecting the color stability. The effects of PEF during electrodialysis with bipolar membranes on cranberry juice deacidification and electrodialytic parameters were evaluated (Pelletier et al., 2017). The rate of deacidification increased by 15%–18% using PEF treatment. About 7%–10% decrease was observed in the energy consumption in comparison to direct current.

The application of PEF affects the activity and stability of enzymes (Terefe et al., 2015). It was noticed that PEF enhanced or reduced the activity of enzymes (Agcam et al., 2014; Ohshima et al., 2007). The application of PEF decreased the activity of papain (Meza-Jiménez et al., 2019). At 13 kV/cm, 288 pulses and 0.2 L/min flow rate, the maximum inactivation of enzyme and a decrease in the sulfhydryl groups (SH) of papain were observed. Relative protein concentration was minimum at 10 kV/cm, 0.4 L/min, and 144 pulses were applied. Zhang et al. (2017) evaluated the effects of PEF treatment (3–21 kV/cm; 5–20 μs pulse width) on enzymatic activity, kinetics, and conformational structure of pectinase. The enzymatic activity was 21.89 ± 1.67% higher at 12 kV/cm electric field strength. PEF treatment improved affinity between enzyme and substrate. PEF also enhanced catalytic efficiency without affecting polypeptide composition of the pectinase molecules. However, the PEF treatment led to an increase in the transfer of the tryptophan micro-environment from the polar surface to the hydrophobic interior and a 10.28% increase in α-helices and a 17.80% reduction in random coils were observed.

The application of PEF (3 kV/cm) enhanced the bioaccumulation of Ca^{2+} ions in the probiotic *Lactobacillus rhamnosus* B 442 strain (Pankiewicz et al., 2020). At 3 kV/cm field strength and 200 μg/mL concentration of calcium in the medium, the highest accumulation of Ca^{2+} ions was observed in cells. Moreover, a higher content of dry matter, fat, protein, and carbohydrates, as well as a lower melting rate was observed in the ice cream prepared using PEF-modified *L. rhamnosus* B 442 bacteria for milk fermentation. The trehalose content in probiotics is positively correlated to the survival of bacterial population during food processing such as spray drying. The intracellular trehalose content was increased to ~100 mM in *L. plantarum* WCFS1 with the application of PEF (square wave form pulse; 7.5 kV/cm; 0.1 ms pulse width) (Vaessen et al., 2018). The impregnation of strawberries with cryoprotectants (12 g/100 g trehalose and 0.2 g/100 g acclimated winter wheat extract containing antifreeze proteins) was carried out by PEF (bipolar rectangular pulse; 850 V/cm; 100 μs pulse width) in conjunction with vacuum infusion (Velickova et al., 2018). The PEF pre-treatment prior to vacuum infusion led to an increase in the cell viability and color retention in comparison to the non-PEF-treated samples. However, the drip loss and texture were not improved by the application of PEF. The treatment of ovomucin-depleted egg white hydrolysates with PEF (bipolar square wave pulse; 1.4–1.7 kV/cm; 20 μs pulse width; 300 Hz) and heating (80°C for 10 min) at pH 4 led to an increase in the antioxidant and anti-inflammatory activity (Liu et al., 2019).

5 Conclusion

PEF food processing is a non-thermal technique which is energy efficient and environment friendly. It preserves the fresh-like quality of food and does not adversely affect the flavor and nutritional properties. PEF can also be used on industrial scale because of energy and time efficient process and also be integrated with a continuous-flow system. Low-intensity PEF is useful for extraction, dehydration, and improving mass transfer while high-intensity PEF is used for inactivation of microorganisms to improve the shelf-life of food products. PEF also influences the chemical, textural, and rheological properties of food. The application of PEF with other hurdle techniques results in efficient microbial inactivation and extraction of bioactive compounds.

Although PEF technique has been evaluated for various purposes, some modifications in this technique are necessary to make it commercially sustainable for specific purpose. PEF processing conditions or parameters should be identified for each food with specific properties. The external factors such as cell/tissue structure, size, conductivity, growth stage, orientation, etc., should also be taken into account during evaluation of PEF efficiency. A thorough insight into the exact mechanism of PEF on various physical and chemical processes in the cell, tissue, and suspension must be ascertained that shall pave the way to develop specific PEF food processing equipment.

References

Aadil, R. M., Zeng, X.-A., Ali, A., Zeng, F., Farooq, M. A., Han, Z., et al. (2015). Influence of different pulsed electric field strengths on the quality of the grapefruit juice. *International Journal of Food Science and Technology*, *50*(10), 2290–2296.

Aadil, R. M., Zeng, X.-A., Sun, D.-W., Wang, M.-S., Liu, Z.-W., & Zhang, Z.-H. (2015). Combined effects of sonication and pulsed electric field on selected quality parameters of grapefruit juice. *LWT—Food Science and Technology*, *62*(1 Pt. 2), 890–893.

Agcam, E., Akyildiz, A., & Akdemir Evrendilek, G. (2016). A comparative assessment of long-term storage stability and quality attributes of orange juice in response to pulsed electric fields and heat treatments. *Food and Bioproducts Processing*, *99*, 90–98.

Agcam, E., Akyıldız, A., & Akdemir Evrendilek, G. (2014). Comparison of phenolic compounds of orange juice processed by pulsed electric fields (PEF) and conventional thermal pasteurisation. *Food Chemistry*, *143*, 354–361.

Agcam, E., Akyıldız, A., & Evrendilek, G. A. (2014). Effects of PEF and heat pasteurization on PME activity in orange juice with regard to a new inactivation kinetic model. *Food Chemistry*, *165*, 70–76.

Aguilar-Rosas, S. F., Ballinas-Casarrubias, M. L., Nevarez-Moorillon, G. V., Martin-Belloso, O., & Ortega-Rivas, E. (2007). Thermal and pulsed electric fields pasteurization of apple juice: Effects on physicochemical properties and flavour compounds. *Journal of Food Engineering*, *83*(1), 41–46.

Aguiló-Aguayo, I., Hossain, M. B., Brunton, N., Lyng, J., Valverde, J., & Rai, D. K. (2014). Pulsed electric fields pre-treatment of carrot purees to enhance their polyacetylene and sugar contents. *Innovative Food Science and Emerging Technologies*, *23*, 79–86.

Aguiló-Aguayo, I., Soliva-Fortuny, R., & Martín-Belloso, O. (2010). Color and viscosity of watermelon juice treated by high-intensity pulsed electric fields or heat. *Innovative Food Science and Emerging Technologies*, *11*(2), 299–305.

Aguiló-Aguayo, I., Soliva-Fortuny, R., & Martín-Belloso, O. (2010). Volatile compounds and changes in flavour-related enzymes during cold storage of high-intensity pulsed electric field- and heat-processed tomato juices. *Journal of the Science of Food and Agriculture*, *90*(10), 1597–1604.

Ahmad Shiekh, K., & Benjakul, S. (2020). Melanosis and quality changes during refrigerated storage of Pacific white shrimp treated with Chamuang (*Garcinia cowa* Roxb) leaf extract with the aid of pulsed electric field. *Food Chemistry*, *309*, 125516.

Akaberi, S., Krust, D., Müller, G., Frey, W., & Gusbeth, C. (2020). Impact of incubation conditions on protein and C-Phycocyanin recovery from *Arthrospira platensis* post- pulsed electric field treatment. *Bioresource Technology*, *306*, 123099.

Altuntaş, J., Akdemir Evrendilek, G., Sangun, M. K., & Zhang, H. Q. (2011). Processing of peach nectar by pulsed electric fields with respect to physical and chemical properties and microbial inactivation. *Journal of Food Process Engineering*, *34*(5), 1506–1522.

Álvarez, I., Mañas, P., Condón, S., & Raso, J. (2003). Resistance variation of *Salmonella enterica* serovars to pulsed electric fields treatments. *Journal of Food Science*, *68*(7), 2316–2320.

Amiali, M., Ngadi, M. O., Smith, J. P., & Raghavan, G. S. V. (2007). Synergistic effect of temperature and pulsed electric field on inactivation of *Escherichia coli* O157:H7 and *Salmonella enteritidis* in liquid egg yolk. *Journal of Food Engineering*, *79*(2), 689–694.

Andreou, V., Dimopoulos, G., Dermesonlouoglou, E., & Taoukis, P. (2020). Application of pulsed electric fields to improve product yield and waste valorization in industrial tomato processing. *Journal of Food Engineering*, *270*, 109778.

Arroyo, C., Eslami, S., Brunton, N. P., Arimi, J. M., Noci, F., & Lyng, J. G. (2015). An assessment of the impact of pulsed electric fields processing factors on oxidation, color, texture, and sensory attributes of Turkey breast meat. *Poultry Science*, *94*(5), 1088–1095.

Arroyo, C., Lascorz, D., O'Dowd, L., Noci, F., Arimi, J., & Lyng, J. G. (2015). Effect of pulsed electric field treatments at various stages during conditioning on quality attributes of beef longissimus thoracis et lumborum muscle. *Meat Science*, *99*, 52–59.

Ayhan, Z., Yeom, H. W., Zhang, Q. H., & Min, D. B. (2001). Flavor, color, and vitamin C retention of pulsed electric field processed orange juice in different packaging materials. *Journal of Agricultural and Food Chemistry*, *49*(2), 669–674.

Bansal, V., Sharma, A., Ghanshyam, C., Singla, M. L., & Kim, K. -H. (2015). Influence of pulsed electric field and heat treatment on *Emblica officinalis* juice inoculated with *Zygosaccharomyces bailii*. *Food and Bioproducts Processing, 95*, 146–154.

Barba, F. J., Parniakov, O., Pereira, S. A., Wiktor, A., Grimi, N., Boussetta, N., et al. (2015). Current applications and new opportunities for the use of pulsed electric fields in food science and industry. *Food Research International, 77*, 773–798.

Barbosa-Canovas, G. V., Pothakamury, U. R., Gongora-Nieto, M. M., & Swanson, B. G. (1999). *Preservation of foods with pulsed electric fields*. Elsevier Science.

Bekhit, A. E. -D. A., Suwandy, V., Carne, A., van de Ven, R., & Hopkins, D. L. (2016). Effect of repeated pulsed electric field treatment on the quality of hot-boned beef loins and topsides. *Meat Science, 111*, 139–146.

Bekhit, A. E. -D. A., van de Ven, R., Suwandy, V., Fahri, F., & Hopkins, D. L. (2014). Effect of pulsed electric field treatment on cold-boned muscles of different potential tenderness. *Food and Bioprocess Technology, 7*(11), 3136–3146.

Ben Ammar, J., Lanoisellé, J. -L., Lebovka, N. I., Van Hecke, E., & Vorobiev, E. (2011). Impact of a pulsed electric field on damage of plant tissues: Effects of cell size and tissue electrical conductivity. *Journal of Food Science, 76*(1), E90–E97.

Bhat, Z. F., Morton, J. D., Mason, S. L., & Bekhit, A. E. -D. A. (2019). Current and future prospects for the use of pulsed electric field in the meat industry. *Critical Reviews in Food Science and Nutrition, 59*(10), 1660–1674.

Bhat, Z. F., Morton, J. D., Mason, S. L., & Bekhit, A. E. -D. A. (2019). Does pulsed electric field have a potential to improve the quality of beef from older animals and how? *Innovative Food Science and Emerging Technologies, 56*, 102194.

Bhat, Z. F., Morton, J. D., Mason, S. L., & Bekhit, A. E. -D. A. (2020). The application of pulsed electric field as a sodium reducing strategy for meat products. *Food Chemistry, 306*, 125622.

Bhat, Z. F., Morton, J. D., Mason, S. L., Mungure, T. E., Jayawardena, S. R., & Bekhit, A. E. -D. A. (2019). Effect of pulsed electric field on calpain activity and proteolysis of venison. *Innovative Food Science and Emerging Technologies, 52*, 131–135.

Bi, X., Liu, F., Rao, L., Li, J., Liu, B., Liao, X., et al. (2013). Effects of electric field strength and pulse rise time on physicochemical and sensory properties of apple juice by pulsed electric field. *Innovative Food Science and Emerging Technologies, 17*, 85–92.

Carbonell-Capella, J. M., Buniowska, M., Cortes, C., Zulueta, A., Frigola, A., & Esteve, M. J. (2017). Influence of pulsed electric field processing on the quality of fruit juice beverages sweetened with *Stevia rebaudiana*. *Food and Bioproducts Processing, 101*, 214–222.

Chian, F. M., Kaur, L., Oey, I., Astruc, T., Hodgkinson, S., & Boland, M. (2019). Effect of pulsed electric fields (PEF) on the ultrastructure and in vitro protein digestibility of bovine longissimus thoracis. *LWT—Food Science and Technology, 103*, 253–259.

Chittapun, S., Jonjaroen, V., Khumrangsee, K., & Charoenrat, T. (2020). C-phycocyanin extraction from two freshwater cyanobacteria by freeze thaw and pulsed electric field techniques to improve extraction efficiency and purity. *Algal Research, 46*, 101789.

Cholet, C., Delsart, C., Petrel, M., Gontier, E., Grimi, N., L'Hyvernay, A., et al. (2014). Structural and biochemical changes induced by pulsed electric field treatments on cabernet sauvignon grape berry skins: Impact on cell wall total tannins and polysaccharides. *Journal of Agricultural and Food Chemistry, 62*(13), 2925–2934.

Chotphruethipong, L., Aluko, R. E., & Benjakul, S. (2019). Enhanced Asian sea bass skin defatting using porcine lipase with the aid of pulsed electric field pretreatment and vacuum impregnation. *Process Biochemistry, 86*, 58–64.

Clemente, I., Condón-Abanto, S., Pedrós-Garrido, S., Whyte, P., & Lyng, J. G. (2020). Efficacy of pulsed electric fields and antimicrobial compounds used alone and in combination for the inactivation of *Campylobacter jejuni* in liquids and raw chicken. *Food Control, 107*, 106491.

Corrales, M., Toepfl, S., Butz, P., Knorr, D., & Tauscher, B. (2008). Extraction of anthocyanins from grape by-products assisted by ultrasonics, high hydrostatic pressure or pulsed electric fields: A comparison. *Innovative Food Science and Emerging Technologies*, *9*(1), 85–91.

Cregenzán-Alberti, O., Halpin, R. M., Whyte, P., Lyng, J. G., & Noci, F. (2015). Study of the suitability of the central composite design to predict the inactivation kinetics by pulsed electric fields (PEF) in *Escherichia coli*, *Staphylococcus aureus*, and *Pseudomonas fluorescens* in milk. *Food and Bioproducts Processing*, *95*, 313–322.

Cserhalmi, Z., Meszaros, L., Sass-Kiss, A., & Toth-Markus, M. (2004). International Congress on Engineering and Food, 11th, March 2004, Montpellier, France. 2004, pp. 7–11.

Dadi, D. W., Emire, S. A., Hagos, A. D., & Eun, J. B. (2019). Effect of ultrasound-assisted extraction of *Moringa stenopetala* leaves on bioactive compounds and their antioxidant activity. *Food Technology and Biotechnology*, *57*(1), 77–86.

de Carvalho, R. J., de Souza, G. T., Pagán, E., García-Gonzalo, D., Magnani, M., & Pagán, R. (2018). Nanoemulsions of *Mentha piperita* L. essential oil in combination with mild heat, pulsed electric fields (PEF) and high hydrostatic pressure (HHP) as an alternative to inactivate *Escherichia coli* O157: H7 in fruit juices. *Innovative Food Science and Emerging Technologies*, *48*, 219–227.

Debao, N., Wang, Q. -Y., Ren, E. -F., Zeng, X. -A., Wang, L. -H., He, T. -F., et al. (2019). Multi-target antibacterial mechanism of eugenol and its combined inactivation with pulsed electric fields in a hurdle strategy on *Escherichia coli*. *Food Control*, *106*, 106742.

Delso, C., Martínez, J. M., Cebrián, G., Álvarez, I., & Raso, J. (2020). Understanding the occurrence of tailing in survival curves of *Salmonella Typhimurium* treated by pulsed electric fields. *Bioelectrochemistry*, *135*, 107580.

Dermesonlouoglou, E., Chalkia, A., Dimopoulos, G., & Taoukis, P. (2018). Combined effect of pulsed electric field and osmotic dehydration pre-treatments on mass transfer and quality of air dried goji berry. *Innovative Food Science and Emerging Technologies*, *49*, 106–115.

Dong, M., Xu, Y., Zhang, Y., Han, M., Wang, P., Xu, X., et al. (2020). Physicochemical and structural properties of myofibrillar proteins isolated from pale, soft, exudative (PSE)-like chicken breast meat: Effects of pulsed electric field (PEF). *Innovative Food Science and Emerging Technologies*, *59*, 102277.

Donsì, F., Ferrari, G., Fruilo, M., & Pataro, G. (2011). Pulsed electric fields—assisted vinification. *Procedia Food Science*, *1*, 780–785.

Duque, S. M. M., Leong, S. Y., Agyei, D., Singh, J., Larsen, N., & Oey, I. (2020). Modifications in the physicochemical properties of flour "fractions" after pulsed electric fields treatment of thermally processed oat. *Innovative Food Science and Emerging Technologies*, *64*, 102406.

Duque, S. M. M., Leong, S. Y., Agyei, D., Singh, J., Larsen, N., & Oey, I. (2020). Understanding the impact of pulsed electric fields treatment on the thermal and pasting properties of raw and thermally processed oat flours. *Food Research International*, *129*, 108839.

Dziadek, K., Kopeć, A., Dróżdż, T., Kiełbasa, P., Ostafin, M., Bulski, K., et al. (2019). Effect of pulsed electric field treatment on shelf life and nutritional value of apple juice. *Journal of Food Science and Technology*, *56*(3), 1184–1191.

El Kantar, S., Boussetta, N., Lebovka, N., Foucart, F., Rajha, H. N., Maroun, R. G., et al. (2018). Pulsed electric field treatment of citrus fruits: Improvement of juice and polyphenols extraction. *Innovative Food Science and Emerging Technologies*, *46*, 153–161.

Elez-Martínez, P., & Martín-Belloso, O. (2007). Effects of high intensity pulsed electric field processing conditions on vitamin C and antioxidant capacity of orange juice and gazpacho, a cold vegetable soup. *Food Chemistry*, *102*(1), 201–209.

Elez-Martínez, P., Soliva-Fortuny, R. C., & Martín-Belloso, O. (2005). Comparative study on shelf life of orange juice processed by high intensity pulsed electric fields or heat treatment. *European Food Research and Technology*, *222*(3), 321.

Evrendilek, G. A. (2016). Change regime of aroma active compounds in response to pulsed electric field treatment time, sour cherry juice apricot and peach nectars, and physical and sensory properties. *Innovative Food Science and Emerging Technologies*, *33*, 195–205.

Evrendilek, G. A. (2017). Impacts of pulsed electric field and heat treatment on quality and sensory properties and microbial inactivation of pomegranate juice. *Food Science and Technology International*, *23*(8), 668–680.

Evrendilek, G. A., Tok, F. M., Soylu, E. M., & Soylu, S. (2008). Inactivation of *Penicillum expansum* in sour cherry juice, peach and apricot nectars by pulsed electric fields. *Food Microbiology*, *25*(5), 662–667.

Evrendilek, G. A., & Zhang, Q. H. (2005). Effects of pulse polarity and pulse delaying time on pulsed electric fields-induced pasteurization of *E. coli* O157:H7. *Journal of Food Engineering*, *68*(2), 271–276.

Evrendilek, G. A., Zhang, Q. H., & Richter, E. R. (1999). Inactivation of *Escherichia coli* O157:H7 and *Escherichia* coli 8739 in apple juice by pulsed electric fields. *Journal of Food Protection*, *62*(7), 793–796.

Evrendilek, G. A., Zhang, Q. H., & Richter, E. R. (2004). Application of pulsed electric fields to skim milk inoculated with *Staphylococcus aureus*. *Biosystems Engineering*, *87*(2), 137–144.

Faridnia, F., Bekhit, A. E. -D. A., Niven, B., & Oey, I. (2014). Impact of pulsed electric fields and post-mortem vacuum ageing on beef longissimus thoracis muscles. *International Journal of Food Science and Technology*, *49*(11), 2339–2347.

Faridnia, F., Bremer, P., Burritt, D. J., & Oey, I. (2016). In *Effects of pulsed electric fields on selected quality attributes of beef outside flat (*Biceps femoris*)* (pp. 51–54). Singapore: Springer Singapore.

Faridnia, F., Ma, Q. L., Bremer, P. J., Burritt, D. J., Hamid, N., & Oey, I. (2015). Effect of freezing as pre-treatment prior to pulsed electric field processing on quality traits of beef muscles. *Innovative Food Science and Emerging Technologies*, *29*, 31–40.

Fauster, T., Giancaterino, M., Pittia, P., & Jaeger, H. (2020). Effect of pulsed electric field pretreatment on shrinkage, rehydration capacity and texture of freeze-dried plant materials. *LWT—Food Science and Technology*, *121*, 108937.

Fauster, T., Schlossnikl, D., Rath, F., Ostermeier, R., Teufel, F., Toepfl, S., et al. (2018). Impact of pulsed electric field (PEF) pretreatment on process performance of industrial French fries production. *Journal of Food Engineering*, *235*, 16–22.

García-Parra, J., González-Cebrino, F., Delgado-Adámez, J., Cava, R., Martín-Belloso, O., Elez-Martínez, P., et al. (2018). Application of innovative technologies, moderate-intensity pulsed electric fields and high-pressure thermal treatment, to preserve and/or improve the bioactive compounds content of pumpkin. *Innovative Food Science and Emerging Technologies*, *45*, 53–61.

Genovese, J., Tappi, S., Luo, W., Tylewicz, U., Marzocchi, S., Marziali, S., et al. (2019). Important factors to consider for acrylamide mitigation in potato crisps using pulsed electric fields. *Innovative Food Science and Emerging Technologies*, *55*, 18–26.

Gómez, B., Munekata, P. E. S., Gavahian, M., Barba, F. J., Martí-Quijal, F. J., Bolumar, T., et al. (2019). Application of pulsed electric fields in meat and fish processing industries: An overview. *Food Research International*, *123*, 95–105.

González-Casado, S., Martín-Belloso, O., Elez-Martínez, P., & Soliva-Fortuny, R. (2018). Enhancing the carotenoid content of tomato fruit with pulsed electric field treatments: Effects on respiratory activity and quality attributes. *Postharvest Biology and Technology*, *137*, 113–118.

Grahl, T., & Markl, H. (1996). Killing of microorganisms by pulsed electric fields. *Applied Microbiology and Biotechnology*, *45*(1–2), 148–157.

Grimi, N., Lebovka, N. I., Vorobiev, E., & Vaxelaire, J. (2009). Effect of a pulsed electric field treatment on expression behavior and juice quality of chardonnay grape. *Food Biophysics*, *4*(3), 191–198.

Guderjan, M., Elez-Martínez, P., & Knorr, D. (2007). Application of pulsed electric fields at oil yield and content of functional food ingredients at the production of rapeseed oil. *Innovative Food Science and Emerging Technologies*, *8*(1), 55–62.

Guo, M., Jin, T. Z., Geveke, D. J., Fan, X., Sites, J. E., & Wang, L. (2014). Evaluation of microbial stability, bioactive compounds, physicochemical properties, and consumer acceptance of pomegranate juice processed in a commercial scale pulsed electric field system. *Food and Bioprocess Technology*, *7*(7), 2112–2120.

Heinz, V., Alvarez, I., Angersbach, A., & Knorr, D. (2001). Preservation of liquid foods by high intensity pulsed electric fields—basic concepts for process design. *Trends in Food Science and Technology*, *12*(3), 103–111.

Hossain, M. B., Aguiló-Aguayo, I., Lyng, J. G., Brunton, N. P., & Rai, D. K. (2015). Effect of pulsed electric field and pulsed light pre-treatment on the extraction of steroidal alkaloids from potato peels. *Innovative Food Science and Emerging Technologies*, *29*, 9–14.

Huang, W., Feng, Z., Aila, R., Hou, Y., Carne, A., & Bekhit, A. E. -D. A. (2019). Effect of pulsed electric fields (PEF) on physico-chemical properties, β-carotene and antioxidant activity of air-dried apricots. *Food Chemistry*, *291*, 253–262.

Hülsheger, H., Potel, J., & Niemann, E. G. (1983). Electric field effects on bacteria and yeast cells. *Radiation and Environmental Biophysics*, *22*(2), 149–162.

Ibey, B. L., Ullery, J. C., Pakhomova, O. N., Roth, C. C., Semenov, I., Beier, H. T., et al. (2014). Bipolar nanosecond electric pulses are less efficient at electropermeabilization and killing cells than monopolar pulses. *Biochemical and Biophysical Research Communications*, *443*(2), 568–573.

Jaeschke, D. P., Mercali, G. D., Marczak, L. D. F., Müller, G., Frey, W., & Gusbeth, C. (2019). Extraction of valuable compounds from *Arthrospira platensis* using pulsed electric field treatment. *Bioresource Technology*, *283*, 207–212.

Janositz, A., Noack, A. K., & Knorr, D. (2011). Pulsed electric fields and their impact on the diffusion characteristics of potato slices. *LWT—Food Science and Technology*, *44*(9), 1939–1945.

Jia, M., Howard Zhang, Q., & Min, D. B. (1999). Pulsed electric field processing effects on flavor compounds and microorganisms of orange juice. *Food Chemistry*, *65*(4), 445–451.

Jin, T. Z., Guo, M., & Yang, R. (2014). Combination of pulsed electric field processing and antimicrobial bottle for extending microbiological shelf-life of pomegranate juice. *Innovative Food Science and Emerging Technologies*, *26*, 153–158.

Jin, T. Z., Yu, Y., & Gurtler, J. B. (2017). Effects of pulsed electric field processing on microbial survival, quality change and nutritional characteristics of blueberries. *LWT—Food Science and Technology*, *77*, 517–524.

Käferböck, A., Smetana, S., de Vos, R., Schwarz, C., Toepfl, S., & Parniakov, O. (2020). Sustainable extraction of valuable components from *Spirulina* assisted by pulsed electric fields technology. *Algal Research*, *48*, 101914.

Kandušer, M., & Miklavčič, D. (2008). Electroporation in biological cell and tissue: An overview. In *Electrotechnologies for extraction from food plants and biomaterials* (pp. 1–37). New York, NY: Springer New York.

Kantono, K., Hamid, N., Oey, I., Wang, S., Xu, Y., Ma, Q., et al. (2019). Physicochemical and sensory properties of beef muscles after pulsed electric field processing. *Food Research International, 121*, 1–11.

Khan, A. A., Randhawa, M. A., Carne, A., Mohamed Ahmed, I. A., Barr, D., Reid, M., et al. (2017). Effect of low and high pulsed electric field on the quality and nutritional minerals in cold boned beef M. longissimus et lumborum. *Innovative Food Science and Emerging Technologies, 41*, 135–143.

Kim, Y. -N., Kwon, H. -J., & Lee, D. -U. (2019). Effects of pulsed electric field (PEF) treatment on physicochemical properties of Panax ginseng. *Innovative Food Science and Emerging Technologies, 58*, 102232.

Kinosita, K., & Tsong, T. Y. (1977). Voltage-induced pore formation and hemolysis of human erythrocytes. *Biochimica et Biophysica Acta (BBA)—Biomembranes, 471*(2), 227–242.

Kotnik, T., Mir, L. M., Flisar, K., Puc, M., & Miklavčič, D. (2001). Cell membrane electropermeabilization by symmetrical bipolar rectangular pulses: Part I. Increased efficiency of permeabilization. *Bioelectrochemistry, 54*(1), 83–90.

Kotnik, T., Rems, L., Tarek, M., & Miklavčič, D. (2019). Membrane electroporation and electropermeabilization: Mechanisms and models. *Annual Review of Biophysics, 48*(1), 63–91.

Kumari, B., Tiwari, B. K., Walsh, D., Griffin, T. P., Islam, N., Lyng, J. G., et al. (2019). Impact of pulsed electric field pre-treatment on nutritional and polyphenolic contents and bioactivities of light and dark brewer's spent grains. *Innovative Food Science & Emerging Technologies, 54*, 200–210.

Lammerskitten, A., Mykhailyk, V., Wiktor, A., Toepfl, S., Nowacka, M., Bialik, M., et al. (2019). Impact of pulsed electric fields on physical properties of freeze-dried apple tissue. *Innovative Food Science and Emerging Technologies, 57*, 102211.

Leong, S. Y., Treadwell, M., Liu, T., Hochberg, M., Sack, M., Mueller, G., et al. (2020). Influence of pulsed electric fields processing at high-intensity electric field strength on the relationship between anthocyanins composition and colour intensity of Merlot (*Vitis vinifera* L) musts during cold maceration. *Innovative Food Science and Emerging Technologies, 59*, 102243.

Li, Q., Wu, Q. -Y., Jiang, W., Qian, J. -Y., Zhang, L., Wu, M., et al. (2019). Effect of pulsed electric field on structural properties and digestibility of starches with different crystalline type in solid state. *Carbohydrate Polymers, 207*, 362–370.

Liang, Z., Cheng, Z., & Mittal, G. S. (2006). Inactivation of spoilage microorganisms in apple cider using a continuous flow pulsed electric field system. *LWT—Food Science and Technology, 39*(4), 351–357.

Liu, C., Grimi, N., Lebovka, N., & Vorobiev, E. (2018). Effects of preliminary treatment by pulsed electric fields and convective air-drying on characteristics of fried potato. *Innovative Food Science and Emerging Technologies, 47*, 454–460.

Liu, C., Grimi, N., Lebovka, N., & Vorobiev, E. (2018). Effects of pulsed electric fields treatment on vacuum drying of potato tissue. *LWT—Food Science and Technology, 95*, 289–294.

Liu, C., Grimi, N., Lebovka, N., & Vorobiev, E. (2020). Impacts of preliminary vacuum drying and pulsed electric field treatment on characteristics of fried potatoes. *Journal of Food Engineering, 276*, 109898.

Liu, T., Burritt, D. J., Eyres, G. T., & Oey, I. (2018). Pulsed electric field processing reduces the oxalate content of oca (*Oxalis tuberosa*) tubers while retaining starch grains and the general structural integrity of tubers. *Food Chemistry, 245*, 890–898.

Liu, T., Burritt, D. J., & Oey, I. (2019). Understanding the effect of pulsed electric fields on multilayered solid plant foods: Bunching onions (*Allium fistulosum*) as a model system. *Food Research International, 120*, 560–567.

Liu, T., Dodds, E., Leong, S. Y., Eyres, G. T., Burritt, D. J., & Oey, I. (2017). Effect of pulsed electric fields on the structure and frying quality of "kumara" sweet potato tubers. *Innovative Food Science and Emerging Technologies*, *39*, 197–208.

Liu, Y. -F., Oey, I., Bremer, P., Silcock, P., Carne, A., & McConnell, M. (2019). Pulsed electric fields treatment at different pH enhances the antioxidant and anti-inflammatory activity of ovomucin-depleted egg white. *Food Chemistry*, *276*, 164–173.

Liu, Z. -W., Zeng, X. -A., Ngadi, M., & Han, Z. (2017). Effect of cell membrane fatty acid composition of *Escherichia coli* on the resistance to pulsed electric field (PEF) treatment. *LWT—Food Science and Technology*, *76*, 18–25.

López-Gámez, G., Elez-Martínez, P., Martín-Belloso, O., & Soliva-Fortuny, R. (2020). Enhancing phenolic content in carrots by pulsed electric fields during post-treatment time: Effects on cell viability and quality attributes. *Innovative Food Science and Emerging Technologies*, *59*, 102252.

López, N., Puértolas, E., Hernández-Orte, P., Álvarez, I., & Raso, J. (2009). Effect of a pulsed electric field treatment on the anthocyanins composition and other quality parameters of Cabernet Sauvignon freshly fermented model wines obtained after different maceration times. *LWT—Food Science and Technology*, *42*(7), 1225–1231.

Luengo, E., Martínez, J. M., Bordetas, A., Álvarez, I., & Raso, J. (2015). Influence of the treatment medium temperature on lutein extraction assisted by pulsed electric fields from *Chlorella vulgaris*. *Innovative Food Science and Emerging Technologies*, *29*, 15–22.

Luo, Q., Hamid, N., Oey, I., Leong, S. Y., Kantono, K., & Alfaro, A. (2019). Physicochemical changes in New Zealand abalone (*Haliotis iris*) with pulsed electric field (PEF) processing and heat treatments. *LWT—Food Science and Technology*, *115*, 108438.

Martin-Belloso, O., Vega-Mercado, H., Qin, B. L., Chang, F. J., Barbosa-Canovas, G. V., & Swanson, B. C. (1997). Inactivation of *Escherichia coli* suspended in liquid egg using pulsed electric fields. *Journal of Food Processing and Preservation*, *21*(3), 193–208.

Martín-García, B., Tylewicz, U., Verardo, V., Pasini, F., Gómez-Caravaca, A. M., Caboni, M. F., et al. (2020). Pulsed electric field (PEF) as pre-treatment to improve the phenolic compounds recovery from brewers' spent grains. *Innovative Food Science and Emerging Technologies*, *64*, 102402.

Martínez, J. M., Delso, C., Álvarez, I., & Raso, J. (2020). Pulsed electric field-assisted extraction of valuable compounds from microorganisms. *Comprehensive Reviews in Food Science and Food Safety*, *19*(2), 530–552.

Martínez, J. M., Luengo, E., Saldaña, G., Álvarez, I., & Raso, J. (2017). C-phycocyanin extraction assisted by pulsed electric field from *Artrosphira platensis*. *Food Research International*, *99*, 1042–1047.

Maza, M. A., Martínez, J. M., Delso, C., Camargo, A., Raso, J., & Álvarez, I. (2020). PEF-dependency on polyphenol extraction during maceration/fermentation of Grenache grapes. *Innovative Food Science and Emerging Technologies*, *60*, 102303.

McDonnell, C. K., Allen, P., Chardonnereau, F. S., Arimi, J. M., & Lyng, J. G. (2014). The use of pulsed electric fields for accelerating the salting of pork. *LWT—Food Science and Technology*, *59*(2 Part 1), 1054–1060.

Medina-Meza, I. G., & Barbosa-Cánovas, G. V. (2015). Assisted extraction of bioactive compounds from plum and grape peels by ultrasonics and pulsed electric fields. *Journal of Food Engineering*, *166*, 268–275.

Mendes-Oliveira, G., Jin, T. Z., & Campanella, O. H. (2020). Modeling the inactivation of *Escherichia coli* O157:H7 and *Salmonella Typhimurium* in juices by pulsed electric fields: The role of the energy density. *Journal of Food Engineering*, *282*, 110001.

Meza-Jiménez, M. d. L., Pokhrel, P. -R., Robles de la Torre, R. R., Barbosa-Canovas, G. V., & Hernández-Sánchez, H. (2019). Effect of pulsed electric fields on the activity of food-grade papain in a continuous system. *LWT—Food Science and Technology*, *109*, 336–341.

Min, S., Jin, Z. T., Min, S. K., Yeom, H., & Zhang, Q. H. (2003). Commercial-scale pulsed electric field processing of orange juice. *Journal of Food Science*, *68*(4), 1265–1271.

Min, S., & Zhang, Q. H. (2003). Effects of commercial-scale pulsed electric field processing on flavor and color of tomato juice. *Journal of Food Science*, *68*(5), 1600–1606.

Monfort, Saldaña, G., Condón, S., Raso, J., & Álvarez, I. (2012). Inactivation of *Salmonella* spp. in liquid whole egg using pulsed electric fields, heat, and additives. *Food Microbiology*, *30*(2), 393–399.

Monfort, S., Gayán, E., Saldaña, G., Puértolas, E., Condón, S., Raso, J., et al. (2010). Inactivation of *Salmonella Typhimurium* and *Staphylococcus aureus* by pulsed electric fields in liquid whole egg. *Innovative Food Science and Emerging Technologies*, *11*(2), 306–313.

Moody, A., Marx, G., Swanson, B. G., & Bermúdez-Aguirre, D. (2014). A comprehensive study on the inactivation of *Escherichia coli* under nonthermal technologies: High hydrostatic pressure, pulsed electric fields and ultrasound. *Food Control*, *37*, 305–314.

Mosqueda-Melgar, J., Raybaudi-Massilia, R. M., & Martín-Belloso, O. (2007). Influence of treatment time and pulse frequency on *Salmonella Enteritidis*, Escherichia coli, and *Listeria monocytogenes* populations inoculated in melon and watermelon juices treated by pulsed electric fields. *International Journal of Food Microbiology*, *117*(2), 192–200.

Mosqueda-Melgar, J., Raybaudi-Massilia, R. M., & Martín-Belloso, O. (2008). Combination of high-intensity pulsed electric fields with natural antimicrobials to inactivate pathogenic microorganisms and extend the shelf-life of melon and watermelon juices. *Food Microbiology*, *25*(3), 479–491.

Moussa-Ayoub, T. E., Jaeger, H., Youssef, K., Knorr, D., El-Samahy, S., Kroh, L. W., et al. (2016). Technological characteristics and selected bioactive compounds of *Opuntia dillenii* cactus fruit juice following the impact of pulsed electric field pre-treatment. *Food Chemistry*, *210*, 249–261.

Nguyen, P., & Mittal, G. S. (2007). Inactivation of naturally occurring microorganisms in tomato juice using pulsed electric field (PEF) with and without antimicrobials. *Chemical Engineering and Processing: Process Intensification*, *46*(4), 360–365.

Noci, F., Riener, J., Walkling-Ribeiro, M., Cronin, D. A., Morgan, D. J., & Lyng, J. G. (2008). Ultraviolet irradiation and pulsed electric fields (PEF) in a hurdle strategy for the preservation of fresh apple Juice. *Journal of Food Engineering*, *85*(1), 141–146.

Novickij, V., Stanevičienė, R., Staigvila, G., Gruškienė, R., Sereikaitė, J., Girkontaitė, I., et al. (2020). Effects of pulsed electric fields and mild thermal treatment on antimicrobial efficacy of nisin-loaded pectin nanoparticles for food preservation. *LWT—Food Science and Technology*, *120*, 108915.

Nowacka, M., Wiktor, A., Anuszewska, A., Dadan, M., Rybak, K., & Witrowa-Rajchert, D. (2019). The application of unconventional technologies as pulsed electric field, ultrasound and microwave-vacuum drying in the production of dried cranberry snacks. *Ultrasonics Sonochemistry*, *56*, 1–13.

O'Dowd, L. P., Arimi, J. M., Noci, F., Cronin, D. A., & Lyng, J. G. (2013). An assessment of the effect of pulsed electrical fields on tenderness and selected quality attributes of post rigour beef muscle. *Meat Science*, *93*(2), 303–309.

Odriozola-Serrano, I., Soliva-Fortuny, R., & Martín-Belloso, O. (2008). Changes of health-related compounds throughout cold storage of tomato juice stabilized by thermal or high intensity pulsed electric field treatments. *Innovative Food Science and Emerging Technologies*, *9*(3), 272–279.

Ohshima, T., Tamura, T., & Sato, M. (2007). Influence of pulsed electric field on various enzyme activities. *Journal of Electrostatics*, *65*(3), 156–161.

Olsen, N. V., Grunert, K. G., & Sonne, A. -M. (2010). Consumer acceptance of high-pressure processing and pulsed-electric field: a review. *Trends in Food Science and Technology*, *21*(9), 464–472.

Ostermeier, R., Giersemehl, P., Siemer, C., Töpfl, S., & Jäger, H. (2018). Influence of pulsed electric field (PEF) pre-treatment on the convective drying kinetics of onions. *Journal of Food Engineering*, *237*, 110–117.

Pakhomov, A. G., Grigoryev, S., Semenov, I., Casciola, M., Jiang, C., & Xiao, S. (2018). The second phase of bipolar, nanosecond-range electric pulses determines the electroporation efficiency. *Bioelectrochemistry*, *122*, 123–133.

Pankiewicz, U., Góral, M., Kozłowicz, K., & Góral, D. (2020). Application of pulsed electric field in production of ice cream enriched with probiotic bacteria (*L. rhamnosus* B 442) containing intracellular calcium ions. *Journal of Food Engineering*, *275*, 109876.

Pashazadeh, B., Elhamirad, A. H., Hajnajari, H., Sharayei, P., & Armin, M. (2020). Optimization of the pulsed electric field-assisted extraction of functional compounds from cinnamon. *Biocatalysis and Agricultural Biotechnology*, *23*, 101461.

Pataro, G., Carullo, D., Falcone, M., & Ferrari, G. (2020). Recovery of lycopene from industrially derived tomato processing by-products by pulsed electric fields-assisted extraction. *Innovative Food Science and Emerging Technologies*, *63*, 102369.

Pavlin, M., Kotnik, T., Miklavčič, D., Kramar, P., & Maček Lebar, A. (2008). Chapter seven electroporation of planar lipid bilayers and membranes. In A. Leitmannova Liu (Ed.), *Advances in planar lipid bilayers and liposomes* (pp. 165–226). (vol. 6). Academic Press.

Pelletier, S., Serre, É., Mikhaylin, S., & Bazinet, L. (2017). Optimization of cranberry juice deacidification by electrodialysis with bipolar membrane: Impact of pulsed electric field conditions. *Separation and Purification Technology*, *186*, 106–116.

Pothakamury, U. R., Barbosa-Canovas, G. V., Swanson, B. G., & Spence, K. D. (1997). Ultrastructural changes in *Staphylococcus aureus* treated with pulsed electric fields/*Cambios ultraestructurales* en *Staphylococcus aureus* sometida a campos electricos pulsantes. *Food Science and Technology International*, *3*(2), 113–121.

Praporscic, I., Lebovka, N., Vorobiev, E., & Mietton-Peuchot, M. (2007). Pulsed electric field enhanced expression and juice quality of white grapes. *Separation and Purification Technology*, *52*(3), 520–526.

Puértolas, E., Hernández-Orte, P., Sladaña, G., Álvarez, I., & Raso, J. (2010). Improvement of winemaking process using pulsed electric fields at pilot-plant scale. Evolution of chromatic parameters and phenolic content of Cabernet Sauvignon red wines. *Food Research International*, *43*(3), 761–766.

Puértolas, E., López, N., Saldaña, G., Álvarez, I., & Raso, J. (2010). Evaluation of phenolic extraction during fermentation of red grapes treated by a continuous pulsed electric fields process at pilot-plant scale. *Journal of Food Engineering*, *98*(1), 120–125.

Pyatkovskyy, T. I., Shynkaryk, M. V., Mohamed, H. M., Yousef, A. E., & Sastry, S. K. (2018). Effects of combined high pressure (HPP), pulsed electric field (PEF) and sonication treatments on inactivation of *Listeria innocua*. *Journal of Food Engineering*, *233*, 49–56.

Qin, B. -L., Chang, F. -J., Barbosa-CÃ¡novas, G. V., & Swanson, B. G. (1995). Nonthermal inactivation of *Saccharomyces cerevisiae* in apple juice using pulsed electric fields. *LWT—Food Science and Technology*, *28*(6), 564–568.

Rajha, H. N., Abi-Khattar, A. -M., El Kantar, S., Boussetta, N., Lebovka, N., Maroun, R. G., et al. (2019). Comparison of aqueous extraction efficiency and biological activities of polyphenols from pomegranate peels assisted by infrared, ultrasound, pulsed electric fields and high-voltage electrical discharges. *Innovative Food Science and Emerging Technologies*, *58*, 102212.

Redondo, D., Venturini, M. E., Luengo, E., Raso, J., & Arias, E. (2018). Pulsed electric fields as a green technology for the extraction of bioactive compounds from thinned peach by-products. *Innovative Food Science and Emerging Technologies*, *45*, 335–343.

Reina, L. D., Jin, Z. T., Zhang, Q. H., & Yousef, A. E. (1998). Inactivation of *Listeria monocytogenes* in milk by pulsed electric field. *Journal of Food Protection*, *61*(9), 1203–1206.

Ribas-Agustí, A., Martín-Belloso, O., Soliva-Fortuny, R., & Elez-Martínez, P. (2019). Enhancing hydroxycinnamic acids and flavan-3-ol contents by pulsed electric fields without affecting quality attributes of apple. *Food Research International*, *121*, 433–440.

Ricci, A., Parpinello, G. P., & Versari, A. (2018). Recent advances and applications of pulsed electric fields (PEF) to improve polyphenol extraction and color release during red winemaking. *Beverages*, *4*(1), 18.

Riemann, F., Zimmermann, U., & Pilwat, G. (1975). Release and uptake of haemoglobin and ions in red blood cells induced by dielectric breakdown. *Biochimica et Biophysica Acta (BBA)—Biomembranes*, *394*(3), 449–462.

Riener, J., Noci, F., Cronin, D. A., Morgan, D. J., & Lyng, J. G. (2008). Combined effect of temperature and pulsed electric fields on apple juice peroxidase and polyphenoloxidase inactivation. *Food Chemistry*, *109*(2), 402–407.

Riener, J., Noci, F., Cronin, D. A., Morgan, D. J., & Lyng, J. G. (2009). Combined effect of temperature and pulsed electric fields on pectin methyl esterase inactivation in red grapefruit juice (*Citrus paradisi*). *European Food Research and Technology*, *228*(3), 373–379.

Rivas, A., Rodrigo, D., Martínez, A., Barbosa-Cánovas, G. V., & Rodrigo, M. (2006). Effect of PEF and heat pasteurization on the physical–chemical characteristics of blended orange and carrot juice. *LWT—Food Science and Technology*, *39*(10), 1163–1170.

Saldaña, G., Álvarez, I., Condón, S., & Raso, J. (2014). Microbiological aspects related to the feasibility of PEF technology for food pasteurization. *Critical Reviews in Food Science and Nutrition*, *54*(11), 1415–1426.

Saldaña, G., Puértolas, E., López, N., García, D., Álvarez, I., & Raso, J. (2009). Comparing the PEF resistance and occurrence of sublethal injury on different strains of *Escherichia coli*, *Salmonella Typhimurium*, *Listeria monocytogenes* and *Staphylococcus aureus* in media of pH 4 and 7. *Innovative Food Science and Emerging Technologies*, *10*(2), 160–165.

Salehi, F. (2020). Physico-chemical properties of fruit and vegetable juices as affected by pulsed electric field: a review. *International Journal of Food Properties*, *23*(1), 1036–1050.

Sampedro, F., Rodrigo, D., & Martínez, A. (2011). Modelling the effect of pH and pectin concentration on the PEF inactivation of *Salmonella enterica serovar Typhimurium* by using the Monte Carlo simulation. *Food Control*, *22*(3), 420–425.

Saulis, G. (2010). Electroporation of cell membranes: The fundamental effects of pulsed electric fields in food processing. *Food Engineering Reviews*, *2*(2), 52–73.

Schilling, S., Alber, T., Toepfl, S., Neidhart, S., Knorr, D., Schieber, A., et al. (2007). Effects of pulsed electric field treatment of apple mash on juice yield and quality attributes of apple juices. *Innovative Food Science and Emerging Technologies*, *8*(1), 127–134.

Schilling, S., Schmid, S., Jäger, H., Ludwig, M., Dietrich, H., Toepfl, S., et al. (2008). Comparative study of pulsed electric field and thermal processing of apple juice with particular consideration of juice quality and enzyme deactivation. *Journal of Agricultural and Food Chemistry*, *56*(12), 4545–4554.

Schindler, S., Krings, U., Berger, R. G., & Orlien, V. (2010). Aroma development in high pressure treated beef and chicken meat compared to raw and heat treated. *Meat Science*, *86*(2), 317–323.

Schottroff, F., Gratz, M., Krottenthaler, A., Johnson, N. B., Bédard, M. F., & Jaeger, H. (2019). Pulsed electric field preservation of liquid whey protein formulations—Influence of process parameters, pH, and protein content on the inactivation of *Listeria innocua* and the retention of bioactive ingredients. *Journal of Food Engineering*, *243*, 142–152.

Semenoglou, I., Dimopoulos, G., Tsironi, T., & Taoukis, P. (2020). Mathematical modelling of the effect of solution concentration and the combined application of pulsed electric fields on mass transfer during osmotic dehydration of sea bass fillets. *Food and Bioproducts Processing*, *121*, 186–192.

Sensoy, I., Zhang, Q. H., & Sastry, S. K. (1997). Inactivation kinetics of *Salmonella dublin* by pulsed electric field. *Journal of Food Process Engineering*, *20*(5), 367–381.

Seratlić, S., Bugarski, B., Nedović, V., Radulović, Z., Wadsö, L., Dejmek, P., et al. (2013). Behavior of the surviving population of *Lactobacillus plantarum* 564 upon the application of pulsed electric fields. *Innovative Food Science and Emerging Technologies*, *17*, 93–98.

Shorstkii, I., Khudyakov, D., & Mirshekarloo, M. S. (2020). Pulsed electric field assisted sunflower oil pilot production: Impact on oil yield, extraction kinetics and chemical parameters. *Innovative Food Science and Emerging Technologies*, *60*, 102309.

Simonis, P., Kersulis, S., Stankevich, V., Sinkevic, K., Striguniene, K., Ragoza, G., et al. (2019). Pulsed electric field effects on inactivation of microorganisms in acid whey. *International Journal of Food Microbiology*, *291*, 128–134.

Sobrino-López, Á., & Martín-Belloso, O. (2006). Enhancing inactivation of *Staphylococcus aureus* in skim milk by combining high-intensity pulsed electric fields and nisin. *Journal of Food Protection*, *69*(2), 345–353.

Soliva-Fortuny, R., Balasa, A., Knorr, D., & Martín-Belloso, O. (2009). Effects of pulsed electric fields on bioactive compounds in foods: a review. *Trends in Food Science and Technology*, *20*(11), 544–556.

Suwandy, V., Carne, A., van de Ven, R., Bekhit, A. E. -D. A., & Hopkins, D. L. (2015). Effect of pulsed electric field on the proteolysis of cold boned beef *M. Longissimus lumborum* and *M. Semimembranosus*. *Meat Science*, *100*, 222–226.

Suwandy, V., Carne, A., van de Ven, R., Bekhit, A. E. -D. A., & Hopkins, D. L. (2015). Effect of pulsed electric field treatment on hot-boned muscles of different potential tenderness. *Meat Science*, *105*, 25–31.

Suwandy, V., Carne, A., van de Ven, R., Bekhit, A. E. -D. A., & Hopkins, D. L. (2015). Effect of pulsed electric field treatment on the eating and keeping qualities of cold-boned beef loins: Impact of initial pH and fibre orientation. *Food and Bioprocess Technology*, *8*(6), 1355–1365.

Suwandy, V., Carne, A., van de Ven, R., Bekhit, A. E. -D. A., & Hopkins, D. L. (2015). Effect of repeated pulsed electric field treatment on the quality of cold-boned beef loins and topsides. *Food and Bioprocess Technology*, *8*(6), 1218–1228.

Tanino, T., Hirosawa, M., Moteki, R., Matsui, M., & Ohshima, T. (2020). Engineering of pulsed electric field treatment using carbon materials as electrode and application to pasteurization of sake. *Journal of Electrostatics*, *104*, 103424.

Teissie, J., Golzio, M., & Rols, M. P. (2005). Mechanisms of cell membrane electropermeabilization: A mini review of our present (lack of ?) knowledge. *Biochimica et Biophysica Acta (BBA)—General Subjects*, *1724*(3), 270–280.

Terefe, N. S., Buckow, R., & Versteeg, C. (2015). Quality-related enzymes in plant-based products: Effects of novel food processing technologies part 2: Pulsed electric field processing. *Critical Reviews in Food Science and Nutrition*, *55*(1), 1–15.

Timmermans, R. A. H., Mastwijk, H. C., Berendsen, L. B. J. M., Nederhoff, A. L., Matser, A. M., Van Boekel, M. A. J. S., et al. (2019). Moderate intensity pulsed electric fields (PEF) as alternative mild preservation technology for fruit juice. *International Journal of Food Microbiology, 298*, 63–73.

Tobajas, A. P., Agulló-García, A., Cubero, J. L., Colás, C., Segura-Gil, I., Sánchez, L., et al. (2020). Effect of high pressure and pulsed electric field on denaturation and allergenicity of Pru 3 protein from peach. *Food Chemistry, 321*, 126745.

Toepfl, S., Heinz, V., & Knorr, D. (2007). High intensity pulsed electric fields applied for food preservation. *Chemical Engineering and Processing: Process Intensification, 46*(6), 537–546.

Toepfl, S., Mathys, A., Heinz, V., & Knorr, D. (2006). Review: Potential of high hydrostatic pressure and pulsed electric fields for energy efficient and environmentally friendly food processing. *Food Reviews International, 22*(4), 405–423.

Töpfl, S. (2006). Pulsed electric fields (PEF) for permeabilization of cell membranes in food- and bioprocessing—Applications, process and equipment design and cost analysis. Unpublished Doctoral Thesis, Technische Universität Berlin, Berlin.

Torregrosa, F., Esteve, M. J., Frígola, A., & Cortés, C. (2006). Ascorbic acid stability during refrigerated storage of orange–carrot juice treated by high pulsed electric field and comparison with pasteurized juice. *Journal of Food Engineering, 73*(4), 339–345.

Traffano-Schiffo, M. V., Laghi, L., Castro-Giraldez, M., Tylewicz, U., Romani, S., Ragni, L., et al. (2017). Osmotic dehydration of organic kiwifruit pre-treated by pulsed electric fields: Internal transport and transformations analyzed by NMR. *Innovative Food Science and Emerging Technologies, 41*, 259–266.

Tsong, T. Y. (1991). Electroporation of cell membranes. *Biophysical Journal, 60*(2), 297–306.

Tylewicz, U., Tappi, S., Mannozzi, C., Romani, S., Dellarosa, N., Laghi, L., et al. (2017). Effect of pulsed electric field (PEF) pre-treatment coupled with osmotic dehydration on physico-chemical characteristics of organic strawberries. *Journal of Food Engineering, 213*, 2–9.

Ukuku, D., Yuk, H., & Zhang, H. (2010). Behavior of pulsed electric fields injured *Escherichia coli* o157:H7 cells in apple juice amended with pyruvate and catalase. *Journal of Microbial and Biochemical Technology, 2*, 134–138.

Vaessen, E. M. J., den Besten, H. M. W., Patra, T., van Mossevelde, N. T. M., Boom, R. M., & Schutyser, M. A. I. (2018). Pulsed electric field for increasing intracellular trehalose content in *Lactobacillus plantarum* WCFS1. *Innovative Food Science and Emerging Technologies, 47*, 256–261.

Vallverdú-Queralt, A., Odriozola-Serrano, I., Oms-Oliu, G., Lamuela-Raventós, R. M., Elez-Martínez, P., & Martín-Belloso, O. (2013). Impact of high-intensity pulsed electric fields on carotenoids profile of tomato juice made of moderate-intensity pulsed electric field-treated tomatoes. *Food Chemistry, 141*(3), 3131–3138.

Vallverdú-Queralt, A., Oms-Oliu, G., Odriozola-Serrano, I., Lamuela-Raventos, R. M., Martín-Belloso, O., & Elez-Martínez, P. (2012). Effects of pulsed electric fields on the bioactive compound content and antioxidant capacity of tomato fruit. *Journal of Agricultural and Food Chemistry, 60*(12), 3126–3134.

Velickova, E., Tylewicz, U., Dalla Rosa, M., Winkelhausen, E., Kuzmanova, S., & Romani, S. (2018). Effect of pulsed electric field coupled with vacuum infusion on quality parameters of frozen/thawed strawberries. *Journal of Food Engineering, 233*, 57–64.

Walkling-Ribeiro, M., Noci, F., Cronin, D. A., Lyng, J. G., & Morgan, D. J. (2009). Antimicrobial effect and shelf-life extension by combined thermal and pulsed electric field treatment of milk. *Journal of Applied Microbiology, 106*(1), 241–248.

Walkling-Ribeiro, M., Noci, F., Cronin, D. A., Riener, J., Lyng, J. G., & Morgan, D. J. (2008). Reduction of Staphylococcus aureus and quality changes in apple juice processed by ultraviolet irradiation, pre-heating and pulsed electric fields. *Journal of Food Engineering, 89*(3), 267–273.

Wang, L., Boussetta, N., Lebovka, N., & Vorobiev, E. (2020). Cell disintegration of apple peels induced by pulsed electric field and efficiency of bio-compound extraction. *Food and Bioproducts Processing*, *122*, 13–21.

Wang, Q., Li, Y., Sun, D. -W., & Zhu, Z. (2018). Enhancing food processing by pulsed and high voltage electric fields: Principles and applications. *Critical Reviews in Food Science and Nutrition*, *58*(13), 2285–2298.

Wibowo, S., Essel, E. A., De Man, S., Bernaert, N., Van Droogenbroeck, B., Grauwet, T., et al. (2019). Comparing the impact of high pressure, pulsed electric field and thermal pasteurization on quality attributes of cloudy apple juice using targeted and untargeted analyses. *Innovative Food Science and Emerging Technologies*, *54*, 64–77.

Wiktor, A., Dadan, M., Nowacka, M., Rybak, K., & Witrowa-Rajchert, D. (2019). The impact of combination of pulsed electric field and ultrasound treatment on air drying kinetics and quality of carrot tissue. *LWT—Food Science and Technology*, *110*, 71–79.

Wu, C., Wu, Q. -Y., Wu, M., Jiang, W., Qian, J. -Y., Rao, S. -Q., et al. (2019). Effect of pulsed electric field on properties and multi-scale structure of japonica rice starch. *LWT—Food Science and Technology*, *116*, 108515.

Wu, X., Wang, C., & Guo, Y. (2020). Effects of the high-pulsed electric field pretreatment on the mechanical properties of fruits and vegetables. *Journal of Food Engineering*, *274*, 109837.

Yeom, H. W., Zhang, Q. H., & Chism, G. W. (2002). Inactivation of pectin methyl esterase in orange juice by pulsed electric fields. *Journal of Food Science*, *67*(6), 2154–2159.

Yildiz, S., Pokhrel, P. R., Unluturk, S., & Barbosa-Cánovas, G. V. (2019). Identification of equivalent processing conditions for pasteurization of strawberry juice by high pressure, ultrasound, and pulsed electric fields processing. *Innovative Food Science and Emerging Technologies*, *57*, 102195.

Yogesh, K. (2016). Pulsed electric field processing of egg products: a review. *Journal of Food Science and Technology*, *53*(2), 934–945.

Yu, X., Bals, O., Grimi, N., & Vorobiev, E. (2015). A new way for the oil plant biomass valorization: Polyphenols and proteins extraction from rapeseed stems and leaves assisted by pulsed electric fields. *Industrial Crops and Products*, *74*, 309–318.

Yu, X., Gouyo, T., Grimi, N., Bals, O., & Vorobiev, E. (2016). Pulsed electric field pretreatment of rapeseed green biomass (stems) to enhance pressing and extractives recovery. *Bioresource Technology*, *199*, 194–201.

Yu, Y., Jin, T. Z., Fan, X., & Wu, J. (2018). Biochemical degradation and physical migration of polyphenolic compounds in osmotic dehydrated blueberries with pulsed electric field and thermal pretreatments. *Food Chemistry*, *239*, 1219–1225.

Zhang, B., Zeng, X. -A., Lin, W. T., Sun, D. -W., & Cai, J. -L. (2013). Effects of electric field treatments on phenol compounds of brandy aging in oak barrels. *Innovative Food Science and Emerging Technologies*, *20*, 106–114.

Zhang, F., Tian, M., Du, M., & Fang, T. (2017). Enhancing the activity of pectinase using pulsed electric field (PEF) treatment. *Journal of Food Engineering*, *205*, 56–63.

Zhu, N., Zhu, Y., Yu, N., Wei, Y., Zhang, J., Hou, Y., et al. (2019). Evaluation of microbial, physicochemical parameters and flavor of blueberry juice after microchip-pulsed electric field. *Food Chemistry*, *274*, 146–155.

Zimmermann, U., Pilwat, G., & Riemann, F. (1974). Dielectric breakdown of cell membranes. *Biophysical Journal*, *14*(11), 881–899.

Chapter 6

Ultraviolet Light-Assisted Titanium Dioxide Photocatalysis for Food Safety

Zarzynska Joanna Magdalena*, **Bogdan Janusz and Anusz Krzysztof**
Department of Food Hygiene and Public Health Protection, Institute of Veterinary Medicine, WULS-SGGW, Nowoursynowska, Warsaw, Poland
**Corresponding author*

1 Introduction

The term "nanomaterial" is defined in European Commission (EC) recommendation 696/2011, as well as by ISO (TS 80004-1:2010) (Warheit and Brown, 2019; Vimercati et al., 2020). The controlled manipulation at the atomic or molecular level and design of different new materials in which known substances may present new properties (quantum effects) are the basis of nanoscience and nanotechnology. Together with bioengineering they give a huge chance to the food sector for establishing modern materials and technologies, which can improve food safety and quality on different levels of production and storage. We can use the term "nano food," proposed by Joseph and Morrison (2006)—"cultivated, produced, processed or packaged using nanotechnology techniques or tools, or to which manufactured nanomaterials have been added." The core idea of nanofood technology is usage of nanomaterials produced intentionally, manufactured on the basis of scientific findings or engineered. After over 15 years of food nanotechnology, there is a particular interest in establishing a law regulation of engineered nanomaterials (NMs) in the food sector and, first of all, to confirm the safety of those materials for the consumers.

Particle size is the key parameter for NM identification. Successfully designed and manufactured engineered NMs have particles size scale 1–100 nm (Som et al., 2010; EU Regulation 2283/2015). Engineered NMs including also structured materials (like aggregates or agglomerates—secondary particles), which may have a size exceeding 100 nm, but holding characteristic nanoscale properties (like large specific surface area or specific physicochemical properties different from those of conventional original material).

Sustainable food processing and engineering challenges. http://dx.doi.org/10.1016/B978-0-12-822714-5.00006-1

The various business operators found opportunities for using NMs in applicative form in different types of production areas, including sunscreens (cosmetology) or work surfaces coatings (e.g., ceramics fabrication) (Bogdan et al., 2015a). The photocatalytic surface coatings from metal oxide nanoparticles (NPs) caught the attention of the food sector, mostly due to their multiactive properties (antibacterial, antifungal, antiviral activity, and deodorizing feature) (Wang et al., 2014; Bogdan et al., 2015a). Most metal oxides can react with microbes at the room temperature. The technological basis of the coatings is the superficial application of a thin layer of some metals in nano oxides forms. A nanosized powder of titanium dioxide (nano-TiO_2) proved to be very useful in coatings. As it is inorganic NM semiconductor, which runs advanced oxidation processes (AOPs) initiated by ultraviolet (UV) radiation (UV light < 388 nm wavelength, bandgap energy of 3.2 eV) (Mihindukulasuriya and Lim, 2014; Verbruggen et al., 2014). AOPs allow obtaining effective self-disinfecting and self-cleaning abilities of the coated surfaces (Wang et al., 2014). Why TiO_2? TiO_2 has been approved by FDA as a generally recognized as safe (GRAS) for drugs, cosmetics, and food (Bodaghi et al., 2013; Wang et al., 2014; Yemmireddy and Hung, 2015). TiO_2 underlined attributes are: high photochemical stability, low production costs, and insolubility in the most reaction environments. It is valued for high UV light absorption rate, as well as for super hydrophilicity (induced by UV radiation). Additionally, TiO_2 generates durable oxidizing power at ambient temperature and pressure (Yemmireddy and Hung, 2015). From three TiO_2 polymorphic forms, two (anatase and rutile) are widely used in nanotechnology. The issues of TiO_2 toxicity have been researched for an extended period of time. It is considered that microsized TiO_2 ($0.1 < \phi \leq 100\ \mu m$) is harmless to humans and animals. For example micro-TiO_2, as the odorless amorphic white powder is applied in many types of food and beverages processing. It is widely used as a pigment (food additive)—whitener (E-171), to lend brightness to food products or anticaking agent (e.g., in chocolate, mints, coffee cream, cheese, yogurts, pop tarts, dressings, carbonated beverages) (Peter et al., 2015; Sodano et al., 2016). In comparison, nano-TiO_2 particle ($1 < \phi < 100$ nm) is having a much bigger active surface, resulting in a higher absorption rate of UV and highly effective photocatalytic activity. TiO_2, as nanoparticle or nanotube forms, is showing better antimicrobial activity against *E. coli* than microsized TiO_2 (Jing et al., 2013). The International Agency for Research on Cancer (IARC) had classified nano-TiO_2 in nanotube forms to the group 2B compounds, which might be carcinogenic for humans (in this group are chloroform, ochratoxin A or nitrobenzene). The United States Environmental Protection Agency (USEPA) reported that TiO_2 is deposited in surface water, ground, and is freed to the atmosphere from sunscreens, textiles, paints, and food packaging (Bogdan et al., 2015a).

2 Photocatalytic properties of TiO_2

The TiO_2 semiconductor electron structure is related to catalytic properties. The photocatalysis is a heterogeneous reaction on the surface of the catalyst, and involves processes of excitation, redox reaction, and recombination. In TiO_2

particle electron structure, a valence band (VB) is full with electrons, as conduction band (CB) is electron free. The energy difference between those two bands (ΔE), named bandgap, represents the amount of energy necessary to be delivered to excite electron from VB to CB (Bogdan et al., 2015a). The width of energy gap rates 3.0 eV for anatase, and 3.2 eV for rutile (equivalent of electromagnetic radiation photon energy, with wavelength < 400 nm). As a source of energy for excitation, UV radiation is used (UV-A, wavelength 315–400 nm or UV-B, wavelength 280–315 nm) (Lee and Ko, 2013). Irradiation of UV-C is too risky for human health. When electron (e^-) migrates from VB toward CB is leaving en electron-hole, which is positively charged (h^+), so as a result, highly unstable status—excitons are obtained, with specific pair hole + electron, oppositely charged (h^+ and e^-), exhibiting strong redox properties. The excitation process can be repeated, leading to hole + electron pairs continuous formation. Most of the formed hole + electron pairs (~ 90%) after excitation recombine rapidly. Electron holes are responsible for the oxidation process, while electrons determine reduction, and by the interaction with the electron acceptor—adherent oxygen or water molecules or hydroxyl ions ($\cdot OH^-$) leading to ROS generation (Fig. 6.1). The titanium dioxide NPs are known to have sufficient huge number of hydroxyl groups on their surface. The rate of ROS formation strongly depends on the type and source of TiO_2 (important parameters: particle size, crystalline phase, nanostructure surface area, isoelectric point) (Hitkova et al., 2012). Hydroxyl radicals have oxidation potential of 2.80 eV, which is much higher than ozone (2.08 eV) or chlorine dioxide (1.36 eV). The other advantages of $\cdot OH^-$ are: zero wastes generation, no chemical toxicity, and no reaction with production plant equipment (like corrosion).

The emerging free radicals inactivate and oxidize microorganisms or organic compounds (organic pollutants present in the air or wastewater, harmful

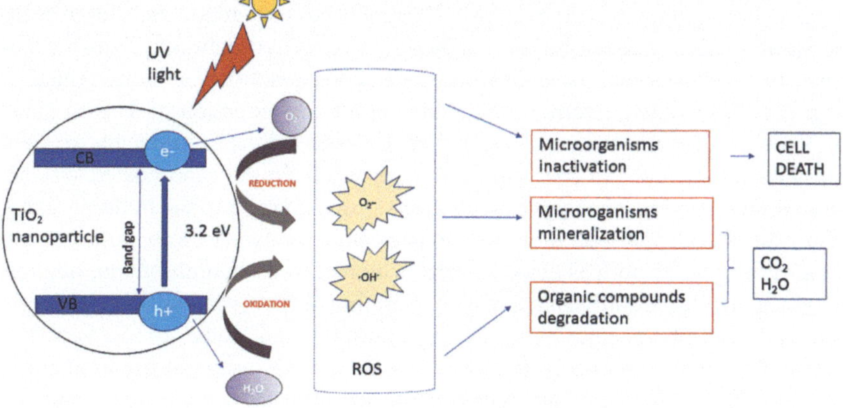

FIGURE 6.1 The nano TiO_2 under UV irradiation shows photocatalytic effect. Redox reactivity of nanoparticle generates ROS, which affect on various microorganisms inactivation (disinfection) and mineralization, as well as organic compounds decomposition.

to humans, like dioxins) to CO_2 and H_2O (Bogdan et al. 2015a)—as it is shown on Fig. 6.1. Oxidized organic compounds are hard to remove from surfaces, air or water, without photocatalysis.

As the main drawbacks of TiO_2 photocatalysis are pointed out wide bandgap and requirements of UV irradiation as a catalysis stimulus. This is causing low utilization of solar radiation (consisting of only ~5% of UV) and efficiency of photocatalysis under visible light is calculated as very low, reaching not even 10%. As a next disadvantage massive recombination of hole + electron pairs is mentioned, which impairs the photocatalytic activity (high quantum yield of photoconversion) (Reddy et al., 2017). TiO_2 has also a low adsorption capacity. Doping with metal cations (e.g., Zn^{2+}, Li^{+}, Nb^{5+}) significantly expands the spectrum of absorption, increases the redox potential, and quantum efficiency (by lowering e–h pairs recombination), although cation doping can result in TiO_2 structure defects. It can be omitted by photocatalyst annealing after doping. For example, the annealing for doped anatase (500°C–600°C) increases the crystallinity and reduces the electron-hole pairs recombination.

Another option is doping by rare-earth metal cations, for example, Ce, which contributes to increasing of PC activity. Ce ions are larger than Ti^{4+}, CeO_2 is located on top of TiO_2 NP surface, as pure cerium dioxide does not have sufficient activity under UV irradiation (Zahornyi, 2018). A disadvantage of cation metal doping is increasing of charge recombination, which leads to decrease of photocatalytic activity, even initiated with UV irradiation. There are some suggestions for using anionic impurities, instead of cationic impurities to gain band gap reduction (Zahornyi, 2018). The doping with nonmetals (e.g., N, C, S, or P) leads to partial replacement of oxygen in the crystalline latticework of semiconductor oxide, hence the band gap narrows and photoconductivity appears under visible light. For example, doping with N narrows the TiO_2 band gap to 2.5 eV (Zahornyi, 2018). The composite TiO_2–N NMs absorb the visible light lower than 550 nm and exhibit photoactivity in NO decomposition reactions.

As was mentioned earlier, nano-TiO_2 shows not only redox abilities. Another useful feature induced by UV irradiation is superhydrophilicity. The mechanism of this phenomenon is also based on redox reactions, as is presented in Fig. 6.2, CB excited electron is responsible for the reduction of Ti ions ($Ti4^{+}$ to $Ti3^{+}$), as electron holes run oxidation of oxide anions to molecular oxygen particles (from $\cdot O^{2-}$ to O_2), which are removed from the nano-coated surface. Remaining oxygen vacancies bind hydroxyl ions, giving surface hydrophilicity. After 30 min of UV radiation, the contact angle between the nano-layer and water drop placed on it is close to zero, which allows obtaining plain, homogenous, invisible film of water, easily evaporating. Superhydrophilicity is reversible and depends on light exposure (Yemmireddy and Hung, 2015). Superhydrophilicity phenomenon is the reason for nano-TiO_2 usage in the production of self-cleaning surfaces and paints or plastics. What is interesting, surfaces covered by a thin layer of TiO_2 are free from pollution even in the dark (Bogdan et al., 2015a). NPs act in pushing water drops or dirt particles off the surfaces,

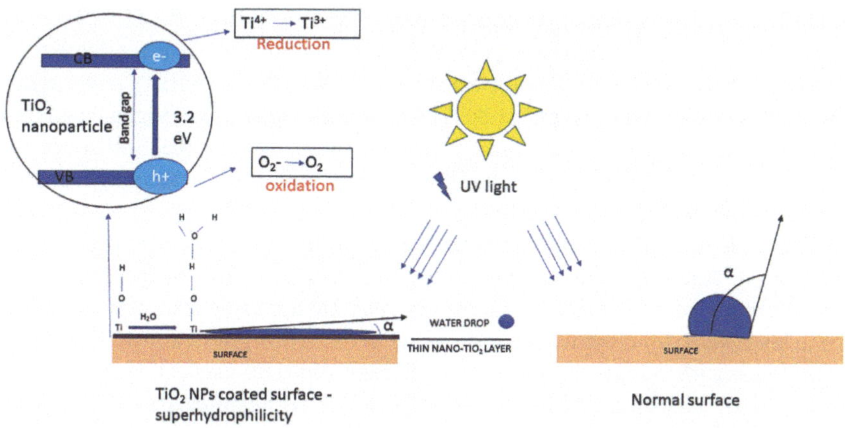

FIGURE 6.2 Surfaces coated with nano TiO_2 thin layer show superhydrophilicity phenomenon under UV irradiation. The mechanism is based on redox reactions. The surface obtains self-cleaning abilities, as water droplets are changing shape, spread out forming thin film, which easily evaporates. In comparison, non-covered surface is more hydrophobic.

which prevent them from getting wet or dirty. By water aerosols spraying on the active surface, water diffuses into the space between surface and dust, in the result eliminating dust—this mechanism is known in plant and animal world surfaces, and it is called superhydrophobicity. The organic compound adsorption to the film surface is giving possibility of transition between hydrophilic and hydrophobic mechanism.

3 TiO_2/UV photocatalysis in the food industry

3.1 Surfaces with antibacterial activity

The surfaces with active photocatalytic coatings exhibit many functional properties, like antistatic effect or deodorization, but the most important attributes for the food industry are self-disinfecting and self-cleaning abilities of nano coatings (Bogdan et al., 2015a). It is a promising technique of microorganisms' elimination in the food processing plants, where regulations restrict the usage of chemical biocides and cleaning substances. It allows reducing the usage of chemical cleaning agents and makes cleaning cycles in food manufacturing plants shorter. Coated surfaces could be tiles, glass, or stainless steel. And the reduction of microbial food contamination is a serious challenge in the food industry, significant for improving food safety. Bacteria have ability to form biofilms in food processing plants, on the working surfaces, and then can be transferred to food product. The occurrence of biofilms in the plant environment can cause post-processing contamination, leading to reducing shelf-life of products, organoleptic changes, or transmission of pathogens causing health issues in consumers. Bacteria biofilms can also influence on effectiveness of technical equipment work and mechanical cleaning is often insufficient.

Microorganisms (bacteria, viruses, yeast, molds) forming biofilms on surfaces coated by a thin layer of nano-TiO_2 after photocatalysis are affected by produced ROS (reactive oxygen species): hydroxyl radicals ($\bullet OH^-$), superoxide anion ($\bullet O_2^-$) and hydrogen peroxide (H_2O_2). ROS inactivate pathogens, and oxidize them to CO_2 and water, making them harmless (similar reaction concerns dead microbes—and it is called mineralization), and in the same time surface obtains self-disinfecting properties.

The susceptibility of infectious pathogens to the photocatalytic process on the basis of scientific findings was sorted in ordered as follows, from the most susceptible: viruses→ prions→ Gram-negative bacteria (covered with complex triple layer wall, lipid membranes are prone to oxidation) → Gram-positive bacteria (thicker peptidoglycan layer is an efficient physical barrier), (van Grieken et al., 2010; Zhu et al., 2018) → yeast→ molds (Bogdan et al., 2015c). The mechanism of viruses' inactivation by AOPs is not fully recognized. Scientists suggest that photocatalysis destroys capsid proteins by generated ROS; subsequently, virus DNA/RNA is fragmented. Non-enveloped viruses are more sensitive to ROS than enveloped ones (as they are separated from environment only by capsid). In enveloped viruses, plasma membrane covering virion acts as a protection layer (the thickness of the layer matters). Among viruses inactivated by TUV process, they are listed as: HSV-1, HBV, A/H5N2, PV, RV-A (Bogdan et al., 2015c). TUV antiviral efficiency can be enhanced in the presence of low halidates concentrations (Br^- or Cl^-) (Zahornyi, 2018). In the study of bacteriophage virus inactivation kinetics in the aqueous medium, a significant reduction of disinfection (to 10 min) was obtained by using 2 M iron sulfate, which promotes hydroxyl radicals generation (Zahornyi, 2018).

The intensity of the antibacterial activity of the TiO_2/UV process depends on factors like: temperature of the environment, humidity, UV light intensity, chemical composition of surface and pollution, concentration, microstructure, and diameter of NPs, density, and susceptibility of microorganisms (Cedrillo-Gonzalez et al., 2016). But some contradictory scientific results can be found in the case of TiO_2 concentration affecting the effectiveness of bacteria eradication (Wang et al., 2014). More powdered nano-TiO_2 is characterized by stronger antibacterial activity ($\phi < 20$ nm), as with this size, the penetration through the damaged cell into the cytosol is possible. With smaller particle size, the surface area increasing (when it is calculated as the surface area to volume ratio) and generates a higher number of ROS (Jing et al., 2011). Bacteria are losing cell membrane integrity due to phospholipids peroxidation (254 nm wavelength UV irradiation), and during this process, carbonyl compounds are formed, some are toxic to bacteria (de Azeredo, 2013). From ROS major attacking, rapidly reacting anti-bacterial agent is $\cdot OH^-$. Other observed signs of cell damage are: plasmolysis, vacuolization, nucleoid condensation, and nucleic acid disruption by $\cdot OH^-$. The hydroxyls radicals can destroy plasmid DNA and induced coenzyme A dimerization inhibits cell respiration by increasing the concentration of dimeric CoA forms—which is underline as a prolonged antibacterial effect

(Zhu et al., 2018). During the oxidation of phospholipids from bacteria cells envelope lot of carbonyl compounds is formed (aldehydes, ketones, carboxylic acids), some of them are potentially toxic to bacteria. Neither in Gram-negative nor Gram-positive bacteria shows visible proteoglycans (PG) degradation—murein shows the highest TUV resistance. In Gram-positive bacteria, PG provide effective physical barrier preventing the molecules entering into the cell. The wall of Gram-positive bacteria is thick (20–80 nm) and surrounded by around 40 layers of PG. Gram-negative bacteria have thin cell wall (2–10 nm) formed of two phospholipid bilayers (outer membrane and plasma membrane) and in between them 2–3 layers of PG are immersed. LPS from outer membrane of Gram-negative bacteria cell wall is easily damaged by oxidation.

Superoxide anion radicals cannot penetrate the cell envelope, because of the negative charge ($\cdot O_2^-$). The high bacteria density (over 10^8 CFU/mL) can weaken the effect of photocatalysis. Various strategies were adopted to overcome the drawbacks, for example, by coupling another semiconductor (both metals and non-metals).

The interest in the photocatalysis process is growing increasingly, as an alternative to traditional disinfection actions also taken in food processing plants or abattoirs. As coating can be applied on walls ceramic tiles (production and clean rooms), as well as on the floors, or stainless tables (as well as on permanent surfaces in frequent contact with workers, also in toilets, to improve the hygienic aspects of production), it is easy to induce photocatalysis by short-term UV radiation. According to scientific findings, 2 min UV exposition results in living bacteria cells reduction by 6–7 logarithmic cycles (effectiveness affected by the type of tiles—shiny or matt). What is the most important for FBOs, the bactericidal effect is highly visible in the eradication of bacteria included in FSC and FHC—*E. coli* (Jing et al., 2011) *S. aureus, and Salmonella Enteritidis*. Wang et al. (2014) tested TiO_2 antimicrobial photocatalytic activity in photoreactor (UV-A light) on bacteria suspension (*M. caseolyticus* and *P. fluorescens*—pathogens commonly responsible for chicken meat spoilage). With an initial population of 10^7 CFU/mL, 120 min of TiO_2 photocatalytic treatment was effective in the complete elimination of both strains. Less time was needed if the initial concentration of bacteria was lower (e.g., 10^5 CFU/mL, 90 min). Based on TEM analysis, the authors proposed three stages of bacterial cell destruction by nano- TiO_2 photocatalysis: (1) photogenerated ROS are getting in contact with bacteria surface; (2) ROS are irreversibly damaging cells' walls and membranes, which results in cytoplasm leakage (K+ and small molecules); (3) the bacteria cells function disrupts, cells lyse, and cellular components degrade, which leads to programmed cell death. Yadav et al. (2016) improved this scheme, and four steps were divided (Fig. 6.3).

As mentioned earlier, mineralization of matter can be obtained thanks to the TiO_2/UV process, which was proved in many experiments. For example, Jacoby et al. (1998), in the test with *E. coli* on the ceramic surface, have observed complete oxidation of 54% of bacteria after 75 min of the photocatalytic process finish.

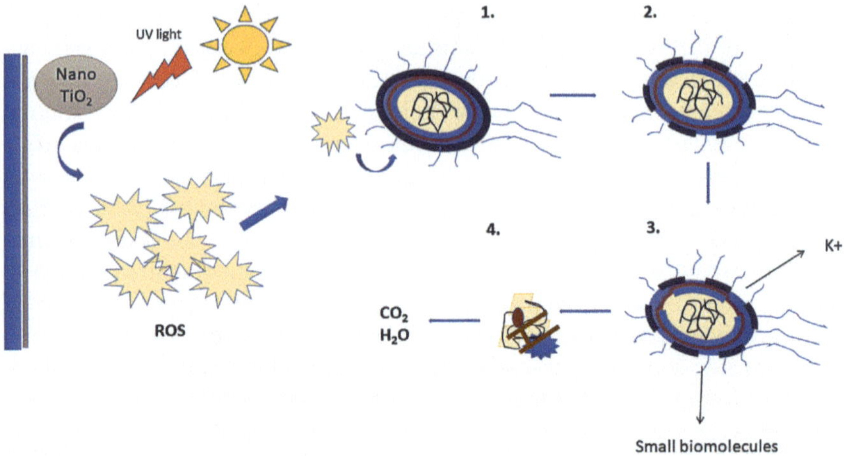

FIGURE 6.3 The mechanism of TUV bacteria degradation—antimicrobial effect depends from amount of generated ROS and for steps of action could be observed. *1*, Photogenerated ROS are getting in contact with bacteria surface; *2*, ROS are damaging cells' walls and membranes, which results in cytoplasm leakage (K+ and small molecules) in step *3*. *4*, The bacteria cells function disrupts, cells lyse and cellular components degrade, which leads to programmed cell death and mineralization. *(Modified from Yadav et al., 2016.)*

Maneerat and Hayata (2006) tested plastic films coated with TiO_2 exposed to UV radiation against mold *Penicillium expansum* on fruits and vegetables (apple, lemons, tomato). The photocatalytic properties of the TiO_2/UV process successfully suppressed the growth of mold. A similar inactivation effect was observed against vegetative forms of *Fusarium verticillioides* or *Apergillus niger*, growing on wet wood coated with a thin layer of TiO_2. But to obtain such effect, prolonged UV-A exposure was needed, taking even 20 days. Such long exposure needs pointing molds' high resistance to photocatalysis. The main factor underlying this fact is high chitin concentration in the molds cell wall (15%). The external layer of mold cells wall consists of galactoproteins, which are less resistant to oxidative injuries than chitin. Spores are not sensitive to active TiO_2. The effectiveness of photocatalysis was also checked on prions and yeasts (UV-A irradiation) (Bogdan et al., 2015b). Tested yeasts—*Candida albicans* and *Saccharomyces cerevisiae* cells were successfully inactivated (nearly 100%) by TiO_2/UV-A catalysis. Vegetative forms are more sensitive then spores. But scientists emphasized that yeasts are having lower sensitivity to AOPs then bacteria, the main factor blocking the oxidizing ROS effect is chitin and β-glucans presented in yeasts wall, which are more resistant to oxidation than PG. Plasma membrane is the most susceptible to oxidative damages. TiO_2 photochemical disinfection was also reported in algae and cancer cells (Yemmireddy and Hung, 2015). Resisting stages, like bacterial endospores, fungal spores, or protozoan cysts, are having increased cell wall thickness (Khezerlou et al., 2018). TiO_2 NPs are showing no toxicity in dark conditions.

In scientific investigations modified-TiO_2 coatings on glass or stainless steel are tested (e.g., Fe_3O_4@TiO_2 or TiO_2-Ag-SiO_2 powder in sol-gel phase with visible light irradiation for tests in fishery plants, or N-TiO_2 coatings) (Zhu et al., 2018; Zahornyi, 2018). In the experiment, the methyl ethyl ketone suspensions with 0.05%–1% wt. TiO_2 NPs were applied to form micro-thick polyethylene films. Also, 1% suspension inactivated *E. coli* (strain ATSS25922) significantly faster under shorter wavelength irradiation (365 nm) (Zahornyi, 2018).

TiO_2/UV process can effectively increase the disinfection of permanently used items, with which employers have frequent contact, and improve hygienic conditions of working area—and in this way, it can improve sanitary standards in the plants, as well as reduce microbiological hazards.

3.2 Active food packaging

Food packaging is one of the most active areas of food business, and technology currently represents the third-largest industry in the world (Mihindukulasuriya and Lim, 2014).

There are many directions of innovations for packaging and a big chance for NM applications. Still, nanotechnology needs optimization in methodology before being widely introduced, as it is costly, and there is no place for untested ideas.

3.2.1 Covers

The combination of food packaging materials with active NPs is a new modern way of surface microbial contamination control, the same as protection, especially for RTE/RTC food. The TiO_2/UV process was successfully applied in the disinfection of food packaging, both plastic bags, and containers (NPs cover of package surface or incorporated as a laminate composite layer). Nanostructured coatings due to inherent advantages over the multilayer films can improve barrier performance and show antimicrobial activity. Eradication of *E. coli* (strain ATCC11775) and *L. monocytogenes* has been proven (Bogdan et al., 2015a). TiO_2/UV process was moreover tested against *E. coli* (strain O157:H7), and photocatalysis effectively decomposed produced toxins (within 4 h) and inactivated living bacteria cells (Mihindukulasuriya and Lim, 2014). Due to the inactivation of *E. coli* TiO_2 powder-coated packaging, film usage is advised for freshly cut products (Cushen et al., 2012). Bodaghi et al. (2013) investigated TiO_2-LDPE film on fresh pears stored under UV light, and 17 days irradiation significantly decreased the number of mesophilic bacteria, as well as yeast cells. Li et al. (2017) pointed out, that nanoTiO_2-LDPE packaging alleviates weight loss, firmness of stored strawberries, and decrease decay. Active packaging can also be used for the preservation of food from animal origin, for example, ripened cheeses. Gumiero et al. (2013) applied high-density PE/calcium carbonate/TiO_2 film, and after UV irradiation observed inhibition of LAB (lactic acid bacteria) and coliforms, and in result better organoleptic properties of cheese.

Zhu et al. (2019) applied 1% TiO_2 NPs suspension (suspended in deionized water, pH 5.8) layer by layer on nonwoven fabric (NWF; characterized with high surface area containing small pores)—in result constructed new type of package, TiO_2 /NWF and used this package for fresh pork loins (heat–sealed) to evaluate effectiveness of TPC against *E. coli*. UV irradiation was applied (254 nm) in time depended experiment (0–60 min). The scientists tested not only the number of microorganisms, but also the organoleptic properties of meat. Disinfection effects were achieved and the efficiency was enhanced by elongation of irradiation time.

Since the pro-ecological approach forcing us to extensive use of paper/paperboard packaging, but in some kind of foods, the applicability is limited by the hydrophilic nature of this material. Scientists and technologists are looking for effective ways of enhancing paper barrier properties. It could be obtained by the incorporation of metal or hydrophobic thermoplastics within paper texture. But another innovative approach proposes the usage of nanoscale coatings. The photocatalytic paper was for the first time prepared in 1995. Traditionally, preparation of photocatalytic paper is carried out by photocatalyst coating size press treatment (e.g., TiO_2). The second approach is based on the deposition of NPs onto individual fibers before sheet formation (Rehim Abdel et al., 2016). For exapmle, Teisala et al. (2012) developed hydrophobic nanostructured TiO_2 coating on paperboard through liquid flame spraying. Youssef et al. (2013) went a step further and created paper substrate from rice straw, then polystyrene nanocomposite with incorporated nano-TiO_2 and nano-Ag was applied, as a coating. Both experiments confirmed the promising value of nanocomposites in the aspects of improved barrier function and antibacterial activity. Novel modified matrices for UV-shielding food packagings are also tested. For example, chitosan-nanoTiO_2 film for red grapes preservation (showing good antimicrobial activity) (Zhang et al., 2017; Lian et al., 2016). As well as chitosan-nanoTiO_2 film supported by graphene oxide (GO), with high bacteriostatic action (tested against biofilm-forming *B. subtilis*) or eco-friendly starch/ TiO_2 matrix (Zhu et al., 2018). Chitosan is well known biodegradable biopolymer, which on its own, can inhibit the growth of many microorganisms. FDA classified chitosan as safe in 2001 (food preservative), and in the food industry has been extensively used for surface coating of meat and fruits (reduce water loss, delay in ripening). Chitosan-TiO_2 nanocomposites possess favorable mechanical, thermal, and photocatalytic properties. Incorporation of TiO_2 nanopowder into chitosan can extend the absorption band over the visible light spectrum (Zhang et al., 2017).

3.2.2 Scavengers

The serious problem of food storage is the admission of oxygen in food packages, which can finish with product shelf-life reduction. Due to oxygen access plenty of degradation reactions within the product can be initiated (e.g., lipid oxidation, rancidity, microbial degradation of proteins), as well as vitamin depletion and organoleptic faults.

Packaging substrates coated with nano-TiO_2 have been tested for their oxygen scavenging abilities. In photoinduced nano-TiO_2, after the promotion of its electrons from the valence to the conduction band, the accumulation of electrons can be observed on the surface of particles. Electron transfer to oxygen is a step in photocatalysis, which determines the rate of reaction. The holes generated by photons of UV light have strong oxidation potential and high reactivity with water (Mihindukulasuriya and Lim, 2014).

Li et al. (2004) demonstrated that glass and acetate films covered with nano-TiO_2 effectively produced a deoxygenated closed environment (similar to head-space in food package). Their results of the oxygen reduction reaction showed the potential applicability of NMs as oxygen scavengers.

The TiO_2/UV process was also tested as the ethylene scavenger. The removal of ethylene vapor is important for the ripening delay of climacteric fruits (e.g., during long-distance shipment) and maintaining the quality of fresh products. TiO_2 is having better scavenging capacity than conventionally used scavengers, as it is not consumed in the reaction (additionally, the energy consumption is low) (Mihindukulasuriya and Lim, 2014). In the experiment of Maneerat and Hayata (2008), polypropylene films coated with TiO_2 (PP covered with ~7 nm particles of TiO_2) were developed to test the removal of ethylene steam in packed horticultural products—and the efficacy was very high. Another group (Ye et al., 2009) experimented with activated felts coated with TiO_2. The ethylene scavenging was confirmed, and as additional findings, the scientists observed that applying a bias voltage to the electrode membrane installation, the rate of ethylene degradation significantly enhanced. In the process, ethylene is converted into CO_2 and H_2O.

Moreover, TiO_2–immobilized photoreactors could be used in horticultural storage facilities, where post-harvested fruits and vegetables are stored. This method can be combined with other physical treatments (e.g., γ irradiation) to enhance scavenging efficiency (Ye et al., 2016). In the experiment of De Chiara et al. (2015) effectiveness of ethylene scavenging during the ripening process of mature green tomatoes was investigated, with the usage of modified TiO_2 (80% + 20% SiO_2) under UV irradiation (3 h). The efficacy of ethylene reduction reached 74%. The limitation of the UV photocatalysis usage is the need of exposition to UV light as an inductor of the process. And UV exposition can affect organoleptic parameters of sensitive fruits/vegetables. The solution of this problem could be a closed photocatalytic air circulation system.

3.2.3 Oxygen indicators

In the packaging with a modified atmosphere (MAP), the oxygen level is reduced, or even the idea is to eliminate oxygen to avoid bacteria growth—depending on product type and its shelf life. To control the maintenance of the targeted oxygen level, producers are using oxygen indicators. In this case, TiO_2/UV indicators can be applied. In such type of indicators, besides of TiO_2 NPs (anatase form), another component could be found. Typical additional components are redox dye (e.g., methylene blue) and an electron donor (e.g., glycerol/glycerine), which are encapsulated in polymer carrier and created as coating layer or film (Mills and Hazafy, 2009; Lawrie et al., 2013; Wen et al., 2018). Indicator irradiated with UV light is activated when excited electrons are accumulating on the surface of TiO_2 semiconductor and are available for dye reduction. When an indicator is subsequently exposed to oxygen, the dye is oxidized and obtains original color, which intensity represents the amount of oxygen inside the packaging. This type of indicator can be used again until the sacrificial electron donor does not undergo depletion. What is useful, when the package is transparent to UV radiation, activation of the indicator can be done after packaging sealing. Mihindukulasuriya and Lim (2014) modified the methodology and created a more sensitive indicator membrane by encapsulating the active factors within electrospun submicron fibers.

4 Potential usage in the food processing plant

4.1 Air and water purification

Nano-TiO_2/UV catalysis has a potential application in keeping air quality in manufacturing sites through air purification. Bioaerosols become currently dominating pollutants in closed rooms, also in processing plants. The components of bioaerosols are volatile organic compounds (VOC) and bacteria concentrated in the ventilation system and spread through it. In the bacteria group, pathogens, such as *M. tuberculosis*, *Str. pneumoniae*, *Str. pyogenes*, *Bordetella pertussis* are also identified (more than 60 bacteria, viruses and fungi are listed as infectious airborne pathogens) (Bogdan et al., 2015a). Traditional technologies used to clean indoor air are not fully effective (like HEPA filters, oznonation, air ionization). To improve air condition systems, UV lamps are used as well. But on the basis of experiments some scientists report, that after removal of UV light source bacteria can regenerate and re-grow (Zahornyi, 2018). Photocatalysis is mentioned as one of the most promising and safe method of indoor air disinfection. According to scientific findings, the TiO_2/UV catalysis process can be useful in VOC decomposition and successful eradication of bacteria (and spores of bacteria, e.g., *E. coli* could be completely mineralized) in a short time, for example, as thin photocatalytic layers in air ventilation systems.

Nano-TiO_2/UV catalysis was reported to be useful in liquid systems, which gives a chance to clean cooling systems in cold stores, as the process is also suitable for the gas phase, although the formation of $\bullet O2^-$ is lesser in this phase. Still, the obtained level exhibits the microorganisms' inactivation potential. One of the examples of photocatalysis modification can be carbon fiber/TiO_2 film, modified with Ag ions (electron acceptor). Ag modification gives the facilitation of bacteria adhesion on the catalyst surface, resulting in microbes' inactivation (Ye et al., 2010).

Another practical application of TiO_2/UV catalysis in hygienic management of food processing plant can be wastewater treatment (also wastewater from the fresh-cut vegetable industry). Disinfection and in the same time mineralization of organic pollutants or decomposition of cyano-bacterias' toxins can be obtained as a result (Wang et al., 2014; Bogdan et al., 2015c; Zhu et al., 2018). AOPs differ from traditionally used physical or chemical water treatments. The main active factor is hydroxyl radicals, very strong oxidants, able to degrade toxins and pollutants without the generation of secondary wastes. Also, some heavy metals can be eliminated, as precipitants (metal hydroxides). TiO_2 is resistant to corrosion. Scientific results on TPC under visible light radiation are promising for the future development of sustainable environmental remediation technologies, for example, with using of renewable sources of energy (Zahornyi, 2018).

Adequately, the TiO_2 suspension/UV-A catalysis process can disinfect drinking water, as well, for example, a significant reduction of coliforms (>91%) after 30 min irradiation has been observed (Zhu et al., 2018). Some microbes, including harmful pathogens (as *Legionella pneumophila*) present limited susceptibility to traditionally used water disinfectants. Additionally, popular conventionally used chemical water treatments (e.g., based on chlorine) generate huge number of harmful to humans and environment by-products, as a result of reaction between organic substances in water and oxidizing disinfectants. Some of those disinfection by-products (DBPs) are classified as teratogens, mutagens, or carcinogens (Bogdan et al., 2015a). For this reason, a need of safe water disinfectant arises. At the same time it should be cheap, non invasive to machines and equipment, as well as easy to handle. And nano TiO_2 fully fulfills those requirements.

A recent study (Monteagudo et al., 2017) showed very useful inactivation of *E. coli* (strain K12) in drinking water. TiO_2 photoreactor exposed to sunlight radiation for 80 min was combined with pasteurization (99.1% inactivation effectiveness). For drinking water disinfection, better antimicrobial effectiveness is obtained by usage TiO_2 suspension (dispersion allows to enlarge the specific surface area) than immobilized particles (immobilization onto glass, polymer films, or activated carbon). Furthermore, the density of suspension is essential—a higher dosage shows more significant photocatalytic activity and generates more ROS (Zhu et al., 2018). The difficulty is the separation of suspension after reaction (membrane filtration is used)—which for now is a limitation of such kind application.

4.2 Fresh vegetable washing

The safety of RTF is a significant concern to consumers, due to frequent outbreaks of foodborne pathogens connected with such kind of products (Pignata et al., 2017). Many efforts are put in minimizing the contamination, at the same time, to protect the gentle products, for example, vegetables. In the systems of fresh vegetables washing, TiO_2 photocatalytic disinfection was tested (carrots, lettuce, onion). For example, the initial number of bacteria in carrot washing water (6.7 logs CFU/mL) was totally eliminated by TPC within the 60 s. Photocatalytic nanotechnology was reported as very effective in the reduction of different bacteria numbers, also for *E. coli*, *S. typhimurium*, *L. monocytogenes*, and *B. cereus* (Zhu et al., 2018).

4.3 Drinks and juice disinfection

In many studies, application of TiO_2 photocatalytic inactivation of microorganisms was tested as a method of improvement of drinks and juices microbiological safety. The treatment was effective against *B. cereus*, *Pseudomonas*, *coliforms*, *S. aureus*, *S. typhimurium*, as well as against yeast and molds, with a result of a significant reduction of microorganisms (Zhu et al., 2018). The limitation in such tests is the dosage of TiO_2 and heat treatment, which can result in organoleptic faults of the products. It must also be pointed out that a high dose of NPs can increase ROS production, which would act not only with pathogens but also with vulnerable oxidized ingredients of drinks (sensory effect). That is why the optimization of the process is essential.

In fruit juices treatment, another example of combined process can be found. TPC is followed by HHP procedure (high hydrostatic pressure; this technique allows inactivation of microbes in a short time), which offers a synergistic practical approach of heat-sensitive food protection methods against bacterial contamination. TPC disrupts cell membranes and destroys genetic material of bacteria, while subsequent HHP treatment allows achieving desired inactivation efficiency level. For example, 20 min TPC with HHP (500 HPa/L min) treatment of apple juice resulted in full eradication of *L. monocytogenes* and *S. aureus* (Yoo et al., 2015). Combined treatment allows skipping limitations of both methods and gives a chance to use lower IR intensity and lower pressure in HHP, which is important for preservation of juice organoleptic properties.

5 Risks connected with nanomaterials—food and consumer safety

Nanotechnology is giving hopes for new innovative approaches in the development of modern packaging and increasing food safety. However, still, profound studies are needed (also toxicological) to ensure the safety of consumers (Fig. 6.4). From a risk assessment perspective for NMs—properties, like biological effects on human health (especially toxicokinetics and interactions between particles and cells), are not strictly related to specific size thresholds, rather are dose-depended. It must be underlined, that structured engineered NMs can have different toxicokinetic behavior, for example, different absorption or metabolism. The general outline for risk assessment of NMs is based on NM composition analysis, then on physicochemical properties check, potential interactions with tissues and cells, and finally on determination of potential exposure levels.

The main risk discussed in the case of NPs presence as packaging components is possible migration into the food products, and potential adverse health

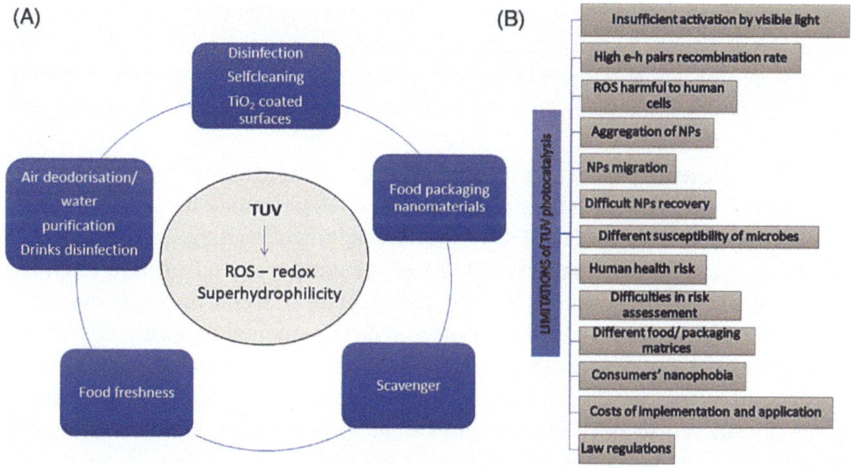

FIGURE 6.4 The current directions of TUV nanotechnology potential applications in food industry (A). Like all innovations, TPC is facing some challenges and limitations (B).

effects as a result (O'Brien and Cummins, 2011; Cushen et al., 2012; Echegoyen and Nerin, 2013). The results of such kind of exposure have not been entirely determined yet. Risk is assessed based on migration test results and depends on the NM toxicity, packaging matrix, and food consumption rate (migration rate increases with decreasing polymer dynamic viscosity and smaller nanoparticle size). The influence of nanoparticle ingestion and their depletion/bioaccumulation within the organism are still not fully understood (Steinhauser and Sayre, 2017). Bioaccumulation criterion is crucial, as NPs elimination from the organisms can take even years. Thus NPs could be a chronic irritant for health. Scientific studies have proven that some of the NPs can be responsible for the induction of intracellular damages (cellular membranes bypassing), pulmonary inflammation, or vascular disease. It is known that NPs can force biological barriers in the organism (e.g., blood-brain, gut) and circulate within the lymphatic system (Sodano et al., 2016). Toxicological mechanisms connected with NMs include generation of ROS, genotoxicity, or protein misfolding. All NPs induce the risk of inflammatory diseases (increased level of inflammatory markers, such as VCAM-1, ICAM-1) (Khezerlou et al. 2018). Ingestion of NMs via food or food contact materials is the most relevant source of NMs oral exposure. EFSA Scientific Committee (2018) distinguished two mods of NMs exposure in food: no presence of NMs in the product on the market/no migration, and in contrary complete NMs transformation in a food product.

The Danish Environmental Protection Agency (Danish EPA, 2015) conducted hazard assessment studies on toxicity and carcinogenicity of TiO_2 and stated that rutile form is better absorbed than anatase form, which is generally considered as more toxic. Anatase TiO_2 is a more efficient photocatalyst, and consequently, most industrial products contain this NP form. It is assumed that high photoreactivity of anatase is based on a higher Fermi level, which reduces the ability to absorb oxygen. And thus increasing the number of hydroxyl groups on TiO_2 NP surface. There are also some findings, that a mixture of both forms (70%–75% anatase, 25%–30% rutile) shows more active photocatalysis (Zahornyi, 2018).

TiO_2 represents one of the most studied engineered NMs in regards to environmental fate. The most dominant processes determining the environmental deposition of photocatalytic TiO_2 NMs are aggregation (e.g., pronounced in saltwater, due to higher ionic strength), sedimentation, and adsorption/retention in soil. Water aggregation can lead to sedimentation. Relative mass transfer is observed in the wastes—also with potential environmental exposure. EPA findings confirmed that photocatalytic TiO_2 from various paintings or coatings can be released to soil via air deposition. Redox reactions as a result of photocatalyis are considered to have low impact on the environment. Short-term aquatic toxicity data are available for daphnia, fish, and algae and long-term studies allowed to obtain results for daphnia and algae. NOEC = 0.89 mg/L and PNEC =18 μg/L (NOEC—no observed effect concentration; PNEC—predicted no effect concentration). On the basis of PEC/PNEC (<1) low aquatic environmental risk was calculated for photocatalytic TiO_2 (Danish EPA, 2015).

But the toxicity mechanism of TiO_2 nanoparticles from nanofood industry is not fully understood. The issue of industrial TiO_2 as a health and environmental hazard was raised a long time in many studies (Bogdan et al., 2015a). As nano-TiO_2 from sunscreen cosmetics with UV filters is classified as inductor of cancer diseases, and nano-TiO_2 is present in many kinds of cosmetics (e.g., antiperspirants, body lotions, face creams), the EU law regulation 1223/2009 allows the maximum amount of nano-TiO_2 on the level of 25% of the content. At the same time, in response to consumers' fears, manufacturers started to work on safer nano-TiO_2 formulas. The products with nano-TiO_2 particles enfolded with a thin layer of ethylene glycol were launched. Subsequent subjection of those coated particles to high-temperature heating (300°C), result in ethylene glycol carbonization and severe blocking of photocatalytic properties of TiO_2, as well as reducing of ROS production (Mihranyan et al., 2012).

E Scientific Committee (2018) for risk assessment takes into consideration characteristics of NM, which may affect toxicity, and on the list of indicators of potential toxicity are placed features characterizing TiO_2, as: altered hydrophilicity, increased reactivity compared to equivalent non-NM (for TiO_2 it is catalytic reactivity), antimicrobial activity, quantum effects (like altered redox properties). The list mentions interaction with biomolecules as well, and it should be taken in consideration not directly for TiO_2, but for ROS as products of photocatalysis.

TiO_2 NPs have generally low toxicity, with LC 50 greater than 100 μg/mL when applied in dispersion together with serum proteins and measured in *in vitro* cell cultures. However irradiated with UV generate free radicals, which affect biological systems. On the other hand, TiO_2 phototoxicity showed promising results in anti-cancer therapy (e.g., breast cancer) (Garvas et al., 2015).

Scientists pointing out technical problems affecting the whole range of risk assessment, mainly the variability of NPs and food matrices, also traditional toxicity research approach is not entirely valid for NMs and needs modification (Cushen et al., 2012). It is hard to distinguish an intentionally added NM in agri/food/feed products from the background levels of those substances present in low concentrations. There are also limited analytic methods dedicated to the characterization of NMs in relevant food matrices. When NMs are used in the manufacturing of food coating materials (FMC), there is a need of characterization of the NMs, their presence in FMC (as surface coating or in the matrix) and possible migrating potential from FMC into food matrix (migration testing, migration modeling in the polymer matrix), as well as the presence in food matrix (e.g., using instrumental ICP-MS detection). Surface NMs can be characterized by size and shape through SEM or TEM microscopy techniques (other applied techniques are FTIR or XRD, X-ray diffraction). Additionally, consideration must be given to the potential release of NMs from FCM under mechanical stress or physical matrix degradation (E Scientific Committee, 2018). In the tests FCM can be, for example, bended, stretched, heated, or scratched. It must be noted, that abrasion is not covered by conventional migration tests.

Insufficient toxicology data is a problematic area in risk assessment, too. Klimisch score criteria can be useful in toxicology studies (Card & Magnuson, 2010). Scientists also confirmed the danger of nano-TiO_2 for the environment based on studies on water purity indicator animal species and soil purity bioindicators (Gottschalk et al., 2009). Same as for industrial nano-TiO_2, nanofood TiO_2 safety reduction of health risk could be obtained by effective immobilization of NPs on surfaces as fixed thin layers resistant to mechanical force. The immobilized NPs won't be transmitted from coatings to the environment, and free NPs won't have contact with consumers' cells. Proven immobilization techniques are, for example, magnetron sputtering, chemical vapor, or sol-gel deposition (Bogdan et al., 2015a).

Consumers' awareness and the presence of neophobias connected with food (e.g., GMO, entomophagy, and artificial meat) (Cushen et al. 2012)—in this case, "nanophobia" toward new unknown technologies should also be taken into account. This can lead to economic problems of innovations (poor acceptance = less demand and fewer chances for business investors), as the process of development, commercialization, and official approval is long and expensive.

6 Nanofood materials—legislation regulatory issues

The main driver of nanotechnology and NMs usage in the food industry sector is the economical value—competitive advantage of innovation launching for the companies, resulting in strengthening their leading position on the market. Other important issues are connected with increased food safety/storage quality, as well as environmental protection through the reduction of the use of risky compounds or dangerous chemicals. Many TPC technologies are on basic TRLs

(Technology Readiness Levels) or are tested in laboratory conditions. For many years while nanotechnology was being introduced, EU law policy tried to work on existing laws dedicated to conventionally obtained foods and materials used in the food sector (Sodano et al. 2016). Nanofood regulations can be divided into horizontal (applicable to all types of products) and vertical (exclusive for agri-food products). Although new knowledge about NMs increased, as well as risks awareness and revealed that current law regulations are incomplete and show an inability to tackle a whole range of NMs risks and emerging new applications. That is why there is still a massive need for regulatory frameworks preventing possible adverse effects for human health and environment (correct risk assessment) and recognizing values affecting nanofood technology.

Regulation (EC) 1907/2006 REACH covers NMs standard information requirements and authorization (RE Committee, 2018), which helps ensure that surface modifications are closely scrutinized. EU regulation forces producers to test active or intelligent packaging for compounds migration. Also, the limits of nanoparticles' maximum levels in food are established. European Food Safety Authority (EFSA) had mandated compounds used in the food packaging; compounds cannot have an adverse effect on consumer health, as well as compounds release or absorption should not mislead the consumers (EU Regulation 10/2011, 450/2009, 1169/2011). Regulation (EC) 1272/2008 covers the aspects of labeling and packaging of chemical substances; NMs are not explicitly pointed. Still, it is important to remember that nanoscale of the content can change properties and classification. In Regulation 1169/2011 clause related to mandatory labeling of NMs was eliminated (to avoid consumers' confusion). Also, Regulation 2283/2015 does not contain any provision of compulsory labeling of manufactured nanomaterials.

Novel food regulation (EU 2015/2283) recognizes the nanofood as novel food and specifies that defined engineered NMs require authorization before usage in food products. This regulation points EFSA as an organ responsible for verifying if novel food consists of engineered NMs, and states that the most up-to-date analytical methods will be used for products safety assessment. IN EU Horizon 2020 program will be implemented Project NPs Pro Safe, presenting the concept that NMs can simultaneously maintain specific physicochemical properties and be safe- with low or absent toxicity for humans and environment (Vimercati et al. 2020).

7 Conclusions

Increasing competition in the food industry and growing consumers' awareness are stimuli for innovations and developments, including advanced technologies, like nanotechnology. But we are still facing the problem with R&D expensiveness, which is also a limitation for many food business operators/food-related industries and can cause more profound market inequalities. It is a must of nanotechnologies optimization and increased access to the industry. At the same time, more efforts should be put following risk assessment & management, as well as in collecting data about toxicity, bioaccumulation possibilities, biokinetics, and specifics of NMs—to ensure consumers' safety. TiO_2/UV photocatalysis has great potential to be incorporated in the food safety area, but still, a lot of tests are needed.

References

Bodaghi, H., Mostofi, Y., Oromiehie, A., Zamani, Z., Ghanbarzadeh, B., Costa, C., et al. (2013). Evaluation of the photocatalytic antimicrobial effects of a TiO_2 nanocomposite food packaging film by in vitro and in vivo tests. *LWT—Food Science and Technology*, *50*(2), 702–706.

Bogdan, J., Jackowska-Tracz, A., Zarzy ska, J., & Pławi ska-Czarnak, J. (2015). Chances and limitations of nanosized titanium dioxide practical application in view of its physicochemical properties. *Nanoscale Research Letters*, *10*, 57. https://doi.org/10.1186/s11671-015-0753-2.

Bogdan, J., Jackowska-Tracz, A., Zarzy ska, J., & Pławi ska-Czarnak, J. (2015). Importance of TiO_2/UV, ZnO/UV and MgO/UV photocatalytic processes in the inactivation of infectious agents. *Medycyna Weterynaryjna*, *71*(8), 472–479.

Bogdan, J., Zarzy ska, J., & Pławi ska-Czarnak, J. (2015). Comparison of infectious agents susceptibility to photocatalytic effects of titanium and zinc oxides: a practical approach. *Nanoscale Research Letters*, *10*, 309. https://doi.org/10.1186/s11671-015-1023-z.

Card, J. W., & Magnuson, B. A. (2010). A method to assess the quality of studies that examine the toxicity of engineered nanomaterials. *International Journal of Toxicology*, *29*(4), 402–410.

Cedrillo-Gonzalez, E. I., Mugonia, C., Montorsib, M., & Siligardi, C. (2016). Evaluation of the correlations between temperature, humidity, incident UV light and the photocatalytic activity of TiO_2 films using a rationale approach. *Applied Surface Science*, *378*, 73–79.

Cushen, M., Kerry, J., Morris, M., Cruz-Romero, M., & Cummins, M. (2012). Nanotechnologies in the food industry—Recent developments, risks and regulation. *Trends in Food Science & Technology*, *24*, 30–46.

Danish EPA (2015). Environmental assessment of nanomaterial use in Denmark, ISBN 978-87-93352-71-1.

de Azeredo, H. (2013). Antimicrobial nanostructures in food packaging. *Trends in Food Science & Technology*, *30*, 56–69.

De Chiara, M. L. V., Pal, S., Licciulli, A., Amodio, M. L., & Colelli, G. (2015). Photocatalytic degradation of ethylene on mesoporous TiO2/SiO2 nanocomposites: Effects on the ripening of mature green tomatoes. *Biosystems Engineering*, *132*, 61–70.

Echegoyen, Y., & Nerin, C. (2013). Nanoparticle release from nanosilver antimicrobial food containers. *Food and Chemical Toxicology*, *62*, 16–22.

EFSA Scientific Committee (2018). Guidance on risk assessment of the application of nanoscience and nanotechnologies in the food and feed chain: Part 1, human and animal health. *EFSA Journal*, 16(7), Article e05327, https://doi.org/10.2903/j.efsa.2018.5327..

Garvas, M., Testen, A., Umek, P., Gloter, A., Koklic, T., & Stancar, J. (2015). Protein corona prevents TiO_2 phototoxicity. *PLoS One*, *10*(6), e0129577. https://doi.org/10.1371/journal.pone.0129577.

Gottschalk, F., Sonderer, T., Scholz, R. W., & Nowack, B. (2009). Modeled environmental concentrations of engineered nanomaterials (TiO2, ZnO, Ag, CNT, fullerenes) for different regions. *Environmental Science & Technology*, *43*(24), 9216–9222.

Gumiero, M., Peressini, D., Pizzariello, A., Sensidoni, A., Iacumin, L., Comi, G., et al. (2013). Effect of TiO2 photocatalytic activity in a HDPE-based food packaging on the structural and microbiological stability of a short-ripened cheese. *Food Chemistry*, *138*(2–3), 1633–1640.

Hitkova, H., Stoyanova, A., Ivanova, N., Sredkova, M., Popova, V., & Iordanova, R. (2012). Study of antibacterial activity of nonhydrolytic synthesized TiO_2 against *E. coli*, *P. aeruginosa* and *S. aureus*. *Journal of Optoelectronic and Biomedical Materials*, *4*(1), 9–17.

Jacoby, W. A., Maness, P. C., Wolfrum, E. J., Blake, D. M., & Fennell, J. A. (1998). Mineralization of bacteria cell mass on a photocatalytic surface in air. *Environmental Science & Technology*, *32*(17), 2650–2653.

Jing, L., Zhou, W., Tian, G., & Fu, H. (2013). Surface tuning for oxide-based nanomaterials as efficient photocatalysts. *Chemical Society Reviews*, *42*, 9509.

Jing, Z., Guo, D., Wang, W., Zhang, S., Qi, W., & Ling, B. (2011). Comparative study of titania nanoparticles and nanotubes as antibacterial agents. *Solid State Sciences*, *13*, 1797–1803.

Joseph, T., & Morrison, M. (2006). Nanoforum report: Nanotechnology in agriculture and food. Available from: ftp.cordis.europa.eu/pub/nanotechnology/doc/nanotechnology_in_agriculture_and_food.pdf/.

Khezerlou, A., Alizadeh-Sanib, M., Azizi-Lalabadib, M., & Ehsani, A. (2018). Nanoparticles and their antimicrobial properties against pathogens including bacteria, fungi, parasites and viruses. *Microbial Pathogenesis*, *123*, 505–526.

Lawrie, K., Mills, A., & Hazafy, D. (2013). Simple inkjet-printed, UV-activated oxygen indicator. *Sensors and Actuators*, *B176*, 1154–1159.

Lee, J. E., & Ko, G. (2013). Norovirus and MS2 inactivation kinetics of UV-A and UV-B with and without TiO_2. *Water Research*, *47*(15), 5607–5613.

Li, D., Ye, Q., Jiang, L., & Luo, Z. (2017). Effects of nano-TiO2-LDPE packaging on postharvest quality and antioxidant capacity of strawberry (*Fragaria ananassa* Duch.) stored at refrigeration temperature. *Journal of the Science of Food and Agriculture*, *97*(4), 1116–1123.

Li, X. -e., Green, A. N. M., Haque, S. A., Mills, A., & Durrant, J. R. (2004). Light-driven oxygen scavenging by titania/polymer nanocomposite films. *Journal of Photochemistry and Photobiology A: Chemistry*, *162*, 253–259.

Lian, Z., Zhang, Y., & Zhao, Y. (2016). Nano-TiO2 particles and high hydrostatic pressure treatment for improving functionality of polyvinyl alcohol and chitosan composite films and nano-TiO_2 migration from film matrix in food simulants. *Innovative Food Science and Emerging Technologies*, *33*, 145–153.

Maneerat, C., & Hayata, Y. (2006). Antifungal activity of TiO_2 photocatalysis against *Penicillium expansum* in vitro and in fruit tests. *International Journal of Food Microbiology*, *107*, 99–103.

Maneerat, C., & Hayata, Y. (2008). Gas-phase photocatalytic oxidation of ethylene with TiO_2 coated packaging film for horticultural products. *Transactions of the ASABE*, *51*, 163–168.

Mihindukulasuriya, S. D. F., & Lim, L. -T. (2014). Nanotechnology development in food packaging: A review. *Trends in Food Science & Technology*, *40*, 149–167.

Mihranyan, A., Ferraz, N., & Strømme, M. (2012). Current status and future prospects of nanotechnology in cosmetics. *Progress in Materials Science*, *57*(5), 875–910.

Mills, A., & Hazafy, D. (2009). Nanocrystalline SnO_2-based, UVB-activated, colourimetric oxygen indicator. *Sensors and Actuators B: Chemical*, *136*, 344–349.

Monteagudo, J. M., Duran, A., Martin, I. S., & Acevedo, A. M. (2017). A novel combined solar pasteurizer/TiO_2 continuous-flow reactor for decontamination and disinfection of drinking water. *Chemosphere*, *168*, 1447–1456.

O'Brien, N., & Cummins, E. (2011). A risk assessment framework for assessing metallic nanomaterials of environmental concern: aquatic exposure and behavior. *Risk Analysis*, *31*(5), 706–726.

Peter, A., Tegla, D., Giurgiulescu, L., Cozmuta, A. M., Nicula, C., Cozmuta, L. M., et al. (2015). Development of Ag/TiO_2-SiO_2-coated food packaging film and its role in preservation of green lettuce during storage. *Carpathian Journal of Food Science and Technology*, *7*(4), 88–96.

Pignata, C., D'Angelo, D., Fea, E., & Gilli, G. (2017). A review on microbiological decontamination of fresh produce with nonthermal plasma. *Journal of Applied Microbiology*, *122*(6), 1438–1455.

REACH Committee (2018). Summary of the Meeting of the REACH Committee. April 2018. (2018) 0357. Brussels. CMTC, pp. 25–26.

Reddy, P. V. L., Kavitha, B., Reddy, P. A. K., & Kim, K. -H. (2017). TiO_2-based photocatalytic disinfection of microbes in aqueous media: A review. *Environmental Research*, *154*, 296–303.

Rehim Abdel, M. H., El-Samahya, M. A., Abdelrahman, A., & Mohram, M. E. (2016). Photocatalytic activity and antimicrobial properties of paper sheets modified with TiO_2/Sodium alginate nanocomposites. *Carbohydrate Polymers*, *148*, 194–199.

Sodano, V., Gorgitano, M. T., Quaglietta, M., & Verneau, F. (2016). Regulating food nanotechnologies in the European Union: Open issues and political challenges. *Trends in Food Science & Technology*, *54*, 216–226.

Som, C., Berges, M., Chaudhry, Q., Dusinska, M., Fernandes, T. F., Olsen, S. I., et al. (2010). The importance of life cycle concepts for the development of safe nanoproducts. *Toxicology*, *269*(2–3), 160–169.

Steinhauser, K. G., & Sayre, P. G. (2017). Reliability of methods and data for regulatory assessment of nanomaterial risks. *NanoImpact*, *7*, 66–74.

Teisala, H., Tuominen, M., Aromaa, M., Makela, J. M., Stepien, M., Saarinen, J. J., et al. (2012). Development of superhydrophobic coating on paperboard surface using the liquid flame spray. *Surface and Coatings Technology*, *205*, 436–445.

van Grieken, R., Marugan, J., Pablos, C., Furones, L., & Lopez, A. (2010). Comparison between the photocatalytic inactivation of Gram-positive *E. faecalis* and Gram negative *E. coli* faecal contamination indicator microorganisms. *Applied Catalysis B: Environmental*, *100*(1–2), 212–220.

Verbruggen, S. W., Keulemans, M., Filippousi, M., Flahaut, D., Van Tendeloo, G., Lacombe, S., et al. (2014). Plasmonic gold-silver alloy on TiO_2 photocatalysis with tunable visible light activity. *Applied Catalysis B: Environmental*, *156*, 116–121.

Vimercati, L., Cavone, D., Caputi, A., DeMaria, L., Tria, M., Prato, E., et al. (2020). Nanoparticles: An experimental study of zinc nanoparticles toxicity on marine crustaceans. General overview on the health implications in humans. *Frontiers in Public Health*, 8, 192. https://doi.org/10.3389/fpubh.2020.00192.

Wang, J., Zhuang, H., Hinton, A., Jr., Bowker, B., & Zhang, J. (2014). Photocatalytic disinfection of spoilage bacteria *Pseudomonas fluorescens* and *Macrococcus caseolyticus* by nano-TiO_2. *LWT—Food Science and Technology*, *59*, 1009–1017.

Warheit, D. B., & Brown, S. C. (2019). What is the impact of surface modifications and particle size on commercial titanium dioxide particle samples?—A review of in vivo pulmonary and oral toxicity studies—Revised 11-6-2018. *Toxicology Letters*, *302*, 42–59.

Wen, J., Huang, S., Sun, Y., Chen, Z., Wang, Y., Li, H., et al. (2018). Titanium dioxide nanotube-based oxygen indicator for modified atmosphere packaging: Efficiency and accuracy. *Materials*, *11*, 2410. https://doi.org/10.3390/ma11122410.

Yadav, H. M., Kim, J. S., & Pawar, S. H. (2016). Developments in photocatalytic antibacterial activity of nano TiO_2: A review. *Korean Journal of Chemical Engineering*, *33*(7), 1989–1998.

Ye, S. -Y., Fan, M. -L., Song, X. -L., & Luo, S. -C. (2010). Enhanced photocatalytic disinfection of *P. expansum* in cold storage using a TiO_2/ACF film. *International Journal of Food Microbiology*, *136*(3), 332–339.

Ye, S. -y., Liang, J. -l., Song, X. -l., Luo, S. -c., & Liang, J. -y. (2016). Modeling intrinsic kinetics in a reactor of corona discharge coupled with TiO_2-activated carbon fibre for ethylene degradation and ozone regulation. *Biosystems Engineering*, *150*, 123–130.

Ye, S. -y., Tian, Q. -m., Song, X. -l., & Luo, S. -c (2009). Photoelectrocatalytic degradation of ethylene by a combination of TiO_2 and activated carbon felts. *Journal of Photochemistry and Photobiology A: Chemistry*, *208*, 27–35.

Yemmireddy, V. K., & Hung, Y. -C. (2015). Selection of photocatalytic bactericidal titanium dioxide (TiO_2) nanoparticles for food safety applications. *LWT—Food Science and Technology*, *61*, 1–6.

Yoo, S., Ghafoor, K., Kim, J. U., Kim, S., Jung, B., Lee, D. U., et al. (2015). Inactivation of Escherichia coli O157:H7 on orange fruit surfaces and in juice using photocatalysis and high hydrostatic pressure. *Journal of Food Protection*, *78*(6), 1098–1105.

Youssef, A. M., Kamel, S., & El-Samahy, M. A. (2013). Morphological and antibacterial properties of modified paper by PS nanocomposites for packaging applications. *Carbohydrate Polymers*, *98*, 1166–1172.

Zahornyi, M. (2018). Functional nanocomposites based of titanium dioxide. LAP Lambert Academic Publishing RU. ISBN: 978-613-8-27486-5.

Zhang, H., Xiao, G., Wang, Y., Zhao, Y., Su, H., & Tan, T. (2017). Preparation of chitosan-TiO_2 composite film with efficient antimicrobial activities under visible light for food packaging applications. *Carbohydrate Polymers*, *169*, 101–107.

Zhu, Z., Cai, H., & Sun, D. -W. (2018). Titanium dioxide (TiO_2) photocatalysis technology for nonthermal inactivation of microorganisms in foods. *Trends in Food Science & Technology*, *75*, 23–35.

Zhu, Z., Cai, H., Sun, D. -W., & Wang, H. -W. (2019). Photocatalytic effects on the quality of park packed in the package combined with TiO2 coated nonwoven fabrics. Journal of Food Process Engineering, 9(42), e12993.

Chapter 7

Bioplastic for Sustainable Food Packaging

Nathdanai Harnkarnsujarit*, Phanwipa Wongphan, Tanaporn Chatkitanan, Yeyen Laorenza and Atcharawan Srisa
Department of Packaging and Materials Technology, Faculty of Agro-Industry, Kasetsart University, Bangkok, Thailand
**Corresponding author*

1 Introduction

Packaging functions include conservation, containment, convenience, and communication. New packaging is generally developed based on these four characteristics. Data indicate that food packaging contributes the highest amount of waste products, with increasing concern regarding the impact of packaging waste on the environment. Policies to reduce packaging waste have been instituted globally including recycling, reduction, and use of environmentally-friendly packaging materials.

Packaging design requires consideration of waste reduction and suitability for waste management. Conventional plastic produced from petroleum-based polymers takes over 100 years to decompose in a landfill and introduces toxic chemicals into the ground. The European Bioplastics Association and Nova-Institute collected market data on bioplastic production. They stated that global production capacity of bioplastic is expected to increase from 2.11 million tons in 2019 to 2.43 million tons in 2024. Asia is the largest hub for bioplastic production, corresponding to 45% of the global output in 2019 followed by Europe (25%), North America (18%), and South America (12%). Starch blends show the most significant global production capacities of 21.3%, followed by PLA and PBAT with 13.9% and 13.4%, respectively. Among other applications, flexible packaging comprises the largest amount of global bioplastic production at 663,000 tons followed by rigid packaging and textiles at 476,000 and 237,000 tons, respectively (European Bioplastics, 2020). Consequently, use of bioplastic is attracting increasing interest from food and packaging manufacturers as an alternative to replace petroleum based-plastic.

Conservation and protection are primary packaging functions that directly impact food quality. Packaging structural design protects foods from exposure

Sustainable food processing and engineering challenges. http://dx.doi.org/10.1016/B978-0-12-822714-5.00007-3

to external factors accelerating deteriorations, including water vapor, oxygen, volatile elements, light, contamination, and mechanical forces as shock, vibration, and impact due to handling during transportation and storage. Package permeability is a major factor affecting the shelf-life of most foods due to mass transport, water vapor, and oxygen. Low permeability is required to increase shelf-life for moisture and oxygen-sensitive food products. Packaging materials used in the food industry include glass, metal, paper, and plastics. Glass and metals are known as impermeable materials giving the least permeation, while paper and porous structures are known as fully permeable materials. Plastic has wide-ranging properties depending on several factors including crystallinity, molecular weight, chemical composition, and additives. High barrier petroleum-based plastics are available in the market including ethylene vinyl alcohol, polyamides, and other laminated and coated materials. However, bioplastics materials mostly have lower barrier properties than petroleum-based plastics with consequent limitations for use with long-shelf-life food products. Therefore, numerous studies have investigated bioplastic packaging to improve barrier properties and hence shelf-life of foods. Bioplastic packaging materials, applications and development in the food industry are discussed in this chapter.

2 Types and forms of bioplastic food packaging

2.1 Bio-based and biodegradable polymers

Bioplastic materials are made from a renewable resource (bio-base) and are biodegradable (Fig. 7.1). Bio-based materials include those made from natural, organic, or renewable resources, including plants and animals such as starch, protein, cellulose, and seaweed polysaccharides. Bio-based polymers can also be produced from plant oils, for example, palm seed, soybean, and linseed oil or fermentation products including polylactic acid (PLA), polyhydroxyalkanoate (PHA), and polyhydroxybutyrate (PHB). Polymerization of monomers from a

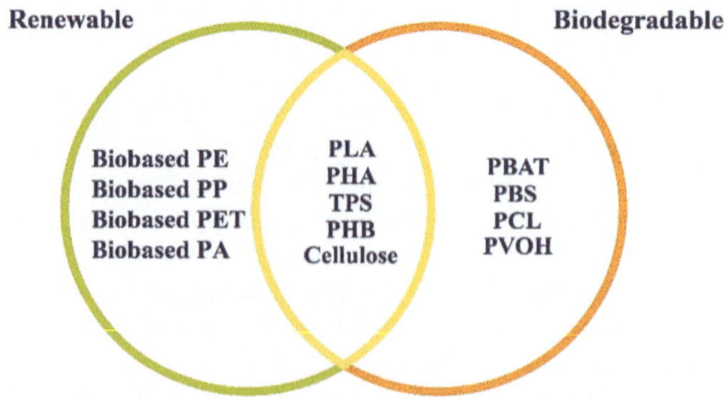

FIGURE 7.1 Bioplastic materials are derived from bio-based renewable resources with biodegradable characteristics.

renewable resource produces bio-based polymers that are not necessarily biodegradable. For example, sugarcane, corn, and cassava can be used to produce bio-based polyethylene (PE), bio-based polypropylene (PP), bio-based polyethylene terephthalate (PET), and bio-based polyamide (PA) that have similar chemical structures and characteristics to their petroleum-based non-renewable resource. These bioplastic materials are considered as bio-based but non-biodegradable materials. Numerous renewable materials also have biodegradation characteristics, for example, PLA, PHA, thermoplastic starch (TPS), PHB, protein, cellulose, and other polysaccharides. Biodegradable materials are classified as those in which chemical structures change rapidly on contact with the biological environment due to hydrolysis and enzymatic activity (Peppas & Brannon-Peppas, 2001; Ross et al., 2017). Polymer scission causes chain cleavage. This decreases molecular weight and produces an increased number of oligomers and other lower molecular weight substances. Biodegradation rates and products depend on several factors in various environmental conditions (Peppas & Brannon-Peppas, 2001). Bioplastic can be derived from petroleum or fossil resources that are synthesized and polymerized forming biodegradable polymers, including polybutylene adipate terephthalate (PBAT), polybutylene succinate (PBS), and polycaprolactone (PCL). Polyvinyl alcohol (PVOH) is a highly hydrophilic material and readily dissolves in water, making disintegration easily in the environment and, therefore, it is commonly considered as a biodegradable bioplastic material.

Standards for biodegradable and compostable plastics have been developed by global organizations, including the American Society for Testing and Materials (ASTM), the International Standards Organization (ISO), the International Committee for Standardization (CEN), the Japanese Institute for Standardization (JIS), and the British Plastics Federation (Table 7.1). Biodegradable packaging undergoes chemical, biodegradability, disintegration, and ecotoxicity testing for products derived from compostable materials, physical characteristics, and compositions under specific environmental conditions. Biodegradation standards of plastic materials specify the conversion of plastic materials to CO_2, water, and biomass in a disposal environment over a defined timespan. Biodegradation standards established for disposal environments include industrial compost, marine environment, anaerobic digestion, landfill, and home compost. These international standards have been established in two major categories, namely testing methods and performance requirements (specifications). For example, ASTM D6400 specifies requirements for biodegradation of compostable plastics that are designed to biodegrade into CO_2, water, and biomass in an industrial compost environment. Compostable plastics are those that pass the ASTM D6400 biodegradation requirements of greater than 90% carbon conversion to CO_2 after 180 days while exposed primarily to hot composting conditions of 58°C and 50% RH (Greene, 2014). ASTM D5338 describes procedures, equipment, materials, and conditions to measure the degree and rate of biodegradation of plastic materials under aerobic thermophilic

TABLE 7.1 Global standards related to biodegradable polymer testing.

Agency/type	Code	Standard	Scope
US Bio-based Standard	ASTM D6866-10	Standard Test Methods for Determining the Bio-based Content of Solid, Liquid, and Gaseous Samples Using Radiocarbon Analysis	Procedures, equipment, materials, and conditions to measure the 14-carbon content of the plastic sample through radiocarbon analysis
US Biodegradation Standards for Industrial Compost Environment	ASTM D6400-04	Standard Specification For Compostable Plastics	Performance requirements for biodegradation of compostable plastic materials that are designed to biodegrade into CO_2, water, and biomass in an industrial compost environment
	ASTM D6868-03	Standard Specification for Biodegradable Plastics Used as Coatings on Paper and Other Compostable Substrates	Performance requirements for biodegradation of materials and products including packaging, films, and coatings that are attached either through lamination or extrusion directly onto the paper to compostable substrate plastic materials
	ASTM D5338-11	Standard Test Method for Determining Aerobic Biodegradation of Plastic Materials Under Controlled Composting Conditions, Incorporating Thermophilic Temperatures	Procedures, equipment, materials, and conditions to measure the degree and rate of biodegradation of plastic materials under aerobic thermophilic composting conditions
International Biodegradation Standards for Industrial Compost Environment	EN 13432-2000	Packaging Requirements for Packaging Recoverable Through Composting and Biodegradation Test Scheme and Evaluation Criteria for the Final Acceptance of Packaging	Performance requirements for biodegradation of plastic materials and products, including packaging, films, and other products

	ISO 17088 (EN 13432)	Plastics-Evaluation of Compostability-Test Scheme and Specification	Performance requirements for biodegradation of plastic materials and products including packaging, films, and other products (very similar to EN13432-2000)
	ISO 14855-2 (EN 14046)	Evaluation of the Ultimate Aerobic Biodegradability and Disintegration of Packaging Materials Under Controlled Composting Conditions. A Method by Analysis of Released Carbon Dioxide-Part 2: Gravimetric Measurement of Carbon Dioxide Evolved in a Laboratory-Scale Test	Test method to evaluate the ultimate aerobic biodegradability of packaging materials based on organic compounds under controlled composting conditions by measurement of released carbon dioxide
	ISO 16929 (EN14045:2003)	Plastics-Determination of the degree of disintegration of plastic materials under simulated composting conditions in a pilot-scale test	Method for the evaluation of the disintegration of plastic materials while under industrial composting conditions
	ISO 20200 (EN14806:2005)	Plastics-Determination of the Degree of Disintegration of Plastic Materials Under Simulated Composting Conditions in a Laboratory-Scale Test	Method for the evaluation of the disintegration of plastic materials while under industrial composting conditions
	AS4736-2006	Biodegradable Plastics-Biodegradable Plastics Suitable for Composting and Other Microbial Treatment	Australian standard includes test methods that enable validation of biodegradation of degradable plastics and designation as "compostable" in Australia. The AS4736 standard is based on EN 13432-2000 standard for the determination of ultimate aerobic biodegradability and disintegration of plastic materials under controlled composting conditions

(Continued)

TABLE 7.1 Global standards related to biodegradable polymer testing. (*Cont.*)

Agency/type	Code	Standard	Scope
US Biodegradation Standards for Marine Environment	ASTM D-7081-05	Non-Floating Biodegradable Plastic in the Marine Environment	Performance requirements for biodegradation of plastic materials and products, including packaging, films, and coatings The marine environment includes conditions of aerobic marine waters or anaerobic marine sediments or both Specifies that the degradation of the plastic materials will not cause any harm to sea life or habitat
	ASTM D6691-09	Standard Test Method for Determining Aerobic Biodegradation of Plastic Materials in the Marine Environment by a Defined Microbial Consortium or Natural Seawater Inoculum	Testing procedures that best simulate the marine environment and a method by which to measure biodegradation Determine the degree and rate of aerobic biodegradation of plastic materials exposed to the indigenous population of existing seawater or synthesized seawater with a pre-grown population of at least 10 aerobic marine microorganisms of known genera
International Aqueous Biodegradation Standards	ISO 14852-1999 (EN14047)	Determination of Ultimate Aerobic Biodegradability of Plastic Materials in an Aqueous Medium-Method by Analysis of Evolved Carbon Dioxide	Method for the evaluation of the biodegradation of plastic materials while exposed to the aqueous mineral medium. Biodegradation is determined by measuring evolved carbon dioxide from the biodegrading sample

	ISO 14851 (EN14048).	Determination of Ultimate Aerobic Biodegradability of Plastic Materials in an Aqueous Medium-Method by Measuring the Oxygen Demand in a Closed Respirometer	Method for the evaluation of the biodegradation of plastic materials while exposed to aqueous mineral medium Biodegradation is determined by measuring oxygen demand produced from the biodegrading sample
US Biodegradation Standards for Anaerobic Digestion	ASTM D5511-02	Standard Test Method for Determining Anaerobic Biodegradation of Plastic Materials under High Solids Anaerobic-Digestion Conditions	Procedures, equipment, materials, and conditions to measure the degree and rate of biodegradation of plastic materials under anaerobic thermophilic conditions in an aqueous environment
	ISO 14853:2005	Determination of Ultimate Anaerobic Biodegradation of Plastic Materials in an Aqueous System. Method of Biogas Production	Method for determining the ultimate biodegradability of plastics by anaerobic microorganisms
US Biodegradation Standards for Active Landfill	ASTM D5526-11	Determining Anaerobic Biodegradation of Plastic Materials Under Accelerated Landfill Conditions	Degree and rate of anaerobic biodegradation of plastic materials under anaerobic mesophilic conditions in an accelerated landfill test environment
	ASTM D7475-11	ASTM D7475-11 Determining the Aerobic Degradation and Anaerobic Biodegradation of Plastic Materials under Accelerated Bioreactor Landfill Conditions	Modifies ASTM D5526 by adding an aerobic degradation test method to the anaerobic biodegradation test method Degree of aerobic degradation and rate of anaerobic biodegradation of plastic materials under anaerobic mesophilic conditions in an accelerated landfill test environment

composting conditions in the laboratory. There are also standard test methods such as D6691 for determining aerobic biodegradation of plastic materials in the marine environment by a defined microbial or natural seawater inoculum as described in D6691. ASTM D-7081-05 indicates performance requirements for biodegradation of plastic materials, including packaging, films, and coatings. This specification establishes the requirements for biodegradation of plastic materials that have similar rates to known compostable materials under aerobic and anaerobic marine conditions, and also specifies that the degradation of the plastic materials shall not cause any harm to sea life or habitat (Greene, 2014).

2.2 Bioplastic materials

Numerous bioplastic materials have been investigated in terms of production and disposal; however, only a few are used in the commercial market due to high cost and limited material performance. Major attempts to reduce production and material cost, enhance manufacturing performance, material properties, and stability are increasing, with strong support for the commercialization of bioplastic packaging for food products. Several packaging properties, including physical, mechanical, barrier, and stability depend on the chemical structures of monomers, molecular weight, and the arrangement of polymers controlled by intra- and inter-molecular interactions. Major bioplastic monomers are shown in Fig. 7.2. Some bio-based materials form very rigid polymer networks and require plasticization to enhance flexibility for packaging production, for example, starch, protein, and non-starch polysaccharides. Types and concentration of plasticizers also affect the properties and stability of packaging materials. Additives and fillers improve material performance as well as aiding the manufacturing process. Nieto (2016) summarized the progress of bio-based polymer productions from renewable resources as (1) application of natural or partially modified bio-based polymers, for example, protein, starch, and non-starch polysaccharides, (2) production of bio-based monomers via fermentation and other chemical reactions followed by polymerization, for example, PLA, PBS, and bio-based PE, and (3) direct harvesting of bio-based polymers via microorganism metabolites, for example, PHA.

2.2.1 *Poly(lactic acid)*

Poly(lactic acid) or poly(lactide), so-called "PLA," is a biocompatible and biodegradable polymer serving a variety of packaging applications. Lactic acid is produced from the fermentation of carbohydrate (starch and cellulose) from biomass resources, for example, corn and cassava, to dextrose that is further converted into lactic acid. Lactide is the cyclic dimer of lactic acid that is present in L-lactide and D-lactide stereoisomers. Poly(lactic acid) is produced by direct polycondensation of lactic acid, while poly(lactide) is produced from ring-opening polymerization of lactide that comes from lactic acid. Though they come from a different process, both terms are often used interchangeably and

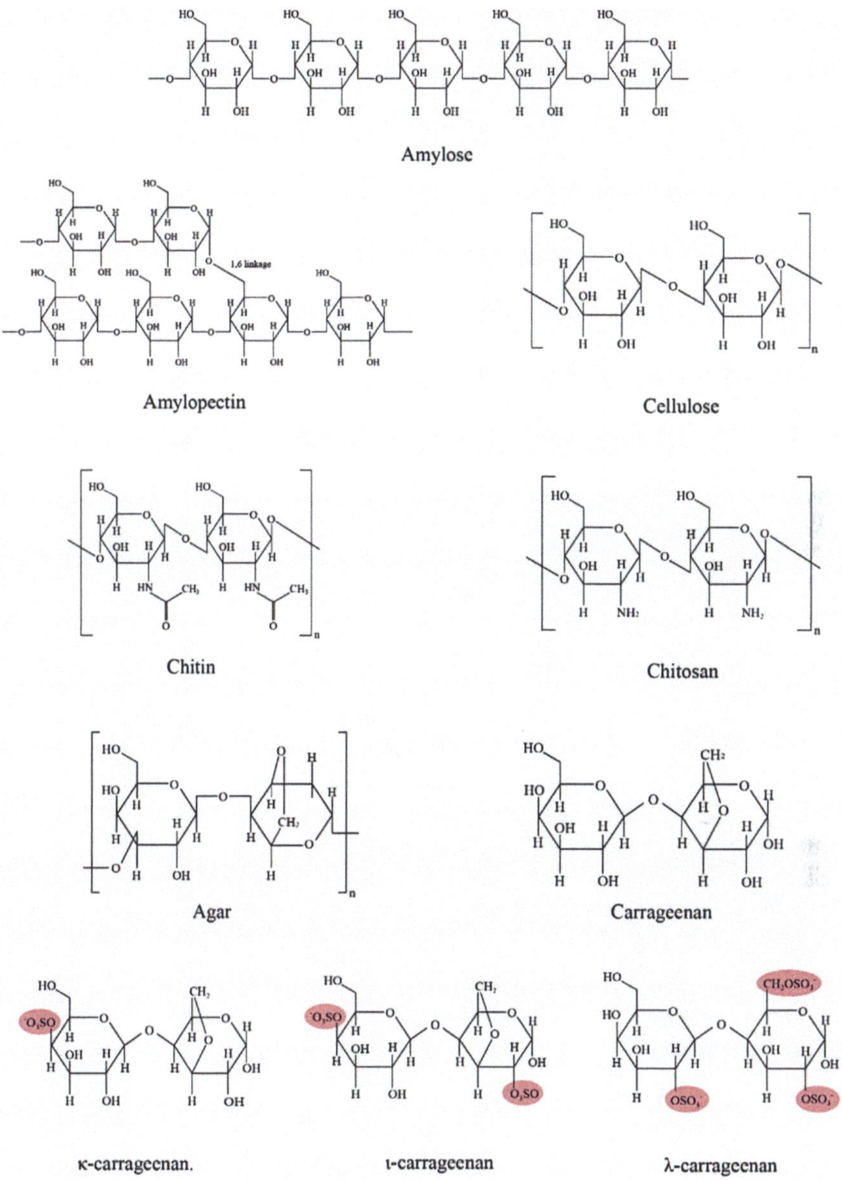

FIGURE 7.2 Structures of several bioplastic materials.

(Continued)

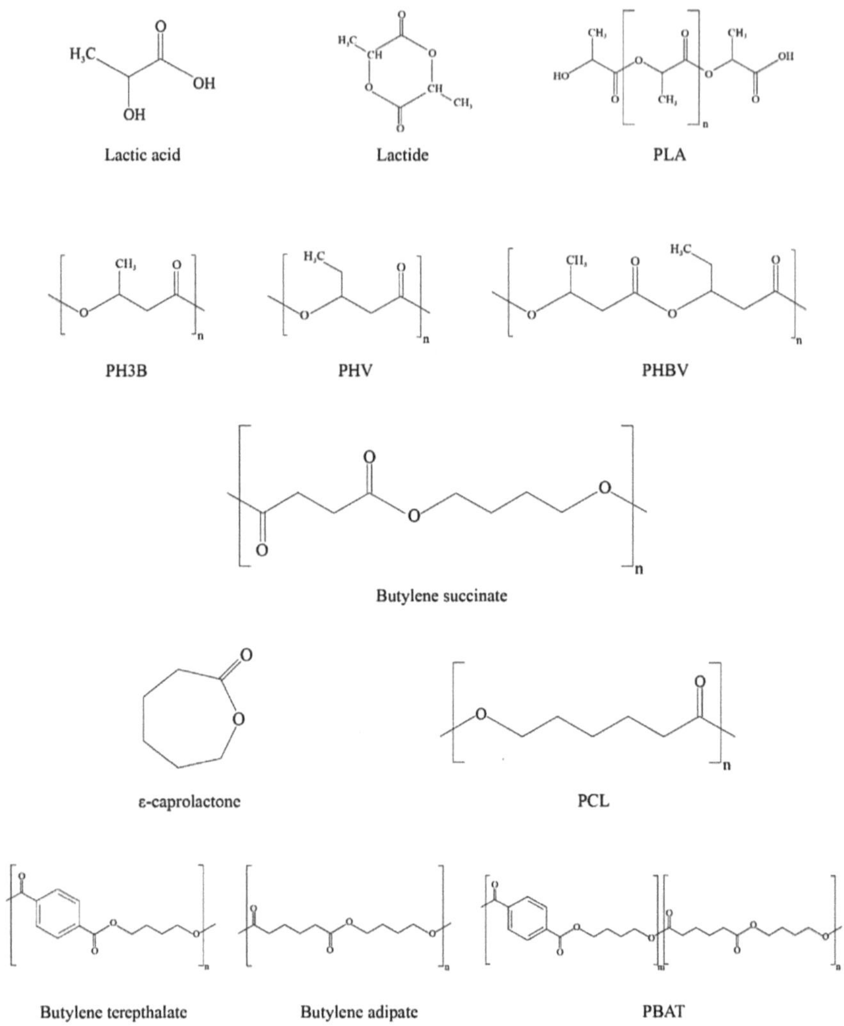

FIGURE 7.2 ***(Cont.)***

called “PLA,” referring to the polymer produced from lactic acid (Sin & Tueen, 2019). The direct condensation of lactic acid to form PLA involves interaction between hydroxyl and carboxylic acid groups with the elimination of water. This results in a low molar mass that is brittle and requires external coupling agents to increase chain length. Conversely, ring-opening polymerization of cyclic lactide is a more effective pathway to produce high molar mass PLA. This process is suitable for large-scale commercial production and effectively controls the chemistry and consequent properties of PLA materials (Ross et al., 2017).

Polymerization of correct L-lactide enantiomer produces poly(L-lactide) or "PLLA," while D-lactide produces poly(D-lactide) or "PDLA." These two forms have semi-crystalline structures. Conversely, polymerization of a D- and L-lactic acid mixture or D, D-lactide/L, L-lactide yields poly(D, L-lactide), or "PDLLA," that has a higher amorphous fraction. PLLA is one of the most useful biodegradable materials among all these PLA stereotypes, partly because its building block L-lactide is produced from the fermentation of starch from agricultural plants. PLLA is more expensive than many petroleum-based polymers but PLLA produced from corn is becoming cheaper as the scale of production is increasing due to higher demand. The abbreviation "PLA" is generally represented for PLLA in industrial materials. PLLA is semi-crystalline (approximately 37% crystallinity) with melting temperature (T_m) of the crystalline fraction at around 170°C-180°C, and glass transition temperature (T_g) of around 60°C-67°C (Ross et al., 2017). Therefore, it is brittle at room temperature and softens at a high temperature that limits its use for in-package hot processed foods and beverages. PLA has low toughness and flexibility, high brittleness and poor thermal stability but gives high clarity characteristics.

PLA is mostly considered as a bioplastic with the highest potential to substitute petroleum-based plastics and low environmental impact, particularly for food packaging purposes. Nevertheless, some drawbacks include high permeability to non-polar gases, for example, oxygen and carbon dioxide, while deformation at high temperatures restricts its industrial applications for processed foods requiring long shelf-life (Rocca-Smith et al., 2019).

2.2.2 Polyhydroxyalkanoate

PHAs are a family of homo- or copolymers of hydroxybutyric acid and hydroxyvaleric acid. The structures differ in pendent groups at the beta-position of the repeating unit. They are commercially produced by microbial fermentation of sugar, glucose, or vegetable oil as a carbon and energy resource. These bacterial metabolites are purified and extracted, while PHA is produced from the renewable resource as PLA. They are highly consumed by several microorganisms giving readily biodegradable property. Major investigations have been conducted on the PHA polymer poly(3-hydroxybutyrate) or "PHB" "P(3HB)." This has a pendant methyl group with a high degree of crystallinity. PHB has similar mechanical and physical properties to PP, including brittleness and stiffness. The melting point of PHB is around 177°C, close to PP, giving thermal stability (Ebnesajjad, 2012). Another PHA family is poly(3-hydroxy valerate) or "PHV." The pendant methyl group is replaced by the ethyl group becoming 3-hydroxy valerate (HV unit), giving PHV more flexibility and less crystallinity than PHB. PHB is rigid and, therefore, is often co-polymerized with PHV forming poly(3-hydroxybutyrate-co-3-hydroxy valerate) called "PHBV." The co-polymer introduces HV units to PHB that disrupt the regular PHB structures, reducing stiffness, crystallinity, crystallization rate, T_g, and T_m with increased HV contents (Ross et al., 2017). PHBV is tougher and more flexible with increased HV.

PHB is used in injection molding to produce food packaging using the same equipment as PP package injection. A conventional plastic converting process can process PHBV, including injection molding, extrusion, blow molding, and blown-film extrusion (Ebnesajjad, 2012).

2.2.3 *Polybutylene succinate*

PBS is a polymer of butylene succinate. It consists of succinic acid (diacid), which is esterified with 1,4-butanediol (diol) to form oligomeric butylene succinate with the elimination of water. Further polymerization via trans-esterification with the catalyst under vacuum produces high molecular weight PBS. Poly(butylene succinate-co-butylene adipate) (PBSA) is produced via co-condensation of succinic and adipate acids with 1-4-butanediol, giving modified physical properties and crystalline structure depending on the composition. Showa Highpolymer produces PBS and PBSA from petroleum called Bionolle. This has similar processability to conventional resins, including LDPE, HDPE, and PP. Bionolle can be processed by injection molding, extrusion, and blowing into films that can then be utilized for agricultural purposes, shopping bags, and compost bags (Puchalski et al., 2018). The T_m and T_g values are around 90°C-120°C and -45°C to −10°C, respectively. The homopolymer PBS is highly crystalline and exhibits T_m of 114°C. Addition of adipate giving a co-polyester exhibits intermediate behavior depending on the composition. Increasing adipate content decreases tensile strength, increases elongation at break, and decreases T_g from −31°C to −61°C (Ebnesajjad, 2012).

2.2.4 *Polycaprolactone*

PCL is synthesized from the cyclic caprolactone monomer. Ring-opening polymerization of caprolactone produces semi-crystalline aliphatic polyester. PCL has a low T_g (−60°C) and low melting point at around 60°C giving soft solids at room temperature. However, it has flexibility and outstanding processability and is used as additive resin or plasticizer to improve processing and end-use properties, including toughness, flexibility, compression set, and tear strength of films (Ebnesajjad, 2012). Hydrolysis of ester bonds in physiological conditions, for example, the human body and microbial enzymes, causes bio-degradation of PCL, which is also used as a control-release device for medical purpose as approved by the FDA.

2.2.5 *Polybutylene adipate terephthalate*

PBAT is a synthetic petroleum-based polymer that is entirely biodegradable in soil and aqueous solutions under composting and anaerobic conditions.

Polybutylene adipate (PBA) is synthesized from 1,4-butanediol and adipate acid. PBAT is produced by copolymerization of diols and dicarboxylic acids via polycondensation between butanediol, adipic acid, and terephthalic acid. PBAT has high elongation at break of approximately 700%, and higher than most biodegradable polyesters including PLA and PBS. PBAT is a co-polyester with an aliphatic-aromatic structure that has higher chain stiffness than other aliphatic polyesters, for example, PCL and PBS because of the presence of terephthalic groups in the polymers (Ebnesajjad, 2012). High flexibility is a major advantage of PBAT polymers that are blended to reduce the rigidity of other bioplastic materials including PLA and TPS films.

2.2.6 *Thermoplastic starch*

Starch is synthesized in plants as energy reserve materials. Starch granules are semi-crystalline structures mainly composed of amylose (20%-35%) and amylopectin. The ratios vary depending on the source of origin, giving marked differences in characteristics and functions. Different plant sources and origins also contribute to a wide variety of starch granule sizes, shapes, and structures. Major compositions of starch are D-glucopyranose units that bind via α-1,4- and α-1,6-glycoside linkages between the first carbon atom (C1) of one molecule connecting to the fourth carbon (C4) or sixth carbon (C6) of the second molecule. Amylose is a linear polymer, consisting of α-1,4-linked D-glucopyranose, while amylopectin contains branch structures linked by α-1,6 bonds, giving a substantially larger size than amylose (Mitrus et al., 2009). The melting temperature of pure starch is greatly higher than the decomposition temperature and this limits applications as bioplastic in its natural form. Accordingly, plasticizers are required to reduce the melting point and achieve the ability to process plastic packaging. Plasticization of starch produces thermoplastic starch or "TPS," which softens and can be shaped or molded at high temperature and solidified at room temperature.

Varieties of plasticized native and modified starch, including cassava, corn, and potato, are used to fabricate food packaging. High temperature and pressure in the presence of a plasticizer, for example, water, glycerol, or sorbitol produce TPS. Glycerol (25%-35%) is a major plasticizer for the plasticization of starch producing TPS. Interaction between starch and plasticizer increases free volume and reduces T_g and T_m. Extrusion is the most useful process to produce TPS pellets as heating and shearing disrupt crystalline structures in starch granules and yield amorphous materials. Types and contents of plasticizers strongly affect material properties and stability, including physical, mechanical, and barrier properties. Moreover, starch is highly hygroscopic due to large numbers of hydroxyl groups that readily absorb water from the environment, causing additional plasticization effects. Increasing plasticizers generally leads to reducing tensile strength, increasing elongation, diffusion, and permeability through polymer matrices. High water sensitivity limits processability in some conventional conversion processes, for example, sticking of starch bubbles is usually found in blown-film extrusion but this can be overcome by blending with other less water sensitivity materials, such as PLA and PBAT. Blending with other polymers and additives yields a variety of material properties, including melt flow characteristics and physical, mechanical, and barrier properties of final packaging.

2.2.7 Non-starch polysaccharide

Polysaccharides are naturally produced during the growth cycles of all living organisms, including animals, plants, and microorganisms. The high molecular weight of polysaccharides causes network rigidity and gives high T_g and T_m values that limit the conventional fabrication processes for packaging. Plasticizers are, therefore, required to improve flexibility and processability. Numerous polysaccharides derived from different resources give various sugar monomers, molecular weights, and conformation. Different components of repeating units including anionic and cationic characters increase polarity. The presence of steric groups due to sugar substitution on the linear polymer backbone, side-chain/branch, or modified ester/ether groups in the polymer impact the structures and properties of polysaccharides (Nieto, 2016), while high numbers of free hydroxyl groups contribute to polar structures and water affinity. Large molecular weight substances with long-chain polymeric structures that absorb water, swell and rather disperse in aqueous media are called hydrocolloids.

Cellulose is an abundant plant-derived material that differs from other plant polysaccharides due to its large single repeating units of molecular chains having crystalline fraction. Isolation from plant cell walls produces microfibrils via chemical extraction. Pure cellulose in all forms is highly crystalline, consisting of high molecular mass and is infusible and insoluble (Mitrus et al., 2009). Neat cellulose is not a thermoplastic material and is usually derivatized to achieve processability. Esterification yields cellulose derivatives that are suitable for thermoplastic processing. Cellulose esters, such as cellulose acetate, cellulose acetate propionate, and cellulose acetate butyrate are thermoplastic materials that can be compounded with plasticizers and other additives via conventional extrusion to produce numerous grades of commercial cellulose plastic (Ebnesajjad, 2012).

Chitin and chitosan are animal-derived nitrogenous polysaccharides. Chitin serves as skeletal polysaccharide, making up shell components of crabs, lobsters, shrimps, and insects. Chitin is hydrophobic and insoluble in water and most organic solvents. Chitin can be degraded by chitinase. Deacetylation converts chitin into chitosan with reduced molecular weight, forming a water-soluble compound. Chitosan is soluble in dilute acids, for example, acetic acid and formic acid, while the degree of N-acetylation strongly impacts physico-chemical and biomedical characteristics (Mitrus et al., 2009; Ebnesajjad, 2012). Chitin and chitosan are biocompatible materials and can absorb heavy metals. They display antimicrobial activity and have been used in water treatment and produce natural antimicrobial packaging.

Agar and carrageenan are plant-derived polysaccharides from seaweeds comprised of 3,6 anhydro-glucopyranose units with β-(1,3)- and α-(1,4)-linked galactose residues. They differ in enantiomers at the α-(1,4)-linkage, with L-enantiomer in agar and D-enantiomer in carrageenan giving different properties. For example, carrageenan is more water sensitive than agar, resulting in poor packaging barrier properties. Plasticized agar film is more rigid, with higher water barrier properties than carrageenan films (Rhim, 2012).

2.2.8 Protein

Proteins can be derived from plants and animals, namely milk (casein and whey protein), tissue (collagen and gelatin), and egg that are mostly insoluble and infusible without prior modification. All proteins are copolymers of various amino acids (up to 20 types of monomer), giving different structural arrangements and interaction. Peptide bonds join amino acids via amide linkages. Condensation polymerization of amino acids produces large molecular weight polymers, while interaction of side-chain amino acids can be non-ionized or ionized and polar (with basic or acid groups) or nonpolar causing different conformations in 3-dimension networks as secondary, tertiary, and quaternary structures. However, modified secondary structures are required to achieve desirable physicochemical and mechanical properties for packaging purposes (Hernandez-Izquierdo and Krochta, 2008; Guerrero & de la Caba, 2017). Strong molecular interactions among protein chains inhibit processability due to a high softening temperature that is often above the decomposition temperature. Modified protein structures and plasticization are required to produce protein-based packaging. Protein modification generally starts with the unfolding of native networks known as denaturation induced by thermal methods or chemicals and exposing hydrophobic structures in native protein (Verbeek & van den Berg, 2010). The formation of aggregates through noncovalent interactions, for example, ionic bond, van der Waals force, and hydrophobic interactions, or covalent bonds, for example, disulfide bonds then forms irreversible interactions and hence polymer networks. Several factors, including pH, temperature, time, and additives control protein characteristics and properties (Verbeek & van den Berg, 2010; Ebnesajjad, 2012; Guerrero & de la Caba, 2017).

Gelatin is an animal protein produced by hydrolysis of collagen and mainly derived from bovine, porcine, and fish (Mitrus et al., 2009). Gelatin is highly hygroscopic and water-soluble; therefore, it is used as edible packaging, for example, capsule shells and films rather than bioplastic packaging for foods. Several studies have investigated the development of milk protein, particularly whey protein, into bioplastic packaging. Plant proteins, including corn zein, wheat gluten, soy, and pea protein can potentially produce biodegradable polymers (Hernandez-Izquierdo and Krochta, 2008; Verbeek & van den Berg, 2010; Reddy & Yang, 2013). Soy and pea protein films are rigid and water sensitive due to a high number of polar amino acids, while wheat gluten has unique viscoelastic properties and water insolubility. Zein derived from corn consists of a large amount of hydrophobic amino acid residues giving hydrophobic film characteristics (Guerrero & de la Caba, 2017).

2.2.9 Plant oil

Polymers can be synthesized from vegetable oils, for example, corn oil and soybean oil. Triglyceride contains reactive groups as carbon-carbon double bonds and allylic and ester groups that can be potentially polymerized. However, proper functionalization of the carbon-carbon double bond is necessary before polycondensation to produce high molecular weight polymers. Free radical polymerization or the incorporation of an epoxy group into polymer chains causes ring-opening or condensation polymerization of plant-based oil. Moreover, triglyceride is converted into monoglyceride or diglyceride through the glycerolysis reaction under high temperature (220°C-230°C) with a catalyst. Monoglyceride or diglyceride further reacts with methacrylic anhydride in triglyceride-based monomers (Ebnesajjad, 2012). However, this material still has limited use in commercial food packaging applications.

2.2.10 Renewable conventional plastics

Several conventional plastics have been commercially produced based on renewable resources, including PE, PP, PET, and PA. For example, I'm green PE is produced from sugarcane and has the same properties and applicability as petroleum-based PE. It can also be recycled through conventional recycling systems of PE. PE is synthesized from ethanol, derived from a renewable resource, and becomes ethylene after a dehydration process. The family of PE from I'm green includes HDPE, LDPE, and LLDPE giving wide ranges of rigid and flexible packaging applications, for example, caps and closures, bags, and films. Conversion of bio-ethanol also produces bio-propylene that is further polymerized into bio-based PP polymers. In 2019, commercial-scale bio-based PP was introduced to the market. In Belgium, Borealis started producing bio-PP, replacing fossil materials based on Neste-produced renewable propane (ChemistryViews, 2020). Moreover, the Mitsui Chemical Group also develops plant-based PP for commercialization by fermentation of non-edible plants to produce isopropanol, and this is further dehydrated to become bio-based PP.

Bio-based PET is also commercially available in food packaging as a blow-molded bottle. It is made from bio-based mono ethylene glycol and purified terephthalic acid that are components of conventional PET. Molasse is a by-product of the sugar cane industry that is fermented to form ethanol and further converted into bio-ethylene, bio-ethylene oxide, and then bio-ethylene glycols, respectively. Purified terephthalic acid can also be produced from a renewable source. Bio-based mono ethylene glycol (30%) reacts with either plant- or petroleum-based purified terephthalic acid (70%) via condensation polymerization forming PET. Bio-based PET is usually produced into beverage bottles, and these materials comprise the largest segment of global bioplastic production in rigid packaging (European Bioplastics, 2020).

2.3 Conversion technology and forms of bioplastic packaging

Plastics can be shaped or molded under specific conditions, for example, pressure, temperature, and time. Packaging in the food industry is mainly produced from thermoplastic materials which melt or soften at a high temperature above T_g. Heating is the essential first step to fabricate plastic packaging. Formation of the viscous state allows material flow under pressure. Solid materials become molten and readily flow during heating with different degrees of viscosity depending on applied pressure, material structures, and temperature. Then, the softened materials are shaped under pressure in molds or extruder die. Afterward, cooling is applied until the product reaches a structural strength for suitable handling and utilization. Postprocessing operations, including finishing, printing, and coating, enhance production throughput, appearance, and packaging performance (Pouzada, 2017). Mixing processes of bioplastic materials and additives include (1) dry blending in various types and size of mixers, and (2) melt blending in batch or continuous process involving heating and shearing.

2.3.1 Extrusion

The extrusion process is the most common conventional plastic processing for thermoplastic materials. Extruders mainly consist of hoppers (feeder), cylinder barrels, screws, and die. Raw material in powder form (with liquid mix) or plastic pellets can be divided into single-screw extruders and twin-screw extruders depending on the number of screws in the barrels. Rotation of the interior screws inside the barrels with electric heating causes heating and shearing that promote plasticization and hence melting of the materials. Temperature can be varied in different zones of the cylindrical barrels to achieve desirable mixing, flowing, pressure, and flow rates (Fig. 7.3). Materials are forced to circulate and

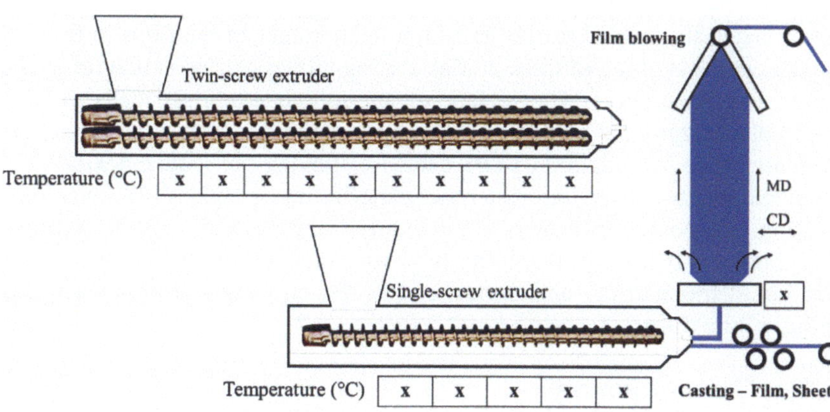

FIGURE 7.3 **Twin-screw and single-screw extrusion process for bioplastic films with film-blowing or casting into films or sheets.** *MD*, Machine direction; *CD*, Cross direction. *(Modified from Huntrakul et al., 2020.)*

flow due to screw rotation inside the helical channel between the cylinder and screw. Heating and shearing also improve the mixing of the molten matrices and promote homogenization of the materials (Pouzada, 2017). At the end of the cylinder barrel is the die that controls the shape and size of the extrudate. The die can be connected with blowing systems for blown-film extrusion or casting materials producing thin films or sheets for further packaging conversion processes.

The single-screw extruder consists of one rotating screw fitted in a cylindrical barrel. Friction between the materials and surface of screw and barrel significantly affects transport and flow. The slip of materials at the barrel wall contributes to a rotation with the screw rather than moving forward. Therefore, frictional forces of materials at the wall and their properties are criteria to select for single or twin-screw extruder. Natural materials containing high water or high-fat content are less suitable for single-screw extruders. Conversely, a twin-screw extruder consists of two screws in an "8" shaped barrel in which one screw protrudes into the channel of another. This effectively prevents rotation of the materials with the screw. The transport mechanism is also less dependent on material characteristics than the single-screw extruder (Janssen, 2009). Better mixing of blended melt materials is achieved, giving more homogeneous blends. Plastic pellets are produced by extrusion and further cut into various forms to reduce size for use in other packaging fabrication cycles. Moreover, the extrudates can be in thin forms (films and sheets) via the casting process or blown into thin wall flexible tubes (films) with desired thickness via the blown-film process (Fig. 7.4). The thickness of the materials can be adjusted by the size of the slit or die and flow rates of the extrudates.

Stability and continuity of the extrudate films govern yields of the manufacturing process. Several commercial grades of bioplastic pellets, for example, PLA and PBAT can be used in blown-extrusion and cast-extrusion. However, problems occur during the production of highly hydrophilic polymeric materials such as starch. The high temperature of the films in front of the die causes sticking of the materials during blown-extrusion that limits the manufacturing process. The surface of amorphous materials tends to stick due to plasticization at high temperature and humidity. Decreasing plasticizer contents in the formulation of TPS decrease sticking but reduce flexibility at room temperature. Blending with other materials is a simple way to overcome sticking. Moreover, addition of some hygroscopic oxide fillers, for example, TiO_2 powders that have the ability to absorb moisture reduces water plasticization of the matrices and, hence, causes less sticking and improves flowability of the films.

2.3.2 Thermoforming

Thermoforming is a packaging conversion process for thermoplastic polymers. The raw materials must be formed into plastic sheets before thermoforming. Cast-extrusion is commonly used to produce plastic sheets from a blend of neat

FIGURE 7.4 Bioplastic processing via (A) blown-extrusion of starch (left) and starch-protein blends (right) into films, and (B) cast-extrusion of starch bioplastic films.

polymeric materials. The preformed sheet is heated to soften the material with deformability for shaping. The temperature range of the materials in the heating process is typically above T_g or slightly below T_m for amorphous and semi-crystalline materials, respectively (Pouzada, 2017). The softened sheet is then clamped and subjected to external force either by vacuum pressing by a plunger, or air compression. Heated molten plastic is forced against a cold mold, forming a packaging structure. Vacuum thermoforming is the most common process to produce food packaging. After shaping, the plastic is cooled and removed from the mold for further cutting and trimming of unwanted parts.

Food trays and cups are commonly produced by the vacuum thermoforming process. The sealing process for food trays and cups with film lids includes heat sealing or adhesive. Permeability of trays and film lids strongly influences the qualities and shelf-life of foods. An example of the development of food trays for freeze-chilled seafood packaging is shown in Fig. 7.5. TPS/PLA trays are compounded and produced for sheet via twin-screw extrusion and vacuum thermoforming into bioplastic trays. Film lids are also developed based on TPS/PLA materials and heat-sealed on the trays. The TPS/PLA trays can be frozen at low temperatures (less than $-60°C$) during freezing with subsequent

FIGURE 7.5 Thermoform trays based on TPS/PLA materials for freeze-chilled seafood packages.

chilling in the fridge at 4°C. The trays can withstand the high humidity of the products (shrimps and fish fillets) during storage for up to 14 days. Slightly increased turbidity was found on the tray bottom. This was exposed to a highly humid area of the fish surface, suggesting recrystallization of polymers, possibly starch components. Moreover, the permeability of gas and volatile components can be limited by the dispersion of zeolites. Such fillers decreased permeability of trimethylamine responsible for the fishy smell of seafood products as determined by gas chromatography and sensory measurement. However, addition of zeolite above 2% possibly resulted in non-homogeneous polymer matrices with several pinholes on film lids and thermoform trays, causing high permeation of water vapor, gas, and volatile compounds.

2.3.3 *Blow molding*

Blow molding is used to produce hollow packaging, namely bottles. The converting process can be extrusion blow molding or injection blow molding. The extrusion blow molding production line consists of an extruder that produces downward-oriented molten tubes or so-called "parison." The parisons move downward, and once the desired length is reached, they are closed with movable molds and cut using a heated tool. The parisons are then blown up, and the materials expand to the mold walls that are at a low temperature. The molds subsequently open when the material cools sufficiently, giving solid objects (Pouzada, 2017). For injection blow molding, the parison is formed by injection that produces tubular materials with a closed bottom and formed ends of the final bottles. The parison is then transferred to another mold to form the final shape of the packaging. Blowing expands the molten materials against the mold walls where it further cools, forming solid objects. Commercial bioplastic bottles are produced from bio-based PET and PLA.

2.3.4 *Injection molding*

Injection molding is a process to fabricate plastic materials, including packaging and closures (Fig. 7.6). Thermoplastic materials are fed into the hopper and

FIGURE 7.6 **Composite packaging based on PP and rice flour (upper row) and PP and rice straw (lower row) at rice materials of 2, 4, 6, 8, and 10 phr (from left to right) plasticized with glycerol and produced by injection molding.**

melted into viscous materials that are then forced into molds. Two significant types of injection molding are ram injection and screw injection. The former uses a piston to push molten plastic forward into the mold, while the latter consists of a small extruder with a screw that moves backward and forward to collect and push the molten plastic into the mold cavities. Screw injection gives faster melting and better mixing and is often used for complicated structures. The molten plastic is forced into mold cavities, forming specific shapes. With the versatility of mold design and manufacturing process, injection molding effectively produces countless varieties of packaging shapes (size, dimension, partition). The mold cavity is filled up with molten plastic and then cooled to become a rigid solid under retained pressure to compensate for shrinkage of the material (Oniszczuk & Janssen, 2009). Plungers eject the solid objects out of the molds.

3 Improved barrier properties of bioplastic packaging

Packaging materials have mechanical, barrier, and optical properties. Barrier properties directly impact the shelf-life of packaged foods. Permeation of water vapor and gas, particularly oxygen, causes deteriorations of quality. Oxygen induced quality loss includes oxidation of lipids, proteins, pigments, and vitamins. Water accelerates physico-chemical changes, including the structural collapse of freeze-dried and dehydrated foods, sticking of food powders, and recrystallization of crystallizable food components, particularly sugars. Moreover, moisture absorption also leads to increased water activity and decreased T_g that increase molecular mobility and accelerate chemical reactions such as non-enzymatic browning, enzyme activity, and other mobility-induced reactions. Permeability of packaging determines the amounts of molecular permeation as a function of time. High permeability of packaging materials typically leads to a shorter shelf-life of packaged foods. Several bioplastic materials have higher permeability to water vapor and gas than conventional petroleum-based plastics. Among others, PLA as the main bioplastic has been investigated globally due to its high potential for commercialization. Improved barrier properties increase the shelf-life of packaged foods, including multilayer, nanofillers, surface treatments, and cross-linking of polymers as:

1. Multilayer packaging materials: coating, lamination, and co-extrusion,
2. Fillers in polymers and adhesives: nanomaterials, for example, nanoclays (inorganic), nanocrystals of biomaterials (organic) as starch, cellulose, chitin, and chitosan derivatives, and
3. Polymer modification: grafting, chemical, and physical cross-linking by irradiation and plasma treatment.

3.1 Multilayer bioplastic packaging

Numerous polymer technologies produce multilayer packaging in the form of films and sheets. Films are flexible, while sheets are further converted into thermoform trays that provide more rigid packaging. Multilayer materials can be achieved either by coating or lamination. The coating is an application of one or more fluid materials on the surface of another solid. Lamination refers to materials that form different layers of solid webs on others using adhesive or tie layers. Co-extrusion is a simple process to prepare conventional multilayer plastic. Delamination in multilayer biopolymer films depends on film compositions and the affinity between each layer. Solid webs with different surface properties and incompatibility are bound together by the adhesive. Wetting of the adhesive on the film surface is influenced by surface energy between the materials. Higher surface energy of the base materials is required to enhance wetting of the adhesive, giving a better adhesion performance, and avoiding delamination. The adhesive also fills any surface irregularities or pinholes to achieve a homogeneous bond between each layer (Rhim et al., 2006). These processes are common for conventional plastics; however, development of multilayer bioplastic packaging is still limited for commercial applications. Several research methods have demonstrated improved barrier properties of bioplastic polymers by multilayer technologies.

Multilayer materials effectively improve strength, gas, and water barriers that are important for the food packaging industry. The structure of multilayer packaging is commonly in the form of a barrier layer in the middle covered by outer structural and printable layers. Rhim et al. (2006) produced a solution-casted multilayer film consisting of soy protein isolate (inner layer) and PLA (outer layer) that gave more than fivefold increased tensile strength compared with neat soy protein isolate film and was comparable to LDPE and HDPE films. The multilayer films were prepared by casting off one outer PLA layer and drying before casting over with a middle layer of soy protein isolate, followed by another PLA outer layer. This method effectively gave strong adhesion between both polymers through hydrogen bonding. The lamination decreased water vapor permeability by 40-fold as compared with soy protein isolate film, while oxygen permeability decreased more than 26-fold as compared with PLA (Rhim et al., 2006). Winotapun et al. (2019) produced multilayer films via blown-extrusion that consisted of LDPE and PLA blended as a core layer with LDPE as outer layers. Different concentrations of PLA (20%-50%) in the core layer formed ribbon-like structures. PLA concentration at 40% and 50% gave oxygen permeability of 20% and 43%, respectively with a long, tortuous path for molecular transport. Moreover, film with 50% PLA effectively limited three volatile compounds from durian, namely diethyl sulfide, ethyl propanoate, and 2-ethyl-1-hexanol. The authors suggested that multilayer films improved the aroma barrier with optimal gas permeability retaining the quality of fresh-cut durian (Winotapun et al., 2019). Rocca-Smith et al. (2019) produced wheat gluten proteins (inner layer) and PLA (outer layer) laminates with improved barrier properties. Different technologies were investigated to develop packaging properties (surface and bulk) and performances (barrier and adhesion), including high-pressure homogenization of film-forming suspensions, corona treatment, wet casting, spin coating, and hot-pressing to shape the multilayers. The authors reported greatly improved oxygen barrier by 2000% and water barrier by 20%. Hot-pressing produced multilayers with enhanced barrier properties by inducing wheat gluten restructuring and increased PLA crystallinity from 28% to 34% (Rocca-Smith et al., 2019).

Coating via the sol-gel process is another flexible method to generate multilayer material with enhanced barriers. Mixtures of various dispersed and continuous phases can be easily applied on the surface of polymers using different techniques, namely dipping, spraying, rolling, and spin-coating. Other additives, for example, antimicrobial, stabilizers, antioxidants, and UV absorbers can also be formulated to enhance the functionality of food packaging by producing nanosized materials and improved compatibility between the interface of organic and inorganic hybrid substances. Iotti et al. (2009) developed organic-inorganic hybrid coatings based on organic PCL, poly(ethylene oxide) or PLA and silica as an inorganic phase. Mixtures were produced for coating PLA films via the sol-gel process giving dense and homogeneous multilayers. The oxygen permeability of the films decreased as the thickness layer increased following non-linear regression (Iotti et al., 2009). Moreover, inorganic vapor-deposited coating is another technology to improve the barrier properties of flexible polymers. This provides transparency, clarity, flexibility, and recyclability. The coating is non-metallic and, therefore, can be used with microwave and metal detectors in the food manufacturing process. It is an alternative to metalized plastic and improves high barrier properties for food packaging. Aluminum oxide (AlO_x), silicon oxide (SiO_x), and carbon are the most common vapor-deposited coatings for conventional plastic. Evaporation and plasma-enhanced chemical vapor deposition (PECVD) processes are used for the deposition of inorganic materials on polymers. PECVD is a plasma polymerization process producing gaseous organosilane and oxygen or dinitrogen oxide. Gaseous silicon dioxide is formed under low-pressure discharge and further condensed on single or both sides of the polymer surface. Silicon oxide vapor-deposited coating has recently been used to improve barrier properties of PLA films. Dong et al. (2017a) successfully used the PECVD process to deposit silicon oxide on PLA films. Young's modulus and tensile strength were improved by 119% and 92%, respectively, while oxygen and carbon dioxide permeability were reduced by 79% and 72%, respectively (at 5°C). The films were used to pack chilled meat giving dynamic equilibrium CO_2 and O_2 of 6%-11% and 8%-13%, respectively that was favorable to preserve protein and lipid and retained product quality (Dong et al., 2017a). Song et al. (2018) also showed that SiO_x coating via PECVD for 60 min effectively improved the barrier of PLA films against oxygen, carbon dioxide, nitrogen, and water vapor by 40.7%, 30.6%, 58.7%, and 53.4% at 25°C, respectively.

3.2 Nanofillers and nanocomposite packaging

Nanocomposite packaging materials have been developed to enhance packaging performance, including layered silicate nanoclays and montmorillonite. Nanofillers refer to materials with at least one dimension in the nanoscale or so-called "nanoparticles." Dispersion of nanofillers into polymer materials produces "nanocomposites," giving proportionally more significant surface area than microscale materials. Smaller size commonly favors interaction between fillers and polymer networks, giving better material properties (De Azeredo, 2009). Nanofillers cause a tortuous path for molecular movement and increase the path length for diffusion.

Inorganic clays, namely montmorillonite, and kaolinite are active materials that improve barrier performance of polymeric packaging. Optimal efficiency of nanofillers is commonly found at low concentrations of around 5%. Compatibility between polymers and nanofillers governs improved properties and performance and, therefore, several attempts to improve compatibility have included chemically modified clays (Arora & Padua, 2010). Rhim (2012) developed bioplastic films containing nanoclays as Cloisite Na+ . Nanoclay (5%) was dispersed in aqueous agar/κ-carrageenan suspension and laminated on preformed PLA films by solution casting. Nanocomposite films showed greater tensile strength, while water vapor permeability decreased due to the presence of dispersed silicate layers with large aspect ratios in the polymer matrix that were impermeable to water vapor. Moreover, the oxygen transmission rate of nanocomposite films decreased 650 times compared with PLA films. Dispersion of metal oxide nanofillers as titanium dioxide (TiO_2) and zinc oxide (ZnO) improved mechanical and barrier properties depending on filler concentration. Moreover, these possibly increased antimicrobial function to the contact surface of food products. Khammeethong et al. (2020) showed that TPS/PBAT films containing TiO_2 and ZnO gave lower microbial counts than pure TPS/PBAT films in pork steaks. Moreover, films wrapping in contact with the meat surface showed efficient antimicrobial function, while not in contact closures showed higher microbial counts. Dispersion of 3% TiO_2 gave higher mechanical and barrier properties than 5%. High contents of fillers can cause clumps and lead to reduced non-homogeneous polymer networks giving lower strength and barrier performance. Fig. 7.7 shows the dispersion of oxide fillers as determined by a scanning electron microscope coupled with an energy dispersive X-ray spectrometer (SEM-EDX).

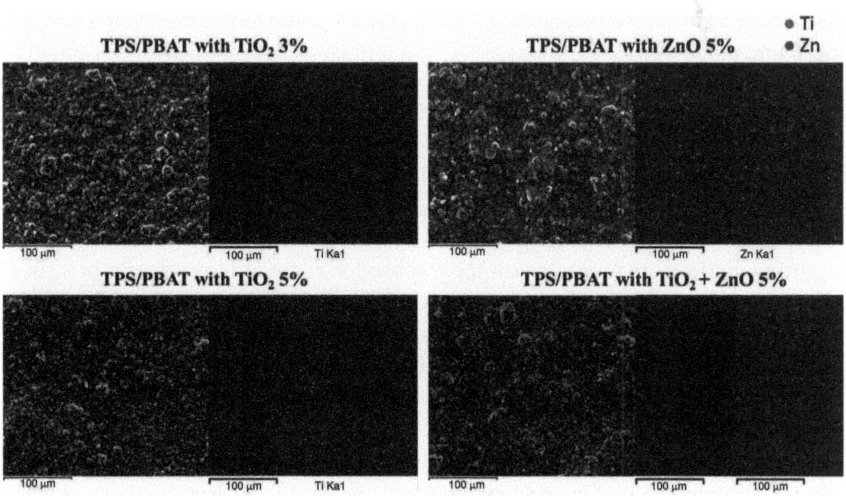

FIGURE 7.7 SEM-EDX showing dispersion of titanium and zinc in TPS/PBAT films containing TiO_2 and ZnO mixtures at 3% and 5%.

Graphene nanoplates are carbon-based nanofillers used to improve packaging properties for conventional and bioplastic materials. Gao et al., (2018a and 2018b) showed that conventional multilayer co-extrusion effectively aligned graphite nanoplatelets in PLA films to produce lightweight food packaging. This method also reinforced efficiency due to improved alignment of graphite nanoplatelets in the films that contributed to a significant 120% reinforcement for mechanical properties. Films with the thinnest filled layer also exhibited optimal water vapor barrier properties due to high crystallinity and high filler orientation. The films mentioned earlier used only 1/10 filler loading and effectively retained similar barrier properties to nano clay dispersed PLA films (H. Gao et al., 2018; Y. Gao, Picot, Tu, Bilotti, & Peijs, 2018). He et al. (2015) also showed that PLA multilayer films containing graphene oxide showed improved surface hydrophobicity, oxygen, and water vapor permeability. PLA multilayer was dipped into aqueous suspensions containing graphene oxide to assemble layer-by-layer film (He et al., 2015). Svagan et al. (2012) also produced multilayer PLA films with high clarity via layer-by-layer technology based on montmorillonite clay and chitosan on extruded PLA film surfaces. Oxygen permeability of the films was significantly reduced by up to 99% with 70 bilayers (Svagan et al., 2012). Satam et al. (2018) produced multilayer spray coating of cationic chitin nanofibers and anionic cellulose nanocrystals on PLA films. The layer-by-layer coating effectively reduced oxygen permeability up to 73% due to densely packed layers of nanomaterials (Satam et al., 2018). Direct addition of nanofillers reduced the strength of packaging structures due to aggregation that caused heterogeneity of the polymer network. Alternatively, nanofillers can be added into the adhesive of the multilayers to improve barrier properties. Reduced permeability by fillers compensates for reduced layers and thickness of materials giving sustainable food packaging. Extensive reviews of nanocomposite packaging were presented by De Azeredo (2009), Arora & Padua (2010), and Duncan (2011).

3.3 Polymer modifications

Plasma or heat is an efficient process to induce oxidation, producing free radical and reactive species that readily promote chain scission and cross-linking of polymer networks. Plasma treatment generates ultraviolet light, high-energy electrons, and reactive species on the exposed surface at atmospheric pressure. Reactive components include reactive oxygen species (ROS) that initiate numerous reactions of biopolymers. Formation of oxygen-containing polar groups such as COH and CO radicals enables further grafting or coating. Plasma is a non-thermal polymer modification that induces the deconstruction of biomacromolecules.

Heidemann et al. (2019) showed that cold plasma treatment under atmospheric air enhanced adhesion between PCL or PLA, and starch layers in multilayer films. The treatment enhanced surface roughness of PCL and PLA while decreasing the contact angle of water droplets. Moreover, multilayer structures showed enhanced delamination resistance with increasing treatment time. Plasma treatment for 10 min gave no delamination and showed the best adhesion. Multilayer films gave a higher water vapor barrier than starch films, while mechanical properties were similar to PCL and PLA films (Heidemann et al., 2019). Cold plasma treatment of zein films displayed a better-ordered porous structure after coating with PLA. The treatment enhanced compatibility between zein and PLA by modification of the protein conformation and increased intermolecular hydrogen bonding. Moreover, cold plasma treatment of zein for 60 s before PLA coating increased surface hydrophobicity (24%), tensile strength (50%), elongation at the break (30%), water vapor barrier (44%), and thermal stability of PLA coated zein films (Chen et al., 2020).

4 Developments of bioplastic food packaging

4.1 Fresh produce

Respiration is a major factor affecting the storage life of fresh fruits and vegetables. Fresh produce exhibits both aerobic and anaerobic respiration. A higher respiration rate gives a lower shelf-life. Aerobic respiration makes use of oxygen and generates carbon dioxide and water, while anaerobic respiration causes fermentation and produces ethanol that adversely affects the organoleptic quality of fresh produce. Concentrations of oxygen and carbon dioxide in the package headspace control the respiration of packaged fresh produce. The absence of oxygen in the package causes anaerobic respiration leading to quality loss of products. Therefore, aerobic respiration at a slower rate is preferred to extend the shelf-life of fruits and vegetables. Higher carbon dioxide and lower oxygen levels often lead to reduced respiration. Aerobic respiration of produce commonly causes reduced oxygen levels, increased carbon dioxide, and moisture contents in the package. Permeation of water vapor and gas are essential to control gas concentration and subsequent respiration of fresh produce. Packaging permeability is, therefore, an important criterion to extend the shelf-life of products. Accumulation of moisture in the package also accelerates the growth of several microflorae leading to faster spoilage. Permeability must be high enough to allow oxygen ingress and removal of excessive moisture and carbon dioxide. Bioplastic packaging has been developed to extend the shelf-life of fruits and vegetables, as shown in Table 7.2. Improved packaging permeability by perforation technology, as commonly used in conventional petroleum-based plastic, has also been applied in bioplastic films. Moreover, antimicrobial agents have also been introduced to packaging to retard the growth of microorganisms. Nevertheless, several research studies showed that addition of substances into the package modified molecular permeation, particularly oxygen and carbon dioxide permeability, that subsequently affected respiration and hence storage-life of fresh produce.

TABLE 7.2 Related researches on developments of bioplastic for fresh produce.

Bioplastic	Treatment	Food product	Packaging	Packaging properties	Food quality	References
PLA	Micro-perforation number - Equilibrium modified atmosphere packaging	Cherry tomato, peach	Film	- Overall transmission rate increased with an increasing number of perforations - Total transmission rate through a pair of perforations decreases with the distance between the two holes	- PLA offered a better optimized, balanced EMA which combined a considerable reduction in weight loss with a slower decay due to pathogen in cherry tomatoes - PLA showed less rotten peaches than PP film	Mistriotis et al. (2016)
PLA	Nanoclays: Na-cloisite 20 and Na-cloisite 30B (1%) Surfactant (Triton X-100)	Fresh-cut melon	Terrine, lid, mesh, and fork		- PLA with nanoclays prevented color change and extended shelf-life	Lorite et al. (2017)
PLA	Bergamot essential oils (9%) Nano-TiO_2 (1%, 2%, and 3%), nano silver (1%)	Mango	Films	- PLA nanocomposite films had lower water vapor permeability than PLA and PLA/bergamot essential oils - PLA films had lower oxygen transmission rate than PLA/bergamot essential oils and PLA nanocomposite	- PLA nanocomposite films gave lower weight loss, retard the changes in color, total acidity, vitamin C, and microbial counts than PLA and PLA/bergamot essential oils - PLA nanocomposite films delayed loss of firmness	Chi et al. (2019)

PLA	Chlorine dioxide, glucono deltalactone	Cherry tomato	Films	- Increase the ratio of sodium chlorite/glucono deltalactone leads to a decreased release in chlorine dioxide (g) - Decreased temperature (22°C-10°C) decreased rate and amount of chlorine dioxide (g) releasing	- Chlorine dioxide gas which generated from the films reduced populations of surface-inoculated *Salmonella* and *Escherichia coli* O157:H7 - An insignificant difference of sensorial quality, color, odor, and firmness to control films	Zhou et al. (2020)
PLA/ P3,4HB	Essential oil: Ginger (4%) and angelica essential oils (4%)	Peach	Films	- Essential oil increased oxygen transmission rate, carbon dioxide transmission rate, water vapor permeability, and elongation but decreased tensile strength of the films due to disrupted H-bonding of the films	- A film with angelica essential oils extended shelf life of peaches > 15 days - The volume fraction of O_2 decreased (14.6%-8.0%), and CO_2 increased (14.2%-21.9%) in the package which inhibited the respiration of the peach - Films had high antioxidant capacities inhibiting polyphenol oxidase and lipoxygenase activities and prevented the destruction of the cell membrane	Jiang et al. (2020b)

(Continued)

TABLE 7.2 Related researches on developments of bioplastic for fresh produce. (*Cont.*)

Bioplastic	Treatment	Food product	Packaging	Packaging properties	Food quality	References
PLA	Biobased: PLA tray and cellulose bag - Conventional: PET tray and PP bag	Tomato	Tray and bag		- Fresh-cut fruits in bio-based compostable package retained greater total solid (+10%-11%), total solubility solid (+4%-10%), and sugars (+5%-12%) than those in a conventional package - Bio-based compostable package better preserves the nutritional value namely lycopene and vitamin C of fruits than conventional package	Patanè et al. (2019)
PLA	Allyl isothiocyanate (5%), tributyl citrate (0%, 2.5%, 5%, 7.5%, and 10%)	Pak choi	Films	- Active compounds gave less strength, less stiffness and more flexibility, lower O_2, and CO_2 transmission rate, and a higher UV absorption rate than the pure PLA films - Allyl isothiocyanate and tributyl citrate decreased T_g and T_m of the films	- Active compounds improved the color stability, inhibited total aerobic bacterial growth of fresh pak choi during storage stored at 4°C and 10°C for 15 days	Gao et al. (2017)

PLA	Thymol and R-(-)-carvone (10%, 15%, and 20%)	Avocado, citrus	Films	- Improved antifungal activity depended on vapor concentration of volatile in the headspace which improved inhibition zone - At 12°C, 20% thymol showed complete growth reduction of avocado isolate, while at 25°C, 15% and 20% thymol showed complete growth reduction of both avocado and citrus isolates - 20% Thymol film had the highest antifungal activity against anthracnose isolates	- 15% R-(-)-carvone and 20% thymol were the most effective at 12°C in suppressing the mycelial growth of the avocado and citrus C	Boonruang et al. (2017)
Pea starch/PLA	Pea starch and PLA bilayers (100/0, 85/15, 75/25, 65/35, 50/50, and 0/100)	Cherry tomato	Bilayer films	- Bilayer films had better toughness, thermal stability, and barrier capacity than PLA films - PLA increased water resistance and tensile strength over monolayer PS film - Incorporation of two layers enhanced the water resistance, thermal properties, and water vapor barrier	- PLA50 bilayer film had excellent oxygen barrier properties that extended the shelf life and delayed the maturation process of cherry tomatoes - Bilayer films reduced the weight loss ratio of cherry tomatoes and extended the retention of organic acids and vitamin C	Zhou et al. (2019)

(Continued)

TABLE 7.2 Related researches on developments of bioplastic for fresh produce. (*Cont.*)

Bioplastic	Treatment	Food product	Packaging	Packaging properties	Food quality	References
PLA	Rice husk ash fillers (0%, 5%, 10%, and 20%)	Banana	Films	- Increasing rice husk ash content led to an increased maximum tensile strength and percentage elongation at break of PLA blend/rice husk ash composite films - Increasing stretching speed of the film process gave higher water vapor transmission rate and oxygen transmission rate values	- PLA blended rice husk ash 20% preserved bananas for up to 14 days, while conventional LLDPE film was able to keep the freshness of bananas for only up to 8 days	Choklob et al. (2019)
PLA	Spray-dried carboxymethyl cellulose from bleached bagasse (0%, 1%, 2%, and 4%)	"Nam Dok Mai" mango	Bags	- The absorption of water vapor of PLA/carboxymethyl cellulose packaging was increased dramatically at 4% (w/w)	- Carboxymethyl cellulose-containing PLA films extended the shelf life of mangoes for 42 days, decreased respiratory rate and physicochemical changes of mangoes as compared to control and unpacked mangoes which had a shelf life of 3 weeks	Kamthai and Magaraphan (2019)

PLA	Carbon nanotubes/chitosan (0%, 1%, 3%, 5%, 7%, and 9%)	Strawberry	Films	- Increased mechanical properties, solubility, and swelling ratio of the films - Water permeability of the fibers decreased with increasing chitosan content	- Antimicrobial activity of the composite fiber films against four microorganisms namely *Escherichia coli, Staphylococcus aureus, Botrytis cinerea,* and *Rhizopus* spp. - Delay the physiological changes and ripening in strawberries and extend their shelf life up to 8 days by the retention of their firmness; decrease in titratable acidity, total soluble solids, and ascorbic acid content; and retardation of decay and weight loss	Liu et al. (2019)
PLA	Allyl isothiocyanate	Tomato	Films	- Films contained antimicrobial activity against inoculated bacteria and fungi	- Reduced numbers of inoculated bacteria and fungi on tomatoes by 2-3 log CFU/g, and inhibited their growth during the 21-day storage - Reduced changes in quality namely color, firmness, total soluble solids, titratable acids, and vitamin C of tomatoes	Gao et al. (2018)

(Continued)

TABLE 7.2 Related researches on developments of bioplastic for fresh produce. (*Cont.*)

Bioplastic	Treatment	Food product	Packaging	Packaging properties	Food quality	References
PLA	Modified atmosphere packaging in PLA namely air (A-PLA) and O_2: N_2: CO_2 of 95:5:0 (O_2-PLA), 1:99:0 (N_2-PLA), 6:82:12 (CO_2-PLA) and PP/LDPE	Fresh-cut celery	Bag	- O_2 and CO_2 permeabilities of PLA were 4 and 5 times lower than PP/PE, respectively. - The steady-state atmosphere was achieved faster in CO_2-PLA (3 days) than A-PLA and N2-PLA bags (7 days) - Levels of $CO_2$and O_2 in the O_2-PLA bags stabilized after 14 days with the A-PP/PE bags reaching steady-state after10 days of storage.	- PLA bags gave higher weight loss after 21 days of storage (~4%) compared to PP/PE bags (<1%) - Celery packaged in A-PLA retained green color during storage which was greener (higher h) than those packaged in A-PP/PE - The high initial concentration of CO_2 in the headspace retained the initial green color of the surface and cut ends while a high O_2 concentration accelerated yellowing and other quality defects - Only O_2-PLA bags inhibited the increase in Listeria populations during the first 10 d of storage	González-Buesa et al. (2014)

PLA	Sodium chlorite and citric acid powder in the ratios of 1:2 and 2:1 (5%-60%)	Grape tomato	Film	- Films were activated by moisture from tomatoes in the package giving release of chlorine dioxide and antimicrobial activity - The increase of reactants in the films produced more chlorine dioxide, while higher PLA content in the films resulted in less release of chlorine dioxide	- Films prevented changes in color and texture during storage - Released ClO_2 package reduced *Salmonella* spp. And *Escherichia coli* O157: H7 inoculated on the tomatoes to undetectable levels (< 5 CFU/cm^2 of tomato) and achieving more than 3 log reduction	Ray et al. (2013)
PBS, PLA, Mater-Bi	PLA, PBS, and LDPE bags (control)	Rambutan	Bag	-	- PBS bags gave the lowest weight loss with yellowish-red skin and fresh green spinterns of rambutan - PBS film effectively extended postharvest life of rambutan for 15 days at 13°C storage	Leabwan et al. (2015)
PBAT/PLA	PBAT/PLA PE	Passion fruit	Films	- PLA/PBAT films can more effectively reduce the rising of ethanol content and delay the total acid, ascorbic acid, and sensory evaluation	- An insignificant difference of weight loss, shrinkage index, firmness, and total sugar of passion fruit packaged with PLA/PBAT films compared with PE films during 21 days at 20°C - PLA/PBAT films are more practical to preserve the overall flavor of passion fruit during storage as determined by sensory evaluation, E-nose, E-tongue, and GC-MS	Zhang et al. (2019a)

(*Continued*)

TABLE 7.2 Related researches on developments of bioplastic for fresh produce. (*Cont.*)

Bioplastic	Treatment	Food product	Packaging	Packaging properties	Food quality	References
PBAT/TPS	PBAT/TPS and cellulose nano whiskers (0%-3%)	Lettuce	Film	- TPS increased the rate of water vapor permeability of the blends - Increased cellulose nano whiskers increased porous surface of the films	- Incorporation of cellulose nano whiskers gave antimicrobial function against molds and yeasts and mesophilic and psychrotrophic bacteria within 10 days	da Silva et al. (2019)
PBAT/TPS	Coconut nanocellulose (0.55%), annatto (0.5%-1%), and citric acid (1%)	Fresh-cut mango	Film	- Nanocrystals reduced water vapor permeability but insignificantly influenced increase stiffness of the films - Tensile strength was dependent on the amount of nanocellulose	- Film maintain good qualities of mango for 14 days	(Leal et al., 2019)
PCL	Rosmarinic acid (6%) and essential oil (6%)	Broccoli	Films	- Tensile strength was decreased	- Bioactive films showed microbial inhibition against *Escherichia coli, Salmonella typhimurium*, of total aerobic microbiota	Takala et al. (2013)
PCL	Silicon oxide	Cherry tomato	Coated film	- PCL/silicon oxide film presented increased elongation at break and appropriate gas permeability and CO_2/O_2 permselectivity	- PCL/ silicon oxide film indicating that delayed ripening, higher nutrient content, and extended shelf life	Yun et al. (2019)

PCL	Different thickness of PCL (27 and 46 μm)	Strawberry	Films	- Different thickness affected CO_2/O_2 permeability ratio and water vapor permeability	- Compared with PA/PE and PE film, PCL film greatly enhanced the shelf life of strawberries to 22 days. - No condensation appeared in PCL packages and the strawberry showed good texture	Yun et al. (2017)
PCL/PLA	Thymol (9%)	Hot pepper	Film	-	- Thymol containing films gave higher vitamin C contents, firmness and lower total phenolic contents, decay speed and microbial growth - After 21 day of storage, the hot pepper packed in film containing thymol was edible, no off-odors, and decay of microorganisms	Qin et al. (2016)
TPS	Titanium dioxide (0.01%, 0.03%, and 0.05 %) and zinc oxide (0.01%, 0.03%, and 0.05 %)	Banana, tomato	Films	- Titanium dioxide and zinc oxide incorporated films exhibited in vitro antimicrobial activity	- TPS films mixed with 0.01 % w/w of titanium dioxide were able to extend the shelf life of bananas from 5-7 days to 14 days, and tomatoes from 7-10 days to 21 days	Tunma (2018)

(Continued)

TABLE 7.2 Related researches on developments of bioplastic for fresh produce. (*Cont.*)

Bioplastic	Treatment	Food product	Packaging	Packaging properties	Food quality	References
PVOH	Nano-silver	Grape	Films	- Inhibition zone diameter of nano-silver on *Aspergillus niger* was 14.4 mm, with minimum bactericidal concentration of 75 ppm - Nano-silver decreased mechanical properties and transparency but enhanced barrier performance	- Nano-silver-containing PVA composite film controlled the mass loss, decay rate, vitamin C and titratable acid content of grapes and had a remarkable effect to prolong grapes' shelf life	Deng et al. (2019)
PVOH	Cellulose nanocrystals (2%, 4%, 6%, and 8%)	Mango	Films	- Thermal stability and tensile properties of the biocomposite films significantly improved by increasing the filler concentration up to 6%	- Mango fruit exhibited a shelf life extension to 19 ± 2 days with proper quality parameters (phenol, flavonoids content and DPPH antioxidant activity), reduced mass loss, increased CO_2, and low O_2 level the package	Perumal et al. (2018)
PVOH	Oregano essential oil (oregano: PVA = 3:5)	Fresh-cut iceberg lettuce	Sachet	- Reduction efficiency against growth of Dickeya chrysanthemi, molds and yeasts, and total mesophilic aerobic bacteria on the surface of iceberg lettuce	- Oregano essential oil released from the sachet inhibited the growth of *D. chrysanthemi*, resulting in 3.9 log CFU/5 pieces reduction over 5 days - Texture and color characteristics of the iceberg lettuce were not affected by the release of oregano	Chang et al. (2017)

Cellulose	Vanillin (0.9 g)	Cantaloupe, pineapple	Film	-	- Vanillin reduced the number of yeasts, which decreased by 4 logs in fresh-cut pineapple on day 6 and inhibited *E. coli* - Vanillin film increased yellowness of pineapple	Sangsuwan et al. (2008)
Cellulose	Guar gum (1%) with silver nanocomposite (20 mL of 0.1 N AgNO3)	Strawberry	Film	- Guar gum with silver nanocomposite improved mechanical strength and antimicrobial effects against Gram-negative, Gram-positive bacteria, and fungi	- Films with guar gum and silver nanocomposite reduced weight loss of strawberry	Kanikireddy et al. (2020)
Cellulose	Silver nanoparticle (1%)	Melon	Absorbent pads	- Release of silver ions controlled the spoilage microorganisms in the cellulose-based absorbent pads	- Silver loaded absorbent pads reduced microbial loads in melon - O_2 and CO_2 contents in the package headspace was not remarkably affected by silver	Fernández et al. (2010)
Poly (lactide-co-glycolide) (PLAG)	Thymol (7.5 g)	Strawberry	Nanofiber film	- Encapsulation in the core-shell nanofiber film gave much slower release of thymol into the atmosphere	- Strawberries wrapped in the control group and pure PLGA nanofibers had decay on the 3rd day but no decay for strawberries wrapped with the thymol/PLGA - Thymol/PLGA nanofiber film reduced microbial growth	Zhang et al. (2019b)

4.2 Bakery

Bakery products have intermediate moisture contents, and water activity is commonly reduced, thereby retarding the growth of several bacteria. The growth rate of microorganisms reduces as water activity decreases. Mold grows at a lower water activity than bacteria that is common in the range of most bakery products. Therefore, mold becomes a significant spoilage microorganism for most bakery products, including buns, bread, cakes, and pizza crust. *Bacillus* spp. is a common spoilage microorganism for bakery products and starch-based foods that produces rope spoilage. Physico-chemical changes in food components also occur in bakery products during storage. Retrogradation of starch causes staling and leads to dehydration of the water previously absorbed by the amorphous phase. This leads to increased hardness and subsequently reduces consumer satisfaction.

Table 7.3 shows several developments of bioplastic to extend shelf-life of bakery products and starch-based foods, for example, pasta. Most researchers attempted to increase shelf-life of bakery products by the addition of antimicrobial agents, including volatile and non-volatile agents. Non-volatile compounds in packaging show antimicrobial action as they directly expose surfaces of food products, while volatile substances give extra benefits as non-direct contact antimicrobial agents. Evaporation of active agents into package headspace and hence food products provides an antimicrobial function without direct contact between package and foods. Essential oils are volatile compounds that show both direct contact and vapor phase antimicrobial function. Nevertheless, the vapor phase commonly shows less antimicrobial capacity than direct contact applications.

Incorporation of essential oils into an active film has been shown to reduce growth of spoilage and pathogenic microorganisms and extend shelf-life. For example, incorporation of cinnamaldehyde into PLA/PBAT films effectively inhibited mold growth in several bakery products (Fig. 7.8). Migration of cinnamaldehyde from films to foods significantly inhibited microbial growth, leading to shelf-life extension. Moreover, the permeability was also modified. Limited oxygen permeation reduces the growth of aerobic microorganisms, while higher water vapor permeability increases water loss, leading to dehydration, reduced water activity, and increased hardness. Films containing volatile antimicrobial agents can have drawbacks on the taint of foods. Therefore, optimal concentrations of volatile compounds giving antimicrobial capacity and desirable organoleptic quality are required.

Srisa (2020) indicated possible mechanisms explaining the effects of films incorporated with volatile essential oil, namely cinnamaldehyde on packaging properties and changes in quality of packaged bread (Fig. 7.9). Essential oils also destabilized the structures and metabolisms of bacteria and molds, leading to cell death (Acosta et al., 2016; Shreaz et al., 2016; Noshirvani et al., 2017a; Wu et al., 2017; Kumar and Kudachikar, 2018), while recrystallization of starch or retrogradation influenced moisture-induced molecular mobility.

TABLE 7.3 Related researches on developments of bioplastic food packaging for bakery products and pasta.

Bioplastic	Fillers	Food product	Packaging	Film properties	Food quality	References
PBAT/rice flour	Potassium sorbate (0%, 1%, 3%, and 5%)	Fresh lasagna pasta	Antimicrobial film	-	- The microbial count was lower in pasta packaged in films containing sorbic acid - Pastas intercalated with films containing 3 and 5 g/100 g of potassium sorbate showed similar characteristics which can be safely consumed for up to 35 days, while films containing no sorbic showed changes in the odor and texture and the appearance of surface slime on day 9	Sousa et al. (2016)
PCL/ chitosan	Grapefruit seed extract (7.0%, 13.0%, 18.4%, 23.1%, and 27.3%)	Salmon and bread	Antimicrobial film	- Addition of grapefruit seed extract increased pits on the film surface and led to the interfacial debonding of polymers, since the cavities between chitosan particles and PCL matrix became more evident in the cross-section. So, the chitosan-PCL interactions were modified when GFSE was present in the films - Grapefruit seed extract decreased resistance to breakage and stretching, however, retaining adequate tensile strength and flexibility for packaging	- Films contains grapefruit seed extract were effective against both *Escherichia coli* and *Pseudomonas aeruginosa* when grapefruit seed extract content increased to 0.5 and 1.0 mL/g, respectively - Films with grapefruit seed extract (≥0.5 mL/g) gave lower *E. coli* in salmon - No visible mold appeared on bread packaged with films containing ≥1.0 mL/g grapefruit seed extract upto 7 days	Wang et al. (2019)

(*Continued*)

TABLE 7.3 Related researches on developments of bioplastic food packaging for bakery products and pasta. (*Cont.*)

Bioplastic	Fillers	Food product	Packaging	Film properties	Food quality	References
Cassava starch	Cinnamon powder and clove powder (0.00%, 0.06%, 0.20%, 0.34%, and 0.40%)	White pan bread slice	Antimicrobial film	- Addition of clove and cinnamon powders (0.20 g/100 g) reduced water vapor permeability, lower tensile strength and elongation at break	- Viable yeast and mould counts of bread slices stored in contact with biodegradable film were similar - Films did not show apparent antimicrobial effect as water activity in read packed in biodegradable films was increased	Kechichian et al. (2010)
Corn starch	ε-Poly-$_L$-lysine (0, 0.2, 0.4, 0.8, 1.6, 3.2, and 6.5 mg ε-PL/cm^2 of biofilm)	Bread slices	Antimicrobial film	- Biofilms with concentrations of ε-Poly-$_L$-lysine less than 1.6 mg/cm^2 showed no fungal inhibition in solid medium against *Aspergillus parasiticus* and *Penicillium expansum*, while the antifungal activity of the films with greater than 1.6 mg/cm^{2}of ε-Poly-$_L$-lysine was dose dependent - Aflatoxins production was significantly inhibited by ε-Poly-$_L$-lysine biopolymer films by 93%-100%	- Films containing 1.6-6.5 mg ε-Poly-$_L$-lysine/cm^2 effectively increased shelf-life of bread inoculated with *A. parasiticus* by 1 day with, while shelf-life of bread with *P. expansum* was increased by 3 day with 6.5 mg ε-Poly-$_L$-lysine/cm^2	Luz et al. (2018)
TPS/ PBAT	Potassium sorbate (0%, 1.5%, 3.0%, and 4.5%)	Fresh pasta sheets	Antimicrobial film	- Films containing potassium sorbate showed lower opacity - Potassium sorbate concentration ≥ 3.0% decreased elongation - Film elongation decreased after 28 days in contact with the fresh pasta, whereas higher potassium sorbate gave greater decrease - Potassium sorbate at 1.5% significantly decreased water vapor permeability after storage due to subsequent recrystallization of the starch, while the water vapor permeability for 3.0% and 4.5% films did not change over storage	- Higher sorbate contents gave lower yeast and mould growth in the packaged pasta - Sorbate possibly prevented darkening of pasta giving higher lightness - Sorbate migration into pasta was less than 0.1% which was less than limit by Brazilian legislation	de Camargo Andrade-Molina et al. (2013)

Triticale flour	-	Dry pasta	Film	- After 45 days of storage the films presented a structure with some few discontinuous zones and few cracks or fissures distributed along the network	- Dry pasta packaged in triticale flour films had similar moisture contents and breaking force to commercial bags - Biodegradable packaging retained low microbial counts during storage - No significant difference in cooking loss or weight gain between pasta packaged in commercial and biodegradable triticale bags during storage time - Decrease in antioxidant activity (ABTS) at 45 days of storage in pasta packaged in biodegradable bags	Aranibar et al. (2020)
Carboxy-methyl cellulose/PVOH	Cinnamon essential oil (0.5%, 1.5%, and 3%)	Sliced wheat bread	Antioxidant/antimicrobial film	- Addition of cinnamon essential oil to the film forming solutions increased total phenolic contents to 6.99 mg gallic acid equivalent/g film and DPPH scavenging to 67% - Films containing 1.5% and 3% cinnamon essential oil were effective against *P. digitatum* and showed complete inhibition in in vitro and in vivo tests	- The higher the CEO content, the greater the shelf life of sliced bread - Films enriched with 1.5% and 3% CEO showed no fungal growth over the 60 days of incubation	Fasihi et al. (2019)

(*Continued*)

TABLE 7.3 Related researches on developments of bioplastic food packaging for bakery products and pasta. (*Cont.*)

Bioplastic	Fillers	Food product	Packaging	Film properties	Food quality	References
Cellulose-derivative polymer	Cinnamaldehyde (0%, 5%, or 10%)	Bread and pastry dough	Antimicrobial film	-	- Films containing 5% and 10% (v/w) cinnamaldehyde effectively inhibited growth of aerobic mesophiles, yeast, and mould in bread - Films were effective in reducing the growth of mesophilic aerobes, *Staphylococcus spp*., yeast and mould in pastry dough. Higher concentration exhibited higher inactivation - Cinnamaldehyde migration from the film (5%) to the pastry dough and bread were 0.005 g/g and 0.0025 g/g, respectively, after 3 days of storage	Lopes et al. (2014).
Chitosan-carboxy-methyl cellulose-oleic acid	Zinc oxide nano particles (0.5%, 1%, and 2%)	Sliced wheat bread	Film	- Addition of oleic acid formed pores, while 2% zinc oxide nano particles provide more compact film structure - Addition of oleic acid and zinc oxide nano particles increased the opacity of the films but decreased water vapor permeability	- Films with nanoparticles exhibited lower firmness over 15 days of storage - Film with 2% zinc oxide nano particles inhibited microorganisms and increased shelf life of sliced wheat bread from 3 to 35 days	Noshirvani et al. (2017b)

Zein/PLA	Carvacrol (5%, 10%, and 20%)	Sliced wheat bread	Electrospun fiber	- Incorporation of carvacrol affected morphology, increased thermal stability in PLA fibers but did not influence the thermal stability - Release of carvacrol increased with increased level of loaded carvacrol - The antioxidant activity of carvacrol loaded zein fibers ranged from 62% to 75%, while antioxidant capacity of PLA fibers varied from 53% to 65% for 5%-20% carvacrol	- Bread packed with PLA and zein fiber at 10% and 20% carvacrol had no mold growth after 7 days of storage at 25°C - The composite fibrous zein and PLA films at 20% carvacrol content inhibited 99.6% and 91.3% growth of mold and yeast	Altan et al. (2018)
Gelatin	Clove, nutmeg, and black pepper oleoresins (10%)	Artisan bread	Antimicrobial film	- Gelatin and microcrystalline cellulose decreased oil and water vapor permeability values by 20%-30% and increased elasticity and seal strength	- Films containing clove and nutmeg oleoresins had similar weight loss, while the samples packaged in films containing black pepper oleoresin had the lowest weight loss values of 13% - No growth of S. aureus and lower growth of mold and yeast in the bread samples packaged in films containing oleoresins - Breads packaged in films with black pepper were scored as soft, while samples packaged in films with nutmeg and clove were perceived as firm and the bread packaged in reference films and without film were perceived as very hard - Films containing black pepper got the highest acceptability during storage period followed by clove and nutmeg package	Figueroa-Lopez et al. (2018)

(Continued)

TABLE 7.3 Related researches on developments of bioplastic food packaging for bakery products and pasta. (*Cont.*)

Bioplastic	Fillers	Food product	Packaging	Film properties	Food quality	References
Cashew gum/ gelatin	Ferulic acid and *Cymbopogon citratus* essential oil (10%)	Bread	Antimicrobial film	- Addition of ferulic acid and *Cymbopogon citratus* essential oil did not significantly affect water vapor permeability or opacity but decreased solubility possibly due to matrix cross-linking - *Cymbopogon citratus* essential oil cross-linked films showed lower rupture stress than non- *Cymbopogon citratus* essential oil films with increased elongation at break from 2% (cross-linked) to 140% (cross-linked/ *Cymbopogon citratus* essential oil)	- Cashew gum and gelatin films incorporated *Cymbopogon citratus* essential oil was more permeable than commercial PE films, giving more hardness of bread - Active films gave shelf-life of 6 days instead of 3 days for commercial PE packaging - Neral and geranial were presented in the crust of the bread, giving antimicrobial action	Oliveira et al. (2020)
Chitosan	Chestnut extract (0.0%, 0.5%, 1.0%, and 1.5%)	Fresh pasta	Antioxidant/ antibacterial film and sachet	- SEM of chitosan incorporated chestnut extract sachet cross-section show a homogenous and evenly distributed structure on the 3rd day when its moisture capacity was highest compared with the initial texture at day 0 during shelf life	- Retrogradation of pasta occurred, giving a hard-like texture after 9 days in chitosan incorporated chestnut extract sachet - Active agents in sachet prevented microbial growth on food surface during the entire 60 days of storage	Kõrge et al. (2020)

Gelatin	Nano chitin and nano ZnO (5%)	Sponge cake	Antimicrobial film	- Addition of nanoparticles reduced water vapor permeability and oxygen permeability of gelatin films	- Antifungal properties of nanoparticles led to less fungal growth on cakes packed in the two layers gelatin nanocomposite and gelatin emulsion films - Cakes packed in the two layers gelatin nanocomposite and gelatin emulsion films showed more organoleptic and texture acceptability	Sahraee et al. (2020)
Chitosan	Potassium sorbate or vanillin (20% and 30%)	Butter cake	Antimicrobial film	- Addition of potassium sorbate resulted in less transparency, lower strength, water vapor permeability, while increased flexibility and oxygen permeability - Addition of vanillin gave more yellow color, reduced film strength and permeability	- Mould was observed on cake wrapped by stretch film and chitosan film on day 4 and 7, respectively, while limited mold counts and no growth on cake wrapped by chitosan film containing potassium sorbate or vanillin during 8 days storage - Cake wrapped with a film containing vanillin became more yellow, however, in the acceptable level	Sangsuwan et al. (2015)

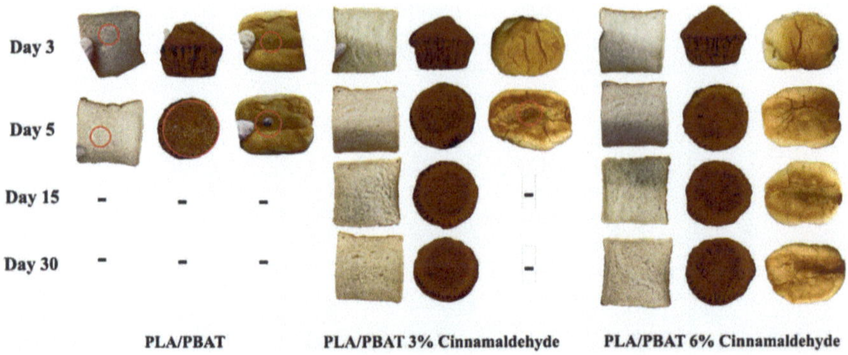

FIGURE 7.8 Appearance of bakery products packed in PLA/PBAT films containing trans-cinnamaldehyde stored for 30 days at 25°C. *Circles* indicate mold growth. *(Modified from Hongkamsang et al., 2020.)*

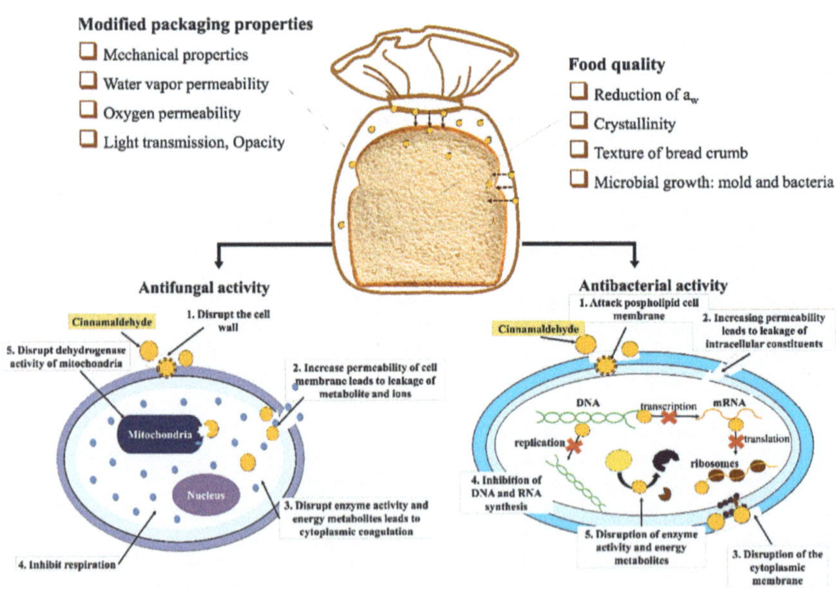

FIGURE 7.9 Interaction between essential oil incorporated packaging, food product quality, and inactivation mechanisms of bacteria and mold. *(Modified from Srisa, 2020.)*

Higher mobility commonly leads to faster recrystallization, giving higher staling and crumb hardness. Moreover, cinnamaldehyde release from the films also enhanced crumb hardness.

4.3 Meat, poultry, and seafood

Meat, poultry, and seafood are highly perishable products. High water activity leads to fast growth of various microorganisms, including spoilage and pathogenic bacteria. Physical changes including discoloration occur due to enzymatic reactions such as polyphenol oxidase which catalyzes melanosis leading to black spots in crustaceans, protein conversion of modified forms of myoglobin giving reduced redness (blooming) concurrent with darkening of red meat, and lipid oxidation inducing darkening of meat color. Microbial contamination from natural flora, handling, and processing primarily limits the shelf-life of chilled meat and seafood products. Packaging technology to extend shelf-life of fresh meat products includes modified atmosphere packaging with high carbon dioxide that inhibits microbial growth and microbial enzyme-induced reactions, for example, melanosis in white shrimp (Kimbuathong et al., 2020). High barrier packaging is commonly used to pack fresh meat in vacuum conditions. This inhibits aerobic microorganism and oxygen-induced deteriorations, particularly oxidation of lipids, proteins, pigments, and vitamins. Bioplastic materials, however, have limited barrier properties in vacuum packaging. Several attempts have been made to improve barrier properties, as mentioned earlier. Bioplastic packaging to extend the shelf-life of meat and seafood products by incorporation of active compounds has also been studied with either antimicrobial or antioxidant functions (Tables 7.4 and 7.5). Natural and synthetic antimicrobial agents have been investigated, including nisin, enzymes, and essential oils that were shown to release from packaging and inactivate microorganisms, giving longer shelf-life. Inorganic fillers and nanofillers have also been investigated in bioplastic materials. These compounds have diverse effects on packaging materials due to non-homogeneity of the matrices, leading to reduced mechanical and barrier properties. Dispersion often caused increased opacity and reduced clarity of the films. Moreover, spoilage of seafood produces volatile biogenic amines that can be detected with a sensor on the food package. Development of bioplastics for seafood packaging has also investigated intelligent packaging that can monitor the quality of packaged products and visually inform consumers.

The strong fishy odor of seafood can be attributed to trimethylamine that has a shallow threshold value of 0.21 ppb. Trimethylamine in fresh seafood increases due to growth of microorganisms and, therefore, is used as a freshness indicator for seafood products. Dried seafood also has a highly intense fishy odor that causes organoleptic dissatisfaction. PLA films can absorb the strong odor of trimethylamine leading to a significant drop in the compounds and consumer detection (Sopajittawattana et al., 2017). Winotapun et al. (2019) also showed that PLA laminate materials improved odor barrier properties against

TABLE 7.4 Related researches on developments of bioplastic food packaging for meat, poultry, and cheese.

Bioplastic	Additive/treatment	Food product	Packaging	Film properties	Food quality	References
PLA	Spice essential oils - *Mentha piperita* essential oil (0.5% and 1%), - *Bunium percium* essential oil (0.5% and 1%)	Ground beef	Film	- Addition of essential oils increased the thickness and opacity of PLA films - Essential oils decreased L* and increased b* of the films - Addition of essential oils significantly lowered the water vapor permeability values	- Both *menthe piperita* essential oil and *bunium percium* essential oil containing films showed productive antibacterial activity against total bacterial count, psychrotropic bacteria, lactic acid bacteria, *pseudomonas* spp., *enterobacteriaceae*, and *staphylococcus aureus* - Combination of both essential oils at 1% extended shelf life from 4 to 7 days	Talebi et al. (2018)
PLA	- Propolis ethanolic extract (0%, 10%, 20%, and 40%)	Dry meat sausage	Film	- Addition of propolis ethanolic extract adversely affected tensile strength, elongation at break, and elastic modulus of PLA films - Phenolic compounds of propolis ethanolic extract were released from PLA films in quantities proportional to propolis ethanolic extract concentration	- Propolis ethanolic extract showed an inhibitory effect against *Staphylococcus aureus* and *Pseudomonas aeruginosa* bacterial species but was more effective against Gram-positive species - Polyethylene glycol (PEG)/calcium carbonate also enhanced inhibitory effect	Safaei and Azad (2020)

PLA	Limonene (0%, 4%, 6%, and 8%)	Fresh beef meat	Film	- Water vapor permeability and oxygen permeability were reduced as the contents of limonene increased	- PLA-limonene at 6% and 8% reduced coliform bacteria, *Pseudomonas* spp. and *Staphylococcus aureus*	Sangkasanya et al. (2018)
PLA	Titanium dioxide (2% and 3%) or silver nanoparticles (1%)	Yunnan cottage cheese	Film	-	- Cheeses packed in PLA/titanium dioxide and PLA/titanium dioxide with silver film provided better retention in pH value, lactic acid bacteria, sensory quality, and antimicrobial activity - Migration of titanium and silver nanoparticles was lower than the limit (10 mg/kg) defined by European Food Safety Authority for food contact materials	Li et al. (2018)
PLA	Nanochitosan (1%) and *Polylophium involucratum* essential oil (0%, 0.5%, 1%, and 1.5%)	Chicken fillets	Film	-	- Wrapped fillets showed lower total volatile base nitrogen, retain peroxide value and reduced microbial counts by approximately 1-3 log CFU/g - PLA films containing 1.5% essential oil prolonged the shelf life of chicken fillet for more than 10 days without any adverse sensorial properties	Javaherzadeh et al. (2020)

(*Continued*)

TABLE 7.4 Related researches on developments of bioplastic food packaging for meat, poultry, and cheese. (*Cont.*)

Bioplastic	Additive/treatment	Food product	Packaging	Film properties	Food quality	References
PLA	*Zataria multiflora* Bioss. essential oil (0%, 0.5%, and 1%), propolis ethanolic extract (0% and 2%), and cellulose nanofiber (0% and 1%)	Cooked sausage	Film	- Addition of *Zataria multiflora* Bioss. essential oil and propolis ethanolic extract improved film flexibility and incorporation of cellulose nanofiber improved tensile strength and modulus - Incorporation of 0.5% *Zataria multiflora* Bioss. essential oil improved water vapor permeability - Films were visually transparent, while composites with propolis ethanolic extract appeared as light red with some particulate areas	- Films containing essential oils gave lower aerobic plate count, lactic acid bacteria and psychotropic counts - PLA with 1% *Zataria multiflora* Bioss. essential oil/propolis ethanolic extract composites increased the shelf life to more than 40 days based on the microbiological and sensory evaluation	Rezaeigolestani et al. (2017)
PLA	Bimetallic silver-copper nanoparticles (2% and 4%) and cinnamon essential oil (25% and 50%)	Chicken meat	Film	- Reinforcement of nanoparticles increased glass transition and improved crystallization temperature indicating decreased polymer chain movement and flexibility - Loading of cinnamon essential oil decreased glass transition, melting, and crystallization temperature - Barrier properties of the composite films were reduced with the increase of cinnamon essential oil loading - Their optical properties were also modified by the addition of both cinnamon essential oil and silver-copper nanoparticles	- Composite films had antimicrobial activity against *Salmonella typhimurium*, *Campylobacter jejuni*, and *Listeria monocytogenes* inoculated in chicken samples - Films loaded with silver-copper nanoparticles and 50% cinnamon essential oil showed maximum antibacterial action during 21 days at the refrigerated condition	Ahmed et al. (2018)

PLA	Essential oils (cinnamon; 0.4, 0.8, 1.2, and 1.6 mL), garlic (1.6 mL), and clove (1.6 mL)	Poultry meat	Film	- Essential oils led to flexible films with significant drop in the glass transition temperature, and mechanical property - Thickness increase with increase essential oil, the maximum thickness was recorded for clove-based film, while garlic oil gave minimum film thickness - Clove and garlic oil showed good compatibility with PLA polymer matrix, while cinnamon oil gave uneven surface because of poor distribution	- Cinnamon and clove oil-based films exhibited a complete zone of inhibition against *C. jejuni* at the maximum concentration (1.6 mL per 2 g PLA/PEG) - Garlic oil gave the lowest antimicrobial activity - Cinnamon and clove showed antimicrobial activity against both *C. jejuni* (7 log reductions) and *S. aureus* (2 log reduction)	Ahmed et al. (2016)
PBS	PBS/PBSA/talc (90/10 and 80/10/10)	Turkey meat	Film	- PBS blends had higher water vapor transmission rate and oxygen transmission rate than PA/PE film - Tensile strength in PBS blends was higher than PA/PE	- Different packaging materials showed no effect to aroma or flavor of packaged meats - Modified film ratios gave different growth of total bacteria, *Psychrophilic* bacteria	Vytejčková et al. (2017)
PCL/ chitosan	Grape seed extract (15%) and nanocellulose (2%)	Chicken breast fillets	Monolayer and bilayer film	-	- Bilayer films with grape seed extract maintained pH, gave lower lipid oxidation of the chicken breast fillets but gave the highest total color change - Total color change was higher for the samples packaged with bilayer films - Film with grape seed extract and nanocellulose gave reduction in total mesophilic aerobic bacteria and total coliform bacteria in chicken breast fillets	Sogut and Seydim (2019)

(Continued)

TABLE 7.4 Related researches on developments of bioplastic food packaging for meat, poultry, and cheese. (*Cont.*)

Bioplastic	Additive/treatment	Food product	Packaging	Film properties	Food quality	References
PHB/PCL	- Organo-clays (Cloistite 30B and 10A; 5% w/w) - Nisin	Ham	Film	- Organo-clays addition promoted a reinforcement effect of the polymer blend, increasing barrier properties, and degradation temperature	- Organo-clays exerted antimicrobial activity against *Lactobacillus plantarum* CRL691 - Nisin adsorption to PHB/PCL film was not affected by clays	Correa et al. (2017)
PCL	Microalgae biopeptide (3%)	Chicken meat cuts	Nanofiber	- SEM showed that diameter of nanofibers containing the biopeptides did not differ to the PCL nanofibers - Film thickness of biopeptide-added PCL was lower than PCL - Tensile strength increased, while elongation and melting temperature decreased for biopeptide-added PCL nanofibers	- PCL/biopeptide nanofibers showed inhibition for DPPH and ABTS radicals. - Nanofibers containing biopeptides effectively controlled lipid oxidation	Gonçalves et al. (2017)

PLLA/PVA/PCL	Tea polyphenol (0.1 g), potassium sorbate (0.1 g), and trehalose (0.1 g)	Fresh pork	Multilayer film	- Film exhibits extremely excellent oxygen barrier property and relatively good water vapor barrier property - No significant change of yield strength and elongation at break of the multilayer film with trehalose, tea polyphenol, and potassium sorbate. - Young's modulus of the mutilayer- tea polyphenol film decreased significantly due to lower crystallinity	- Total viable count and total volatile basic nitrogen of pork in multilayer film were greatly inhibited - Multilayer film with trehalose, tea polyphenol, and potassium sorbate gave lower and more stable pH value, and higher sensory score - Multilayer-potassium sorbate film showed the longest shelf life of 31 days	Dong et al. (2017b)
TPS	Irradiated starch (0, 5, 10, or 20 kGy) Lime juice (4.5 mL)	Minced chicken	Film	- Lime juice reinforced intermolecular binding which increased tensile strength and reduced elongation at break of the films - Films containing lime juice had antioxidant and antibacterial properties	- Films with lime juice extended shelf-life of chilled minced chicken meat from 2 to 12 days - Lipid peroxidation of minced chicken packed in active films was lower	Kanatt (2020)

(*Continued*)

TABLE 7.4 Related researches on developments of bioplastic food packaging for meat, poultry, and cheese. (*Cont.*)

Bioplastic	Additive/treatment	Food product	Packaging	Film properties	Food quality	References
Starch	Cellulose nanocrystal (10%) and two grape pomace extracts (Cabernet Franc; 8%) and (Viognier; 8%)	Ready-to-eat-chicken meat	Film	- Cellulose nanocrystal significantly increased tensile strength and decreased elongation at break and water vapor permeability - Grape pomace extracts increased elongation water vapor permeability - Cellulose nanocrystal and grape pomace extract increased film opacity - Films with Viognier and cellulose nanocrystal extract exhibited more significant release of phenolic compound	- Films incorporating grape pomace extracts exhibited a more substantial inhibitory effect on *S. aureus* ATCC 29213 compared to *L. monocytogenes* ATCC 7644 - Nanocomposite films incorporating grape pomace extract inhibited growth of *L. monocytogenes* inoculated on surface of ready-to-eat deli meat products	Xu et al. (2018)
Starch	*Zataria multiflora* essential oil (0.5% and 1%) and cinnamaldehyde (0.5% and 1%),	Ground beef	Film	- Treated films have identical thickness but exhibited lower tensile strength and higher elongation at break than control - GC-MS analysis of *Zataria multiflora* essential oil showed the main compounds of *Zataria multiflora* essential oil as thymol (37.94%), p-cymene (9.5%), and 3-carene (4.98%)	- Beef wrapped with the films containing nanoemulsion of *Zataria multiflora* essential oil had the best chemical and sensory properties - Best antioxidant effect was obtained for *Zataria multiflora* essential oil - Essential oils delayed the increased pH values of meat	Amiri et al. (2019)

Starch	Eugenol (1%, 3%, and 5 %)	Pork	Film	-	- Mechanism for release of eugenol from films was well-controlled by diffusion and erosion - Film displayed good antibacterial activity against *E. coli, S. aureus*, and *L. monocytogenes* depending on dose of eugenol - Film with 3% eugenol can extend the shelf-life of pork beyond 50%	Cheng et al. (2019)
Starch	Gallic acid, chitosan, and carvacol	Ready- to-eat ham	Film	-	- Gallic acid was the least effective antimicrobial - Starch films with chitosan and carvacrol inhibited *L. monocytogenes* - Starch films loaded with chitosan or chitosan and carvacrol delayed growth of microbiota by 1-2 weeks	Zhao et al. (2019)
Starch	Sodium benzoate (0.001%), and citric acid (30%) and pulsed light treatment	Sliced cheddar cheese	Film	- Starch films containing sodium benzoate increased tensile strength up to 60% after pulsed light treatment - Partially replacing glycerol with citric acid gave up to 8 times higher tensile strength, with no influence by pulsed light	- Cheese packed in antimicrobial starch film-citric acid or antimicrobial starch film-citric acid/sodium benzoate and treated with PL showed an average reduction of 4.5 log CFU/slice after 3 days of storage - Cheese packed in antimicrobial starch film-sodium benzoate or starch films and treated with pulsed light reached a reduction of 2.0 log CFU/slice of *L. innocua* immediately after the treatment, and maintained these levels throughout the storage period	de Moraes et al. (2020)

(*Continued*)

TABLE 7.4 Related researches on developments of bioplastic food packaging for meat, poultry, and cheese. (*Cont.*)

Bioplastic	Additive/treatment	Food product	Packaging	Film properties	Food quality	References
PVOH/PP	Rhubarb ethanolic extracts (0%, 0.5%, 1%, 2%, and 4%) and cinnamon essential oil (0%, 0.04%, 0.08%, 0.16%, and 0.32 %)	Fresh beef	Film	-	- Antimicrobial films gave better quality of packaged beef in comparison to the control film and maintained the pH and total volatile base nitrogen of beef steaks - Films with extract had lower total viable count	Han et al. (2014)
Cellulose	Carboxymethyl cellulose: PVOH (1:1, 1:2, and 2:1) and clove oil (1%, 2%, and 3%)	Ground chicken meat	Film	- Films containing clove oil had slight yellowish color, decreased both tensile strength and puncture force but insignificant changed water vapor transmission rate - Increase in PVOH concentration improved the puncture force and tensile strength, decreased water vapor transmission rate	- Packaged meat in films containing clove oil had lower total viable counts and displayed a shelf life of 12 days, whereas, control samples spoiled within 4 days in refrigerated storage - Films showed efficacy to inhibit inoculum strains namely *Staphylococcus aureus* and *Bacillus cereus* in ground chicken meat	Muppalla et al. (2014)
Cellulose	Emulsion containing oregano, thyme, and cinnamon essential oils (10%)	Minced beef	Pads		- Pads were active against specific meat bacterial species namely *Pseudomonas putida, Pseudomonas fragi, Pseudomonas fluorescens, Enterococcus faecalis, and Lactococcus lactis* and some common foodborne pathogens including *Salmonella enterica, Campylobacter jejuni,* and *Staphylococcus aureus*	Agrimonti et al. (2019)

Cellulose chitosan	*Ziziphora clinopodioides* essential oil (0%, 0.5%, 1%, and 2%), *Ficus carica* extract (0% and 1%), and montmorillonite (0.1%)	Camel meat	Film	- Excellent dispersion of montmorillonite into the chitosan and carboxymethyl cellulose matrices without any aggregation - Thickness values of designated chitosan-nanomontmorillonite and carboxymethyl cellulose-nanomontmorillonite films containing *Ziziphora clinopodioides* essential oil and *ficus carica* extract were insignificantly higher than control	- Microbial counts of meat samples packed in active films were decreased approximately 1-4 log CFU/g compared to control - Nanocomposite films retarded the increases in total volatile base nitrogen, pH, peroxide value, protein carbonyl content and thiobarbituric acid reactive substances - Sensory attributes namely odor, color and overall acceptability were significantly enhanced in meat packed in active films	Khezrian and Shahbazi (2018)
Cellulose	Nisin (156.3, 312.5, 625, 1250, 2500, 5000, and 10,000 IU/mL)	Processed meat	Film	-	- Incorporation of nisin into cellulose films reduced *L. monocytogenes* populations on frankfurters, while total aerobic bacteria were reduced by 3.3 log CFU/g after 14 days of storage	Nguyen et al. (2008)
Cellulose	Okra mucilage (30%, 40%, and 50%), and zinc oxide nanoparticles (20%)	Chicken breast meat	Film	-	- Growth of microorganisms was significantly inhibited by okra mucilage and zinc oxide nanoparticles in the packaging - Increase of lipid oxidation and total volatile base nitrogen were significantly restricted by incorporation of okra mucilage and zinc oxide nanoparticles - Chicken samples packed with okra mucilage and zinc oxide nanoparticles had higher lightness	Mohammadi et al. (2019)

(*Continued*)

TABLE 7.4 Related researches on developments of bioplastic food packaging for meat, poultry, and cheese. (*Cont.*)

Bioplastic	Additive/treatment	Food product	Packaging	Film properties	Food quality	References
Cellulose	Pink pepper essential oils (0%, 2%, 4%, and 6%)	Sliced cheese	Film	-	- Essential oils added to the films gave antimicrobial capacity against *L. monocytogenes* and *S. aureus* - Affinity between non-polar molecules of essential oils and lipid components of cheese allowed migration of essential oils to food, indicating favorable for direct contact application	Dannenberg et al. (2017)

TABLE 7.5 Related researches on developments of bioplastic film packaging for seafood.

Bioplastic	Additives	Food product	Packaging properties	Food quality	References
PLA	Nanochitosan (0.5%, 1, 2%)	Indian white prawn	- Nanochitosan decreased tensile strength, elongation at break and heat sealing strength due to hydrophilic nature of chitosan that is not compatible to hydrophobic nature of PLA giving weakness of interfacial adhesion - Chitosan decreased film transparency - Increased chitosan increased migration which enhanced antimicrobial capacity	- 1% and 2% nanochitosan incorporation effectively reduced the microbial growth, total volatile based nitrogen which enhanced the shelf life of prawn during chilled storage - Gram positive *Listeria monocytogenes* showed higher reduction (2.23%-67.09%) than Gram negative *E. coli* (5.06%-30.46%)	Fathima et al. (2018)
PLA	Thyme, rosemary, and oregano essential oil (10%)	Fish fillet	- Addition of essential oils did not affect permeation, migration, and elongation at break of films - Tensile strength and Young modulus significantly decreased on films containing essential oils.	- Film containing 5% and 10% oregano oil exhibited higher antioxidant activity and reduced oxidation of minced fish - Oregano essential oil gave lowest Thiobarbituric acid value - All essential oils gave a subtle odor/taste to the product	Zeid et al. (2019)
PLA	*Origanum vulgare* essential oil (0.5%, 1%, and 1.5 %)	Refrigerated rainbow trout	- PLA-essential oil films were more flexible than neat PLA films with up to sixfold increase in the elongation at break (2.82%-16.78%) for 0.5% oil - Essential oil decreased glass transition temperature of PLA due to plasticization and modified mechanical properties - Surface roughness as determined by atomic force microscopy increased with essential oils	- Films exhibited antimicrobial activities which were more effective against bacteria in direct contact than a vapor phase - *Staphylococcus aureus* was the most sensitive bacteria to essential oil-containing films, while *Escherichia coli* was the most resistant. - Essential oil with 1.5% effectively reduced the microbial growth of rainbow trout during chilled storage	Javidi et al. (2016)

(Continued)

TABLE 7.5 Related researches on developments of bioplastic film packaging for seafood. (*Cont.*)

Bioplastic	Additives	Food product	Packaging properties	Food quality	References
PLA	ZnO (1.5%) Zatari bois essential oil or *mentha piperita* essential oil (0.5%, 1%, and 1.5%)	*Otolithes ruber* fillets	- Films had antibacterial activity against five common food-borne pathogens namely *Escherichia coli, salmonella enterica, Pseudomonas aeruginosa, Bacillus cerus, and staphylococcus aureus* as determined by disc diffusion method - PLA/ZnO films showed no antimicrobial activity against bacteria	- Films showed antimicrobial capacity against enterobacteria, H_2S-porducing bacteria and lactic acid bacteria - Active films significantly increased the shelf life of fillets from 8 to 16 days. Films decreased lipid oxidation, decreased total volatile base nitrogen - Atomic absorption spectrophotometer showed migration of Zn^{2+} ions from the nanocomposite film to fillet were still far below the migration limit of 40 mg/day on zinc daily consumption	Heydari-Majd et al. (2019)
PBAT	Oregano essential oil (0.0%, 2.5%, 5.0%, 7.5%, and 10%)	Fish fillet	- Essential oils reduced thermal stability of films and gave and rough surfaces in cross-sections - Tensile strength, elongation at break, and elastic modulus were impaired - Higher essential oils gave higher water vapor permeability	- Films reduced total coliforms, *Staphylococcus aureus* and psychrotrophic microorganisms - Higher essential oil concentration gave more potent antioxidant activity of the films	Cardoso et al. (2017)

PCL/whey protein isolate coating	*Urtica dioica* L. extract in whey protein isolate (5 and 7 g/100 g)	Rainbow trout fillets	- Incorporation of plant extracts did not affect uniformity of fibrillar shape and nanofibers presented similar morphological characteristic of PCL	- Coating containing *Utirca dioca* L. water extract exhibited higher antimicrobial efficiency against mesophilic, psychrophilic, and lactic acid bacteria as well as *Enterobacteriaceae* - Coating exhibited reduced growth including *Enterobacteriacea*, psychrophilic, and lactic acid bacteria count, total volatile essential nitrogen and thiobarbituric acid which extended shelf life up to 15 days	Erbay et al. (2017)
PLA/PHB	Fennel oil (6%)	Oyster	- Fennel oil improved elongation at break but did not affect tensile strength PLA and PHB blends showed a good transparency, with addition fennel oil, haze decreased and transmittance increased - Fennel oil greatly enhanced oxygen barriers of PLA-PHB, antioxidant capacity and anti-bacterial activity - Maximum release of fennel oil took place in 65% ethanol suggesting an outstanding applicable for package of oily food	- Films showed ability to reduce about 1 log CFU/g on both aerobic and anaerobic bacteria of oyster - Changes in pH and total volatile based nitrogen were retarded with the addition of fennel oil - Addition of fennel oil into films effectively prolonged shelf life of oyster by 2-3 days	Miao et al. (2019)
PLA/PHB	Mono-caprylin glycerate or glycerol monolaurate (0.5%, 1%, and 2%) and cinnamaldehyde (5%)	Salmon	- Mono-caprylin glycerate and glycerol monolaurate showed higher tensile strength due to plasticizer effect - Films containing mono-caprylin glycerate and glycerol monolaurate showed higher radical scavenging - Cinnamaldehyde increase the water vapor transmission rate due to hydrophilicity of aldehyde group	- Mono-caprylin glycerate gradually stabilized total bacteria count and total volatile based nitrogen of salmon dices due to releasing behavior - Monolaurate as plasticizer can hold the releasing of cinnamaldehyde where cinnamaldehyde acts as antibacterial agent	Ma et al. (2018)

(*Continued*)

TABLE 7.5 Related researches on developments of bioplastic film packaging for seafood. (*Cont.*)

Bioplastic	Additives	Food product	Packaging properties	Food quality	References
PVOH/ starch	Purple sweet potato extracts or red cabbage extracts 0%, 0.5%, and 1%	Shrimp	- Purple potato extracts or red cabbage extracts containing films showed color changes in different buffers, increased thickness, thermal stability, insignificantly affected water vapor permeability - Mechanical properties were improved from 7.3 to 11.3 or 9.1 MPa in tensile strength and from 92% to 249% or 284% in elongation at break with purple potato extracts or red cabbage extracts, respectively	- Developed films presented visual changes corresponding to total volatile based nitrogen variation when shrimps were spoiled and used as an indicator for monitoring shrimp freshness	Zhang et al. (2020)
PVOH/ chitosan	Jambolan anthocyanins (0.1% and 0.3%)	Shrimp	- Scanning electron microscopy and fourier-transform infrared results revealed that anthocyanin extract was compatible with chitosan and PVOH. - Anthocyanin from jambolan extract did not modify the physical integrity but impacted thickness and optical properties of films	- Films containing anthocyanin exhibited color change at different pH values and used to monitor the freshness of shrimp at different temperatures (between −20°C and 20°C)	Merz et al. (2020)
PVOH/ starch	Betalains from red pitaya peel extract (0%, 0.25%, 0.5%, and 1 %)	Shrimp	- Hydrogen bonds formed among starch, PVOH, glycerol, and betacyanins greatly improved the water vapor barrier and mechanical properties of the films - Films exhibited improved light barrier, antioxidant and antimicrobial properties	- Films were sensitive to ammonia and was successfully applied to monitor the freshness of shrimp	Qin et al. (2020)
PVOH/ starch	Roselle anthocyanins (0.03%, 0.06%, and 0.12%)	Fish	- X-ray diffraction and scanning electron microscopy indicated that reduced crystallinity of PVA and improved compatibility between starch and PVA due to roselle anthocyanins - Extract decreased water content and tensile strength, increased elongation at break	- Colorimetric films for real-time monitoring of fish freshness	Zhai et al. (2017)

Carboxy methyl-cellulose/starch	Purple sweet potato anthocyanins (0.1%, 0.3%, 0.5%, 0.7%, and 0.9 %)	Fish	- Extract increased tensile strength but decreased elongation of break and moisture content	- Film color was changed from red to blue and green when exposed to different pH of ammonia	Jiang et al. (2020a)
Carboxy methyl cellulose-Sodium caseinate	Prebiotic: *L. acidophilus, L. reuteri, L. casei, L. rhamnosus, B. bifidum* (9 log CFU/mL)	Trout fillet	- Prebiotic decreased tensile strength, puncture force, puncture deformation, water vapor transmission rate, oxygen permeability, and swelling index compared with film without prebiotic	- Incorporation of probiotic remarkably delayed growth of microbial spoilage Film retarded the chemical spoilage indicators including peroxide value, thiobarbituric acid reactive substance, total volatile based nitrogen, and protein carbonyls - Films supplemented with *L. acidophilus, L. reuteri, L. casei, L. rhamnosus, B. bifidum* extended shelf-life of trout for > 2 weeks	Mozaffar-zogh et al. (2020)
Gelatin	Glycerol and squalene (15%)	Fresh fish crackers	- Gelatin/glycerol/squalene films had improved water vapor and oxygen barrier properties	- Fried fish crackers stored in films containing glycerol/squalene had lower moisture content and water activity than glycerol bags and exhibited negligible changes in color and texture - After 30 days, samples stored in glycerol/squalene films had lower hydroperoxides, secondary lipid oxidation products and less volatile compounds, especially those from autoxidation and non-enzymatic browning reaction -Oxidation and loss of polyunsaturated fatty acids, mainly n-3 and n-6 was reduced - Fresh fish crackers stored in gelatin/glycerol/squalene exhibited negligible changes in color and texture	Ali et al. (2020)

(Continued)

TABLE 7.5 Related researches on developments of bioplastic film packaging for seafood. (*Cont.*)

Bioplastic	Additives	Food product	Packaging properties	Food quality	References
Protein (sun-flower)	Clove essential oil (0.75 mL/g)	Sardine patties	- Clove essential oil modified protein interactions reducing water solubility and glass transition temperature of the films, but it did not modify markedly their moisture content, thickness, color, opacity, water vapor permeability, and mechanical properties	- Films retarded autoxidation and slightly delayed growth of total mesophiles	Salgado et al. (2013)
Agar	Protein hydrolysate (0.5 g/g agar) Clove essential oil (0.5 g/g agar)	Flounder fillets	- Addition of protein hydrolysate or clove essential oil decreased tensile strength, while protein hydrolysate gave higher tensile strength (19.89 MPa) than essential oil (10.16 MPa) - Protein hydrolysate and clove essential oil increased water vapor permeability of the films	- Antimicrobial activity against pathogens such as *Staphylococcus aureus, Yersinia enterocolitica, Aeromonas hydrophila, Debaryomyces hansenii*, and *Listeria innocua*. - Fillets covered with protein hydrolysate and, particularly clove essential oil films, presented lower total volatile bases and pH values, and delayed growth of H_2S-producing microorganism	Da Rocha et al. (2018)
Gelatin/ chitosan	Essential oil: clove, fennel, cypress, lavender, thyme, herb of the cross, pine, and rosemary. (0.75 mL/g polymer)	Cod	- Gelatin-chitosan-clove film is less soluble than the gelatin-clove film which slower release essential oil to maintain sufficient concentration over a more extended period of time	- Films were tested for antimicrobial activity against 18 genera of bacteria - Clove essential oil showed the highest microbial inhibitory effect, followed by rosemary and lavender - Clove and thyme essential oils had the most effective antimicrobial capacity in fish - Film containing essential oil delayed total volatile-based nitrogen formation in cod giving stable pH	Gómez-Estaca et al. (2010)

the strong smell of durian. PLA is, therefore, a potential material for packaging highly odorous food products.

5 Conclusions

Global production and investigation of bioplastic for food packaging are increasing. Different forms of bioplastic have been investigated and commercially produced as films, pouches, trays, and cups. Conventional packaging conversions, including extrusion, thermoforming, blow molding, and injection molding produce commercial bioplastic packaging. Permeability of the packaging directly affects the quality and shelf-life of food products. Bioplastic has high permeability, and this limits applications for products requiring long shelf-life. High barrier improvement of bioplastic is, therefore, required to extend the shelf-life of foods. Multilayer materials produced by coating, lamination, and co-extrusion have shown improved barrier properties. Moreover, inorganic and organic fillers and package treatment, for example, plasma treatment, effectively improved packaging barriers. Research on the development of bioplastic packaging for food products, including fresh produce, bakery, meat, and seafood has mainly focused on shelf-life extension of packaged foods by modified permeability and adding antimicrobial functions as active packaging. Improved packaging properties and active functions to extend shelf-life are required to enhance the performance of bioplastic packaging for sustainable food production.

References

Acosta, S., Chiralt, A., Santamarina, P., Rosello, J., González-Martínez, C., & Cháfer, M. (2016). Antifungal films based on starch-gelatin blend, containing essential oils. *Food Hydrocolloids*, *61*, 233–240.

Agrimonti, C., White, J. C., Tonetti, S., & Marmiroli, N. (2019). Antimicrobial activity of cellulosic pads amended with emulsions of essential oils of oregano, thyme and cinnamon against microorganisms in minced beef meat. *International Journal of Food Microbiology*, *305*, 108246.

Ahmed, J., Arfat, Y. A., Bher, A., Mulla, M., Jacob, H., & Auras, R. (2018). Active chicken meat packaging based on polylactide films and bimetallic Ag-Cu nanoparticles and essential oil. *Journal of Food Science*, *83*(5), 1299–1310.

Ahmed, J., Hiremath, N., & Jacob, H. (2016). Antimicrobial, rheological, and thermal properties of plasticized polylactide films incorporated with essential oils to inhibit *Staphylococcus aureus* and *Campylobacter jejuni*. *Journal of Food Science*, *81*(2), E419–E429.

Ali, A. M. M., de la Caba, K., Prodpran, T., & Benjakul, S. (2020). Quality characteristics of fried fish crackers packaged in gelatin bags: Effect of squalene and storage time. *Food Hydrocolloids*, *99*, 105378.

Altan, A., Aytac, Z., & Uyar, T. (2018). Carvacrol loaded electrospun fibrous films from zein and poly(lactic acid) for active food packaging. *Food Hydrocolloids*, *81*, 48–59.

Amiri, E., Aminzare, M., Azar, H. H., & Mehrasbi, M. R. (2019). Combined antioxidant and sensory effects of corn starch films with nanoemulsion of *Zataria multiflora* essential oil fortified with cinnamaldehyde on fresh ground beef patties. *Meat Science*, *153*, 66–74.

Aranibar, C., Pedrotti, F., Archilla, M., Vazquez, C., Borneo, R., & Aguirre, A. (2020). Storage and preservation of dry pasta into biodegradable packaging made from triticale flour. *Journal of Food Science and Technology*, *57*(2), 693–701.

Arora, A., & Padua, G. W. (2010). Nanocomposites in food packaging. *Journal of Food Science*, *75*(1), R43–R49.

Boonruang, K., Kerddonfag, N., Chinsirikul, W., Mitcham, E. J., & Chonhenchob, V. (2017). Antifungal effect of poly (lactic acid) films containing thymol and R-(-)-carvone against anthracnose pathogens isolated from avocado and citrus. *Food Control*, *78*, 85–93.

Cardoso, L. G., Santos, J. C. P., Camilloto, G. P., Miranda, A. L., Druzian, J. I., & Guimarães, A. G. (2017). Development of active films poly (butylene adipate co-terephthalate)-PBAT incorporated with oregano essential oil and application in fish fillet preservation. *Industrial Crops and Products*, *108*, 388–397.

Chang, Y., Choi, I., Cho, A. R., & Han, J. (2017). Reduction of *Dickeya chrysanthemi* on fresh-cut iceberg lettuce using antimicrobial sachet containing microencapsulated oregano essential oil. *LWT-Food Science and Technology*, *82*, 361–368.

Chen, G., Chen, Y., Jin, N., Li, J., Dong, S., Li, S., et al. (2020). Zein films with porous polylactic acid coatings via cold plasma pre-treatment. *Industrial Crops and Products*, *150*, 112382.

Cheng, J., Wang, H., Kang, S., Xia, L., Jiang, S., Chen, M., et al. (2019). An active packaging film based on yam starch with eugenol and its application for pork preservation. *Food Hydrocolloids*, *96*, 546–554.

Chi, H., Song, S., Luo, M., Zhang, C., Li, W., Li, L., et al. (2019). Effect of PLA nanocomposite films containing bergamot essential oil, TiO2 nanoparticles, and Ag nanoparticles on shelf life of mangoes. *Scientia Horticulturae*, *249*, 192–198.

Choklob, W., Gupta, R. K., & Kajorncheappunngam, S. (2019). PLA blend/RHA permeable composite films for fruit packaging. *Engineering and Applied Science Research*, *46*(3), 210–218.

Correa, J. P., Molina, V., Sanchez, M., Kainz, C., Eisenberg, P., & Massani, M. B. (2017). Improving ham shelf life with a polyhydroxybutyrate/polycaprolactone biodegradable film activated with nisin. *Food Packaging and Shelf Life*, *11*, 31–39.

Da Rocha, M., Alemán, A., Romani, V. P., López-Caballero, M. E., Gómez-Guillén, M. C., Montero, P., et al. (2018). Effects of agar films incorporated with fish protein hydrolysate or clove essential oil on flounder (*Paralichthys orbignyanus*) fillets shelf-life. *Food Hydrocolloids*, *81*, 351–363.

da Silva, J. B. A., Santana, J. S., de Almeida Lucas, A., Passador, F. R., de Sousa Costa, L. A., Pereira, F. V., et al. (2019). PBAT/TPS-nanowhiskers blends preparation and application as food packaging. *Journal of Applied Polymer Science*, *136*(26), 47699.

Dannenberg, G. D. S., Funck, G. D., Cruxen, C. E. D. S., Marques, J. D. L., Silva, W. P. D., & Fiorentini, Â. M. (2017). Essential oil from pink pepper as an antimicrobial component in cellulose acetate film: Potential for application as active packaging for sliced cheese. *LWT-Food Science and Technology*, *81*, 314–318.

De Azeredo, H. M. (2009). Nanocomposites for food packaging applications. *Food Research International*, *42*(9), 1240–1253.

de Camargo Andrade-Molina, T. P., Shirai, M. A., Victória Eiras Grossmann, M., & Yamashita, F. (2013). Active biodegradable packaging for fresh pasta. *LWT-Food Science and Technology*, *54*(1), 25–29.

de Moraes, J. O., Hilton, S. T., & Moraru, C. I. (2020). The effect of pulsed light and starch films with antimicrobials on *Listeria innocua* and the quality of sliced cheddar cheese during refrigerated storage. *Food Control*, *112*, 107134.

Deng, J., Chen, Q. J., Peng, Z. Y., Wang, J. H., Li, W., Ding, Q. M., et al. (2019). Nano-silver-containing polyvinyl alcohol composite film for grape fresh-keeping. *Materials Express*, *9*(9), 985–992.

Dong, T., Song, S., Liang, M., Wang, Y., Qi, X., Zhang, Y., et al. (2017). Gas permeability and permselectivity of poly (l-lactic acid)/SiOx film and its application in equilibrium-modified atmosphere packaging for chilled meat. *Journal of Food Science*, *82*(1), 97–107.

Dong, T., Zhang, Y., Qi, X., Liang, M., Song, S., Liu, L., et al. (2017). Evaluation of the effects of prepared antibacterial multilayer film on the quality and shelf-life stability of chilled meat. *Journal of Food Processing and Preservation*, *41*(5), e13151.

Duncan, T. V. (2011). Applications of nanotechnology in food packaging and food safety: barrier materials, antimicrobials and sensors. *Journal of Colloid and Interface Science*, *363*(1), 1–24.

Ebnesajjad, S. (Ed.). (2012). *Handbook of biopolymers and biodegradable plastics: properties, processing and applications*. William Andrew.

Erbay, E. A., Dağtekin, B. B. G., Türe, M., Yes¸ilsu, A. F., & Torres-Giner, S. (2017). Quality improvement of rainbow trout fillets by whey protein isolate coatings containing electrospun poly (ε-caprolactone) nanofibers with *Urtica dioica* L. extract during storage. *LWT-Food Science and Technology*, *78*, 340–351.

Fasihi, H., Noshirvani, N., Hashemi, M., Fazilati, M., Salavati, H., & Coma, V. (2019). Antioxidant and antimicrobial properties of carbohydrate-based films enriched with cinnamon essential oil by Pickering emulsion method. *Food Packaging and Shelf Life*, *19*, 147–154.

Fathima, P. E., Panda, S. K., Ashraf, P. M., Varghese, T. O., & Bindu, J. (2018). Polylactic acid/chitosan films for packaging of Indian white prawn (*Fenneropenaeus indicus*). *International Journal of Biological Macromolecules*, *117*, 1002–1010.

Fernández, A., Picouet, P., & Lloret, E. (2010). Cellulose-silver nanoparticle hybrid materials to control spoilage-related microflora in absorbent pads located in trays of fresh-cut melon. *International Journal of Food Microbiology*, *142*(1-2), 222–228.

Figueroa-Lopez, K. J., Andrade-Mahecha, M. M., & Torres-Vargas, O. L. (2018). Development of antimicrobial biocomposite films to preserve the quality of bread. *Molecules*, *23*(1), 212.

Gao, H., Fang, X., Chen, H., Qin, Y., Xu, F., & Jin, T. Z. (2017). Physiochemical properties and food application of antimicrobial PLA film. *Food Control*, *73*, 1522–1531.

Gao, H., Wu, W., Chen, H., Qin, Y., Fang, X., & Jin, T. Z. (2018a). Microbial inactivation and quality improvement of tomatoes treated by package film with allyl isothiocyanate vapour. *International Journal of Food Science & Technology*, *53*(8), 1983–1991.

Gao, Y., Picot, O. T., Tu, W., Bilotti, E., & Peijs, T. (2018b). Multilayer coextrusion of graphene polymer nanocomposites with enhanced structural organization and properties. *Journal of Applied Polymer Science*, *135*(13), 46041.

Gómez-Estaca, J., De Lacey, A. L., López-Caballero, M. E., Gómez-Guillén, M. C., & Montero, P. (2010). Biodegradable gelatin-chitosan films incorporated with essential oils as antimicrobial agents for fish preservation. *Food Microbiology*, *27*(7), 889–896.

Gonçalves, C. F., Schmatz, D. A., Uebel, L. D. S., Kuntzler, S. G., Costa, J. A. V., Zimmer, K. R., et al. (2017). Microalgae biopeptides applied in nanofibers for the development of active packaging. *Polímeros*, *27*(4), 290–297.

González-Buesa, J., Page, N., Kaminski, C., Ryser, E. T., Beaudry, R., & Almenar, E. (2014). Effect of non-conventional atmospheres and bio-based packaging on the quality and safety of *Listeria monocytogenes*-inoculated fresh-cut celery (*Apium graveolens* L.) during storage. *Postharvest Biology and Technology*, *93*, 29–37.

Greene, J. P. (2014). *Sustainable plastics: Environmental assessments of biobased, biodegradable, and recycled plastics*. John Wiley & Sons, Inc.

Guerrero, P., & de la Caba, K. (2017). Protein-based films and coatings. In M. A. P. R. Cerqueira, R. N. C. Pereira, O. L. da Silva Ramos, J. A. C. Teixeira, & A. A. Vicente (Eds.), *Edible food packaging: Materials and processing technologies* (pp. 81–120). CRC Press.

Han, C., Wang, J., Li, Y., Lu, F., & Cui, Y. (2014). Antimicrobial-coated polypropylene films with polyvinyl alcohol in packaging of fresh beef. *Meat Science*, *96*(2), 901–907.

He, X., Wu, L. L., Wang, J. J., Zhang, T., Sun, H., & Shuai, N. (2015). Layer-by-layer assembly deposition of graphene oxide on poly (lactic acid) films to improve the barrier properties. *High Performance Polymers*, *27*(3), 318–325.

Heidemann, H. M., Dotto, M. E., Laurindo, J. B., Carciofi, B. A., & Costa, C. (2019). Cold plasma treatment to improve the adhesion of cassava starch films onto PCL and PLA surface. *Colloids and Surfaces A: Physicochemical and Engineering Aspects*, *580*, 123739.

Hernandez-Izquierdo, V. M., & Krochta, J. M. (2008). Thermoplastic processing of proteins for film formation-a review. *Journal of Food Science*, *73*(2), R30–R39.

Heydari-Majd, M., Ghanbarzadeh, B., Shahidi-Noghabi, M., Najafi, M. A., & Hosseini, M. (2019). A new active nanocomposite film based on PLA/ZnO nanoparticle/essential oils for the preservation of refrigerated Otolithes ruber fillets. *Food Packaging and Shelf Life*, *19*, 94–103.

Hongkamsang, K., Pasanga, S., & Harnkarnsujarit, N. (2020). Development of volatile released polymer to enhance functional property of food packaging for bakery products. Bachelor Thesis, Kasetsart University.

https://www.european-bioplastics.org/market/. 2020. (Accessed 30 September 2020).

https://www.chemistryviews.org/details/news/11223436/Certified_Renewable_Polypropylene.html. 2020. (Accessed 30 September 2020).

Huntrakul, K., Yoksan, R., Sane, A., & Harnkarnsujarit, N. (2020). Effects of pea protein on properties of cassava starch edible films produced by blown-film extrusion for oil packaging. *Food Packaging and Shelf Life*, *24*, 100480.

Iotti, M., Fabbri, P., Messori, M., Pilati, F., & Fava, P. (2009). Organic-inorganic hybrid coatings for the modification of barrier properties of poly (lactic acid) films for food packaging applications. *Journal of Polymers and the Environment*, *17*(1), 10–19.

Janssen, L.P.B.M. (2009). Extruders. In L.P.B.M. Janssen and L. Moscicki (Eds.), *Thermoplastic starch: A green material for various industries* (pp. 119-139). WILEY-VCH Verlag GmbH & Co. KGaA, Weinheim, Germany.

Javaherzadeh, R., Bafroee, A. T., & Kanjari, A. (2020). Preservation effect of *Polylophium involucratum* essential oil incorporated poly lactic acid/nanochitosan composite film on shelf life and sensory properties of chicken fillets at refrigeration temperature. *LWT-Food Science and Technology*, *118*, 108783.

Javidi, Z., Hosseini, S. F., & Rezaei, M. (2016). Development of flexible bactericidal films based on poly (lactic acid) and essential oil and its effectiveness to reduce microbial growth of refrigerated rainbow trout. *LWT-Food Science and Technology*, *72*, 251–260.

Jiang, G., Hou, X., Zeng, X., Zhang, C., Wu, H., Shen, G., et al. (2020). Preparation and characterization of indicator films from carboxymethyl-cellulose/starch and purple sweet potato (*Ipomoea batatas* (L.) lam) anthocyanins for monitoring fish freshness. *International Journal of Biological Macromolecules*, *143*, 359–372.

Jiang, J., Gong, L., Dong, Q., Kang, Y., Osako, K., & Li, L. (2020). Characterization of PLA-P3, 4HB active film incorporated with essential oil: Application in peach preservation. *Food Chemistry*, *313*, 126134.

Kamthai, S., & Magaraphan, R. (2019). Development of an active polylactic acid (PLA) packaging film by adding bleached bagasse carboxymethyl cellulose (CMCB) for mango storage life extension. *Packaging Technology and Science*, *32*(2), 103–116.

Kanatt, S. R. (2020). Irradiation as a tool for modifying tapioca starch and development of an active food packaging film with irradiated starch. *Radiation Physics and Chemistry*, 108873.

Kanikireddy, V., Varaprasad, K., Rani, M. S., Venkataswamy, P., Reddy, B. J. M., & Vithal, M. (2020). Biosynthesis of CMC-Guar gum-Ag0 nanocomposites for inactivation of food pathogenic microbes and its effect on the shelf life of strawberries. *Carbohydrate Polymers*, *236*, 116053.

Kechichian, V., Ditchfield, C., Veiga-Santos, P., & Tadini, C. C. (2010). Natural antimicrobial ingredients incorporated in biodegradable films based on cassava starch. *LWT-Food Science and Technology*, *43*(7), 1088–1094.

Khammeethong, T., Khamta, S., & Harnkarnsujarit, N. (2020) Enhanced food packaging performance to extend shelf-life of perishable food product using reinforced oxide filler. Bachelor Thesis, Kasetsart University.

Khezrian, A., & Shahbazi, Y. (2018). Application of nanocompostie chitosan and carboxymethyl cellulose films containing natural preservative compounds in minced camel's meat. *International Journal of Biological Macromolecules*, *106*, 1146–1158.

Kimbuathong, N., Leelaphiwat, P., & Harnkarnsujarit, N. (2020). Inhibition of melanosis and microbial growth in Pacific white shrimp (*Litopenaeus vannamei*) using high CO2 modified atmosphere packaging. *Food Chemistry*, *312*, 126114.

Kõrge, K., Bajić, M., Likozar, B., & Novak, U. (2020). Active chitosan-chestnut extract films used for packaging and storage of fresh pasta. *International Journal of Food Science & Technology*, *55*(8), 3043–3052. https://doi.org/10.1111/ijfs.14569.

Kumar, A., & Kudachikar, V. B. (2018). Antifungal properties of essential oils against anthracnose disease: a critical appraisal. *Journal of Plant Diseases and Protection*, *125*(2), 133–144.

Leabwan, N., Loylerd, K., Enmak, P., Changprasert, S., Korppiboon, A., & Tengrang, S. (2015). Observations of the effect of different bioplastic packaging on prolonging the storage life of rambutan and biodegradability of the packaging. In International symposium on durian and other humid tropical fruits 1186 (pp. 177-184).

Leal, I. L., da Silva Rosa, Y. C., da Silva Penha,, J., Cruz Correia, P. R., da Silva Melo, P., Guimarães, D. H., ... Machado, B. A. S. (2019). Development and application starch films: PBAT with additives for evaluating the shelf life of Tommy Atkins mango in the fresh-cut state. *Journal of Applied Polymer Science*, *136*, 48150. https://doi.org/10.1002/app.48150.

Li, W., Li, L., Zhang, H., Yuan, M., & Qin, Y. (2018). Evaluation of PLA nanocomposite films on physicochemical and microbiological properties of refrigerated cottage cheese. *Journal of Food Processing and Preservation*, *42*(1), e13362.

Liu, Y., Wang, S., Lan, W., & Qin, W. (2019). Fabrication of polylactic acid/carbon nanotubes/chitosan composite fibers by electrospinning for strawberry preservation. *International Journal of Biological Macromolecules*, *121*, 1329–1336.

Lopes, F. A., de Fátima Ferreira Soares, N., de Cássia Pires Lopes, C., da Silva, W. A., Júnior, J. C. B., & Medeiros, E. A. A. (2014). Conservation of bakery products through cinnamaldehyde antimicrobial films. *Packaging Technology and Science*, *27*(4), 293–302.

Lorite, G. S., Rocha, J. M., Miilumäki, N., Saavalainen, P., Selkälä, T., Morales-Cid, G., et al. (2017). Evaluation of physicochemical/microbial properties and life cycle assessment (LCA) of PLA-based nanocomposite active packaging. *LWT-Food Science and Technology*, *75*, 305–315.

Luz, C., Calpe, J., Saladino, F., Luciano, F. B., Fernandez-Franzon, M., Manes, J., et al. (2018). Antimicrobial packaging based on varepsilon-polylysine bioactive film for the control of mycotoxigenic fungi in vitro and in bread. *Journal of Food Processing and Preservation*, *42*(1), e13370.

Ma, Y., Li, L., & Wang, Y. (2018). Development of PLA-PHB-based biodegradable active packaging and its application to salmon. *Packaging Technology and Science*, *31*(11), 739–746.

Merz, B., Capello, C., Leandro, G. C., Moritz, D. E., Monteiro, A. R., & Valencia, G. A. (2020). A novel colorimetric indicator film based on chitosan, polyvinyl alcohol and anthocyanins from jambolan (*Syzygium cumini*) fruit for monitoring shrimp freshness. *International Journal of Biological Macromolecules*, *153*, 625–632.

Miao, L., Walton, W. C., Wang, L., Li, L., & Wang, Y. (2019). Characterization of polylactic acids-polyhydroxybutyrate based packaging film with fennel oil, and its application on oysters. *Food Packaging and Shelf Life*, *22*, 100388.

Mistriotis, A., Briassoulis, D., Giannoulis, A., & D'Aquino, S. (2016). Design of biodegradable bio-based equilibrium modified atmosphere packaging (EMAP) for fresh fruits and vegetables by using micro-perforated poly-lactic acid (PLA) films. *Postharvest Biology and Technology*, *111*, 380–389.

Mitrus, M., Wojtowicz, A., & Moscicki, L. (2009). Biodegradable polymers and their practical utility. In Leon P.B.M. Janssen and Leszek (Ed), *Thermoplastic starch* (pp. 1-33). Moscicki, WILEY-VCH Verlag GmbH & Co. KGaA, Weinheim.

Mohammadi, H., Kamkar, A., Misaghi, A., Zunabovic-Pichler, M., & Fatehi, S. (2019). Nanocomposite films with CMC, okra mucilage, and ZnO nanoparticles: Extending the shelf-life of chicken breast meat. *Food Packaging and Shelf Life*, *21*, 100330.

Mozaffarzogh, M., Misaghi, A., Shahbazi, Y., & Kamkar, A. (2020). Evaluation of probiotic carboxymethyl cellulose-sodium caseinate films and their application in extending shelf life quality of fresh trout fillets. *LWT-Food Science and Technology*, *126*, 109305.

Muppalla, S. R., Kanatt, S. R., Chawla, S. P., & Sharma, A. (2014). Carboxymethyl cellulose-polyvinyl alcohol films with clove oil for active packaging of ground chicken meat. *Food Packaging and Shelf Life*, *2*(2), 51–58.

Nguyen, V. T., Gidley, M. J., & Dykes, G. A. (2008). Potential of a nisin-containing bacterial cellulose film to inhibit *Listeria monocytogenes* on processed meats. *Food Microbiology*, *25*(3), 471–478.

Nieto, M. B. (2016). Edible film and packaging using gum polysaccharides. In M. A. P. R. Cerqueira, R. N. C. Pereira, O. L. da Silva Ramos, J. A. C. Teixeira, & A. A. Vicente (Eds.), *Edible food packaging: Materials and processing technologies* (pp. 9–79). CRC Press.

Noshirvani, N., Ghanbarzadeh, B., Gardrat, C., Rezaei, M. R., Hashemi, M., Le Coz, C., et al. (2017). Cinnamon and ginger essential oils to improve antifungal, physical and mechanical properties of chitosan-carboxymethyl cellulose films. *Food Hydrocolloids*, *70*, 36–45.

Noshirvani, N., Ghanbarzadeh, B., Rezaei Mokarram, R., & Hashemi, M. (2017). Novel active packaging based on carboxymethyl cellulose-chitosan-ZnO NPs nanocomposite for increasing the shelf life of bread. *Food Packaging and Shelf Life*, *11*, 106–114.

Oliveira, M. A., Gonzaga, M. L. C., Bastos, M. S. R., Magalhães, H. C. R., Benevides, S. D., Furtado, R. F., et al. (2020). Packaging with cashew gum/gelatin /essential oil for bread: Release potential of the citral. *Food Packaging and Shelf Life*, *23*, 100431.

Oniszczuk, T., & Janssen, L.P.B.M. (2009). Injection-molding. In L.P.B.M. Janssen and Leszek Moscicki (Eds.), *Thermoplastic starch: A green material for various industries* (pp. 185-196). WILEY-VCH Verlag GmbH & Co. KGaA, Weinheim, Germany.

Patanè, C., Malvuccio, A., Saita, A., Rizzarelli, P., Siracusa, L., Rizzo, V., et al. (2019). Nutritional changes during storage in fresh-cut long storage tomato as affected by biocompostable polylactide and cellulose based packaging. *LWT-Food Science and Technology*, *101*, 618–624.

Peppas, N.A. & Brannon-Peppas, L. (2001). Biodegradable Polymers in Parker, G. (ed.) *Encyclopedia of materials: science and technology.*

Perumal, A. B., Sellamuthu, P. S., Nambiar, R. B., Sadiku, E. R., Phiri, G., & Jayaramudu, J. (2018). Effects of multiscale rice straw (*Oryza sativa*) as reinforcing filler in montmorillonite-polyvinyl alcohol biocomposite packaging film for enhancing the storability of postharvest mango fruit (*Mangifera indica* L.). *Applied Clay Science*, *158*, 1–10.

Pouzada, A. S. (2017). Processing, design, and performance of plastics products. In *Brydson's plastics materials* (pp. 205-246). Butterworth-Heinemann.

Puchalski, M., Szparaga, G., Biela, T., Gutowska, A., Sztajnowski, S., & Kruci ska, I. (2018). Molecular and supramolecular changes in polybutylene succinate (PBS) and polybutylene succinate adipate (PBSA) copolymer during degradation in various environmental conditions. *Polymers*, *10*(3), 251.

Qin, Y., Liu, Y., Zhang, X., & Liu, J. (2020). Development of active and intelligent packaging by incorporating betalains from red pitaya (*Hylocereus polyrhizus*) peel into starch/polyvinyl alcohol films. *Food Hydrocolloids*, *100*, 105410.

Qin, Y., Zhuang, Y., Wu, Y., & Li, L. (2016). Quality evaluation of hot peppers stored in biodegradable poly (lactic acid)-based active packaging. *Scientia Horticulturae*, *202*, 1–8.

Ray, S., Jin, T., Fan, X., Liu, L., & Yam, K. L. (2013). Development of chlorine dioxide releasing film and its application in decontaminating fresh produce. *Journal of Food Science*, *78*(2), M276–M284.

Reddy, N., & Yang, Y. (2013). Thermoplastic films from plant proteins. *Journal of Applied Polymer Science*, *130*(2), 729–738.

Rezaeigolestani, M., Misaghi, A., Khanjari, A., Basti, A. A., Abdulkhani, A., & Fayazfar, S. (2017). Antimicrobial evaluation of novel poly-lactic acid based nanocomposites incorporated with bioactive compounds in-vitro and in refrigerated vacuum-packed cooked sausages. *International Journal of Food Microbiology*, *260*, 1–10.

Rhim, J. W. (2012). Physical-mechanical properties of agar/κ-carrageenan blend film and derived clay nanocomposite film. *Journal of Food Science*, *77*(12), N66–N73.

Rhim, J. W., Mohanty, K. A., Singh, S. P., & Ng, P. K. (2006). Preparation and properties of biodegradable multilayer films based on soy protein isolate and poly (lactide). *Industrial & Engineering Chemistry Research*, *45*(9), 3059–3066.

Rocca-Smith, J. R., Pasquarelli, R., Lagorce-Tachon, A., Rousseau, J., Fontaine, S., Aguié-Béghin, V., et al. (2019). Toward sustainable PLA-based multilayer complexes with improved barrier properties. *ACS Sustainable Chemistry & Engineering*, *7*(4), 3759–3771.

Ross, G., Ross, S., & Tighe, B. J. (2017). Bioplastics: new routes, new products. In *Brydson's plastics materials* (pp. 631–652). Butterworth-Heinemann.

Safaei, M., & Azad, R. R. (2020). Preparation and characterization of poly-lactic acid based films containing propolis ethanolic extract to be used in dry meat sausage packaging. *Journal of Food Science and Technology*, *57*(4), 1242–1250.

Sahraee, S., Milani, J. M., Ghanbarzadeh, B., & Hamishehkar, H. (2020). Development of emulsion films based on bovine gelatin-nano chitin-nano ZnO for cake packaging. *Food Science & Nutrition*, *8*(2), 1303–1312.

Salgado, P. R., López-Caballero, M. E., Gómez-Guillén, M. C., Mauri, A. N., & Montero, M. P. (2013). Sunflower protein films incorporated with clove essential oil have potential application for the preservation of fish patties. *Food Hydrocolloids*, *33*(1), 74–84.

Sangkasanya, S., Naknaen, P., & Petchwattana, N. (2018). Feasibility application of biodegradable active films produced by poly (lactic acid) incorporating with limonene as an antimicrobial packaging for fresh beef meat. In *Proceedings of the 2018 8th International Conference on Bioscience, Biochemistry and Bioinformatics* (pp. 60-64).

Sangsuwan, J., Rattanapanone, N., & Pongsirikul, I. (2015). Development of active chitosan films incorporating potassium sorbate or vanillin to extend the shelf life of butter cake. *International Journal of Food Science & Technology*, *50*(2), 323–330.

Sangsuwan, J., Rattanapanone, N., & Rachtanapun, P. (2008). Effect of chitosan/methyl cellulose films on microbial and quality characteristics of fresh-cut cantaloupe and pineapple. *Postharvest Biology and Technology*, *49*(3), 403–410.

Satam, C. C., Irvin, C. W., Lang, A. W., Jallorina, J. C. R., Shofner, M. L., Reynolds, J. R., et al. (2018). Spray-coated multilayer cellulose nanocrystal-chitin nanofiber films for barrier applications. *ACS Sustainable Chemistry & Engineering*, *6*(8), 10637–10644.

Shreaz, S., Wani, W. A., Behbehani, J. M., Raja, V., Irshad, M., Karched, M., Ali, I., et al. (2016). Cinnamaldehyde and its derivatives, a novel class of antifungal agents. *Fitoterapia*, *112*, 116–131.

Sin, L. T., & Tueen, B. S. (2019). *Polylactic acid: a practical guide for the processing, manufacturing, and applications of PLA*. William Andrew.

Sogut, E., & Seydim, A. C. (2019). The effects of chitosan-and polycaprolactone-based bilayer films incorporated with grape seed extract and nanocellulose on the quality of chicken breast fillets. *LWT-Food Science and Technology*, *101*, 799–805.

Song, S., Wang, Y., Liang, M., Qi, X., Yang, J., Jin, Y., et al. (2018). Mechanical and gas barrier properties of poly (L-lactic acid) by plasma-enhanced chemical vapor deposition of SiO x. *Polymer-Plastics Technology and Engineering*, *57*(6), 581–590.

Sopajittawattana, T., Pijitt, P., Assavavichai, A. & Harnkarnsujarit, N. (2017). Application of activated charcoal in active packaging for seafood odor scavenging. *Proceeding of the 4th Packaging and Materials Innovation Symposium (PMIS 2017)*. Bangkok, Thailand.

Sousa, G. M., Yamashita, F., & Soares Júnior, M. S. (2016). Application of biodegradable films made from rice flour, poly(butylene adipate-co-terphthalate), glycerol and potassium sorbate in the preservation of fresh food pastas. *LWT-Food Science and Technology*, *65*, 39–45.

Srisa, A. (2020). Antifungal film based on cinnamaldehyde incorporated polylactic acid and poly(butylene adipate-co-terephthalate) blends (PLA/PBAT) for shelf-life extension of bakery products. Master Thesis, Kasetsart University.

Svagan, A. J., Åkesson, A., Cárdenas, M., Bulut, S., Knudsen, J. C., Risbo, J., et al. (2012). Transparent films based on PLA and montmorillonite with tunable oxygen barrier properties. *Biomacromolecules*, *13*(2), 397–405.

Takala, P. N., Salmieri, S., Boumail, A., Khan, R. A., Vu, K. D., Chauve, G., et al. (2013). Antimicrobial effect and physicochemical properties of bioactive trilayer polycaprolactone/methylcellulose-based films on the growth of foodborne pathogens and total microbiota in fresh broccoli. *Journal of Food Engineering*, *116*(3), 648–655.

Talebi, F., Misaghi, A., Khanjari, A., Kamkar, A., Gandomi, H., & Rezaeigolestani, M. (2018). Incorporation of spice essential oils into poly-lactic acid film matrix with the aim of extending microbiological and sensorial shelf life of ground beef. *LWT-Food Science and Technology*, *96*, 482–490.

Tunma, S. (2018). Starch based nanocomposites in active packaging for extended shelf life of fresh fruits. *Walailak Journal of Science and Technology (WJST)*, *15*(4), 273–281.

Verbeek, C. J., & van den Berg, L. E. (2010). Extrusion processing and properties of protein-based thermoplastics. *Macromolecular Materials and Engineering*, *295*(1), 10–21.

Vytejčková, S., Vápenka, L., Hradecký, J., Dobiáš, J., Hajšlová, J., Loriot, C., et al. (2017). Testing of polybutylene succinate based films for poultry meat packaging. *Polymer Testing*, *60*, 357–364.

Wang, K., Lim, P. N., Tong, S. Y., & Thian, E. S. (2019). Development of grapefruit seed extract-loaded poly(ε-caprolactone)/chitosan films for antimicrobial food packaging. *Food Packaging and Shelf Life*, *22*, 100396. https://doi.org/10.1016/j.fpsl.2019.100396.

Winotapun, C., Phattarateera, S., Aontee, A., Junsook, N., Daud, W., Kerddonfag, N., et al. (2019). Development of multilayer films with improved aroma barrier properties for durian packaging application. *Packaging Technology and Science*, *32*(8), 405–418.

Wu, J., Sun, X., Guo, X., Ge, S., & Zhang, Q. (2017). Physicochemical properties, antimicrobial activity and oil release of fish gelatin films incorporated with cinnamon essential oil. *Aquaculture and Fisheries*, *2*(4), 185–192.

Xu, Y., Rehmani, N., Alsubaie, L., Kim, C., Sismour, E., & Scales, A. (2018). Tapioca starch active nanocomposite films and their antimicrobial effectiveness on ready-to-eat chicken meat. *Food Packaging and Shelf Life*, *16*, 86–91.

Yun, X., Qi, X., Zhang, Y., Song, S., & Dong, T. (2019). Application of SiO x-coated poly (ε-caprolactone) film for preservation of cherry tomato. Polymers and Polymer Composites, 0967391119874558.

Yun, X., Wang, Y., Li, M., Jin, Y., Han, Y., & Dong, T. (2017). Application of permselective poly (ε-caprolactone) film for equilibrium-modified atmosphere packaging of strawberry in cold storage. *Journal of Food Processing and Preservation*, *41*(6), e13247.

Zeid, A., Karabagias, I. K., Nassif, M., & Kontominas, M. G. (2019). Preparation and evaluation of antioxidant packaging films made of polylactic acid containing thyme, rosemary, and oregano essential oils. *Journal of Food Processing and Preservation*, *43*(10), e14102.

Zhai, X., Shi, J., Zou, X., Wang, S., Jiang, C., Zhang, J., et al. (2017). Novel colorimetric films based on starch/polyvinyl alcohol incorporated with roselle anthocyanins for fish freshness monitoring. *Food Hydrocolloids*, *69*, 308–317.

Zhang, K., Huang, T. S., Yan, H., Hu, X., & Ren, T. (2020). Novel pH-sensitive films based on starch/polyvinyl alcohol and food anthocyanins as a visual indicator of shrimp deterioration. *International Journal of Biological Macromolecules*, *145*, 768–776.

Zhang, R., Lan, W., Ding, J., Ahmed, S., Qin, W., He, L., et al. (2019). Effect of PLA/PBAT antibacterial film on storage quality of passion fruit during the shelf-life. *Molecules*, *24*(18), 3378.

Zhang, Y., Zhang, Y., Zhu, Z., Jiao, X., Shang, Y., & Wen, Y. (2019). Encapsulation of thymol in biodegradable nanofiber via coaxial eletrospinning and applications in fruit preservation. *Journal of Agricultural and Food Chemistry*, *67*(6), 1736–1741.

Zhao, Y., Teixeira, J. S., Saldaña, M. D., & Gänzle, M. G. (2019). Antimicrobial activity of bioactive starch packaging films against *Listeria monocytogenes* and reconstituted meat microbiota on ham. *International Journal of Food Microbiology*, *305*, 108253.

Zhou, S., Jin, T., Sheen, S., Zhao, G., Liu, L., Juneja, V., et al. (2020). Development of sodium chlorite and glucono delta-lactone incorporated PLA film for microbial inactivation on fresh tomato. *Food Research International*, *132*, 109067.

Zhou, X., Yang, R., Wang, B., & Chen, K. (2019). Development and characterization of bilayer films based on pea starch/polylactic acid and use in the cherry tomatoes packaging. *Carbohydrate Polymers*, *222*, 114912.

Chapter 8

Intelligent Packaging

Sustainable food processing and engineering challenges

Mariusz Tichoniuk*, **Marta Biegańska and Ryszard Cierpiszewski**
Department of Industrial Product Quality and Packaging Development, Institute of Quality Science, Poznań University of Economics and Business, Poznań, Poland
**Corresponding author*

1 Introduction

Food processing and packaging in line with sustainable attitude should ensure that the resources required for foodstuff production and supply are used most efficiently (Dobon et al., 2011). Although the food product sustainability is widely described logically, food manufacturers and distributors require clear solutions and practical examples of implementing this philosophy in the area of their business operations (Akkerman et al., 2010; Holden et al., 2018; Woodhouse et al., 2018; Cabot et al., 2019). Intelligent packaging (IP) provides many different opportunities for the implementation of the principles of sustainable development into food processing, but often their application also involves additional resources and engineering challenges (Wang, 2019).

Agricultural commodities require sustainable processing and prevention so that the resources used for cultivation or breeding them will not be wasted during food production and distribution. For this reason, among other things, food waste becomes an essential aspect of economics and ecology (Poyatos-Racionero et al., 2018). Food and Agriculture Organization of the United Nations (FAO) reports that one-third of all food produced worldwide for human consumption is lost or wasted each year. It generates losses in the global economy of nearly $ 1,000 billion a year and represents 7% of greenhouse gas emissions (Broekmeulen and van Donelaar, 2019; Schmidt Rivera et al., 2019). The more detailed data provided by FAO reveals that global food waste concerns the processing and distribution of various types of agricultural commodities. This fact leads to unacceptably high levels of loss of cereals (30% of the harvest), oilseeds (20%), root crops, fruits and vegetables (40%–50 %), meat and dairy products (20%), and fish (35%) (FAO, 2015). The type of food waste and loss strongly depends on the individual situation and local conditions in a particular region or culture.

Sustainable food processing and engineering challenges. http://dx.doi.org/10.1016/B978-0-12-822714-5.00008-5

The organizational and technical limitations during the harvest, storage and transport, food processing, packaging, and distribution are the main reasons for food loss in low-income countries. Small and medium fish and agricultural processing plants require special attention here. In the medium- and high-income countries, the most significant impact on the amount of food wasted have consumer behavior as well as legal and organizational requirements defining the priorities in a given sector. For example, improperly planned agricultural subsidies can lead to the overproduction of arable crops, at least some of which are lost or wasted. Incorrectly defined legal provisions regarding the safety and quality of food may force the participants of the food supply chain to the disposal of food products that are still safe for consumers.

On the other hand, proper planning of purchases by food consumers and consumption of food supplies before their expiry date would help to avoid unnecessary food waste (FAO, 2015). Due to various factors that can be controlled by carefully selected packaging in the supply chain, around 88 million tons of food are wasted annually in the European Union (Buisman et al., 2019). The stages where food losses arise due to improper storage, transport, or processing are primary production (11% of all wasted food), food processing (19%), retail and wholesale (5%), food service and catering (12%), and household consumption (54%) (ActInPack, 2018). The last indication shows how important it is to involve the final consumer in the sustainable approach in food distribution and consumption. The primary sources in food waste in retail refer to the expired shelf life of foods, an oversupply of products, and packaging appearance that is unattractive to consumers (Müller and Schmid, 2019).

It is also worth mentioning in the view of sustainability that according to Adekomaya et al., 40% of food products being produced require refrigeration and cold chain logistics (transport of food) uses 15% of world fossil fuel energy. Food preservation, especially of perishables, consumes energy for keeping it under refrigerated conditions. Improving the food transport system could lead to the reduction of fossil fuel consumption, decrease of carbon dioxide emission, and slow down global warming (Adekomaya et al., 2016).

An efficient flow of information about a packaged food product and its environment is one of the most critical aspects of sustainable development in the food supply chain (Dobon et al., 2011; Heising et al., 2017). The main functionality of packaging in the food distribution chain is related to the product protection against contamination, external environment, and mechanical damage during storage and transportation. The packaging also provides logistic utilities and convenience in product application by each participant of the distribution chain by ensuring unwanted losses and enable efficient tampering of the loading units (Wang, 2019). Also, packaging carries information about a product that enables tracking in the supply chain and provides necessary information about food such as product content, nutritional value, best before date, and recommendations on consumption and storage conditions. The communication function of packaging is also associated with marketing tasks that help to sell a desired amount of products and indirectly reduce the waste of unused products (Kalpana et al., 2019). The effectiveness of packaging applied in the food supply chain becomes more important about cost reduction, increase companies' competitiveness, and especially saving limited resources, including reduction of food waste. Conventional packaging has limited potential ineffective transfer of information about the product and its surroundings because they are not able to provide a direct response to variable conditions in the food supply chain or changes of food product itself (Poyatos-Racionero et al., 2018). Traditional packaging provides only mechanical support for the packaged product and protects it from external conditions (e.g., water vapor, oxygen, light, insects, microbial contamination, and dirt). As such, it has only a passive role and allows only protection of the product from the outside environment. In order to change that passive role of packaging into an active one, that enables extending shelf life, the interaction between packaging and product is required (Lee and Rahman, 2014; Ghaani et al., 2016).

So-called smart materials and solutions extend the information function of conventional packaging enabling continuous collecting and processing information on packed products (e.g., freshness, ripeness) or their surrounding (e.g., temperature, humidity). Many review papers report on the types of intelligent packaging and their potential for use in the food distribution chain (Heising et al., 2017; Sohail et al., 2018; Kalpana et al., 2019; Müller and Schmid, 2019). The primary purpose of the application of intelligent packaging for any product focuses on one of two reasons (Lydekaityte and Tambo, 2020). The main goal is to increase the functionality of the product or the efficiency of its use by overcoming limitations in monitoring the current state of the product and its surroundings. The second aspect is to involve the consumer and other participants in the distribution chain in the active flow of information about the packed product and its properties. IP systems like the most popular time-temperature indicators (TTIs) or integrity indicators, as well as still developed food freshness indicators, can respond directly to significantly changed conditions which affect the quality and safety of packaged food (Shaefer and Cheung, 2018; Tichoniuk, 2019). These groups of intelligent packaging can be applied to determine the real end-of-life point for food products considering the conditions within the distribution chain regardless of the initially specified expiration date. It makes the transfer from the so-called fixed shelf life (FSL) attitude to the dynamic shelf life (DSL) approach possible. The indicators implemented in IP react on actual packed product changes and its surrounding, which frequently results in its deterioration. The adjustable expiration date in the DSL system could help to avoid wasting food still suitable for consumption (Herbon et al., 2012; Albrecht et al., 2019). Much better communication and more efficient traceability in the supply chain could be provided by new data carriers such as Radio Frequency Identification (RFID) systems that can be coupled with sensors for ongoing monitoring of the condition of products and their surrounding (Müller and Schmid, 2019). Efficient communication about the state of food could reduce the probability of food waste at any stage of the supply chain (Ghaani et al., 2016; Cerqueira et al., 2018). The chapter presents a review of the challenges and opportunities in the application of intelligent packaging systems for sustainable food processing and distribution.

2 Definitions and types of intelligent packaging

IP is one of the newest technologies in the field of food packaging. Although it is still evolving and is not entirely commercially viable, this technology has great potential for improving food safety and traceability, and for maintaining their quality. It is also essential to improve the convenience of the consumer using the packed goods (Ghaani et al., 2016). With the growing interest in these packaging systems, many definitions of these types of solutions have appeared. Sometimes, instead of the term intelligent packaging for describing new packaging technologies, various terms were used, for example, "interactive," "smart," "clever," or "communicative." These phrases are sometimes given without definition or explanation. The introduction of other terms, however, causes further complications and confusing terminology (Yam et al., 2005; Dainelli et al., 2008). In scientific literature, the definition given by Yam and co-authors is often accepted. They define IP as a packaging system that can perform specific functions, for example, recording, tracking, detection, communication to facilitate decision making aimed at extending the shelf life of food, improving its quality, providing information, and increasing consumer safety (Yam et al., 2005).

The definition of IP was also established in the framework regulations (EC) No 1935/2004 of the European Parliament and of the Council of 27 October 2004 on materials and articles intended to come into contact with food and COMMISSION REGULATION (EC) No 450/2009 of 29 May 2009 on active and intelligent materials and articles intended to come into contact with food. According to these regulations, "intelligent materials and articles" mean materials and articles which monitor the condition of packaged food or the environment surrounding the food. It is clear from this term that IP is intentionally designed to collect and display information in the food supply chain, and not to extend the shelf life directly. IP system monitors the internal or external environment of the packaging, collects information about temperature, integrity, freshness of packaged food, record changes, and then provides information to manufactures, retailers, and consumers (Yam et al., 2005; Chowdhury and Morey, 2019). Such a technique can be used, for example, in the meat industry (Fang et al., 2017).

The use of IP in poultry packaging can improve the safety and quality of poultry products, as well as maintain the development of the poultry industry. By monitoring the conditions inside the packaging of poultry products, as well as the environment surrounding them, such products have great potential in improving food safety and maximizing customer satisfaction (Chowdhury and

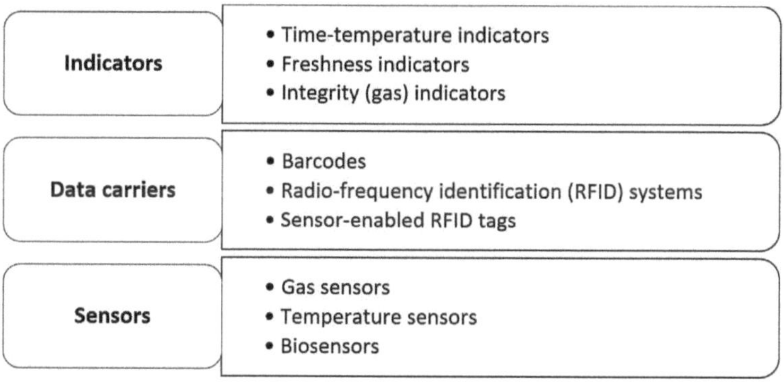

FIGURE 8.1 Basic types of intelligent packaging systems used in the food supply chain. *(Source: Own work based on Ghaani et al., 2016; Cerqueira et al., 2018; Sohail et al., 2018; Tichoniuk, 2019.)*

Morey, 2019. The IP system is usually based on attaching an external device to the final packaging, which is in the form of small, inexpensive labels or tags that can acquire, store and convey information about the functions and properties of packaged food. Based on these functions, IP technologies can be divided into three broad categories: indicators, data carriers, and sensors (Fang et al., 2017; Lee et al., 2015). The scheme for such classification is given later, but it should be remembered that there are no sharp boundaries between individual types of IP systems (Fig. 8.1).

The most directly acting IP systems are indicators. They provide information related to the presence or absence of some reaction, its intensity, or the concentration of target compounds influencing the condition of packaged products (Ahmed et al., 2018). Most often, such information is communicated by visual changes, for example, color change or dye diffusion (O'Grady and Kerry, 2008). The indicators can be introduced inside or outside the package depending on the parameter monitored (Ghaani et al., 2016). The most applicable systems in the food supply chain are time-temperature indicators (TTIs), integrity (gas) indicators, and freshness indicators. Table 8.1 shows the area of IP application and gives examples of commercially available IP systems.

TTIs are simple, inexpensive labels attached to the package. There are three types of temperature indicators: critical temperature indicators (CTI), critical temperature/time integrators (CTTI), time or temperature integrators, or indicators (TTI) (Taoukis, 2001). The first group, that is, critical temperature indicators, shows if the ambient temperature has exceeded the permitted temperature. At the same time, the excess temperature lasted for some time (from several minutes to several hours), that is, for a period sufficient to change selected parameters of the indicator. The second type of indicator shows whether the ambient temperature has exceeded the critical (permitted) temperature and the whole

TABLE 8.1 Commercially available intelligent packaging indicators.

Type of indicators	Area of application	Examples of commercial indicators
Time-temperature indicators (TTIs)	Meat, poultry, fishery products (especially at cooling or freezing temperature) Beverages with recommended temperature of consumption	Monitor Mark (3M Company) On Vu (Freshpoint/BASF) VITSAB (Vitsab Internat.) Fresh Check (Temptime Co) TopCryo (TRACEO)
Integrity (gas) indicators	All packaged foods—especially in aseptic packaging or with modified atmosphere system	Ageless Eye (Mitsubishi Gas Chemical Co) Insignia after opening freshness timer/pull-timer (Insignia Technologies)
Freshness indicators	Perishable food products, for example, meat, fishery products, dairy products, vegetables, and fruits	Freshpoint (Freshpoint Quality Assurance Ltd) Fresh Tag (COX Technologies) ripeSense (ripeSense) SensorQ (DSM NV & Food Quality Sensor Intern. Inc.)

Own work based on Poyatos-Racionero et al., 2018; Cerqueira et al., 2018; Tichoniuk, 2019.

history of its changes above the critical temperature. Therefore, they indicate the amount of energy supplied to the indicator, which is equivalent to a temperature exceeding the assumed critical temperature. The third type of TTI gives the ability to track the influence of any temperature on a product throughout its shelf life. They monitor the full history of temperature from the moment they are activated (Taoukis, 2001). The application of TTIs in the food supply chain can help avoid adverse storage and transport conditions for food, thereby reducing its waste.

During transport and storage of food, the temperature is mainly controlled by data loggers, most often measuring the temperature of the food environment. Smart wireless autonomous microsystems such as RFID systems have been developed to monitor temperature within the supply chain. Although this technology has many advantages, the temperature can only be monitored during transport and storage from manufacturers to retailers. Once the consumer has purchased the goods, temperature measurement in this way is no longer possible. However, most often breaks in the cold chain occur on the way from the seller to consumer. Therefore, systems that allow temperature monitoring also at this stage are of great importance (Kreyenschmidt et al., 2010).

3 Time and temperature indicators (TTIs)

Indicators are devices that show the presence or absence of a target substance. This fact is usually indicated through a characteristic change, like a change in color (Lee and Rahman, 2014; Ghaani et al., 2016; Müller and Schmid, 2019). Temperature plays a significant role in shelf life and quality of food, especially when it needs to be stored in chilled or frozen conditions. Temperature fluctuations over time contribute to the loss of nutritional and sensory qualities of perishable foods (Lee and Rahman, 2014; Ahmed et al., 2018; Gao et al., 2020). In order to help aid this problem and enable monitoring of the temperature fluctuations within the supply chain, TTIs were developed.

TTIs are the most popular IP systems applied for food products because of their simple, user-friendly forms, and relatively low costs of the application. They usually take the form of labels providing easy to read the information (e.g., by changing the color) in case of the excess of temperature conditions that could reduce the quality and safety of food (Pereira et al., 2015; Wang et al., 2015). They are based on a different mechanism, such as:

- mechanical,
- chemical,
- electrochemical,
- enzymatic,
- microbiological reactions that are irreversible and easy to read through a visible color change of the label or part of it.

TTIs show either cumulative information on the entire temperature history or a partial-history showing when the temperature to which the product was exposed exceeded a predetermined threshold. The visible TTI's color change is irreversible, and the rate of that change increases at higher temperatures to reflect the deterioration of the product (Taoukis, 2001; Taoukis and Labuza, 2003; Lee and Rahman, 2014; Ghaani et al., 2016; Müller and Schmid, 2019; Gao et al., 2020). The length of time from activation of an indicator to its termination reflected by a color change corresponds to the shelf life of the packaged food. Time and temperature play a vital role in the dynamics affecting food spoilage, thus using TTI labels in the supply chain can help to monitor and control those factors and reduce food waste and inform consumers about the remaining shelf life of products (Taoukis, 2010; Gao et al., 2020). Food waste can be reduced by better matching the shelf life with the actual state of food. They affect microbiological and sensory quality, among other conditions, for the production, processing, and storage of food products. For a better estimate of the shelf life, it should be possible to predict microbial growth in a given product accurately. Processor defined shelf life may not accurately determine this period; as a rule, it is intentionally underestimated, to address potential abuse within the supply chain and to ensure its quality and security, which is important, especially when we look at shelf life from the consumers' perspective (Lin et al., 2016). As the TTIs responses are visible color changes, they are easy to observe even by end-consumers, and with the development of optics and image processing, those TTIs color measurements can be done using smartphones. This in turn can facilitate their proper use by consumers (like color-blind people), but also within the entire distribution chain and help improve online workflow and dynamic shelf-life calculation (Gao et al., 2020). Besides, a reliable and easy to use and read TTI should have a relatively flexible range of configuration and provide a reading for a useful time. Such an indicator must be resistant to ambient conditions, nontoxic, and for economic reasons should have a long shelf life before its activation. Moreover, due to heterogeneous structures of international food chains, it must be easily adapted to them, and be cost-effective and easy to use and handle/read (Taoukis, 2001; Kreyenschmidt et al., 2010).

3.1 Essential technologies of TTI construction

During the last 20 years, commercial applications of different TTIs have been increasing. A large number of them have been discussed and evaluated in the literature. Although, there are many TTIs described in the literature and available commercially, they can all be divided into three main groups: the ones based on a polymerization reaction, diffusion-based, and enzymatic indicators (Taoukis, 2001; Chun et al., 2013; Lee and Rahman, 2014; Ghaani et al., 2016; Müller and Schmid, 2019; Gao et al., 2020). An essential aspect for the practical application of TTIs is their proper response to temperature changes as well as compliance of its indications/readings with the product's deterioration kinetics (its actual state at the moment of reading) (Albrecht et al., 2019).

The activation energy is the minimum energy level at which a chemical reaction occurs. Therefore the activation energy of TTI and a given food product must coincide. If this condition is met, TTIs can be used on a larger scale to assess the quality of food products. Moreover, the product shelf life also has to be known to select suitable TTIs, because the response time should correspond to the shelf life of the product (Lu et al., 2013; Ahmed et al., 2018). Different methods of kinetic modeling are used to guarantee the best-suited TTI indications for changes in food quality. The most popular is the Arrhenius model, but it is not without flaws, because there are deviations from the linearity of data and insufficient data, which in turn translates into the poor matching of TTI indications. These limitations have influenced the development of new models, such as the fuzzy reasoning-based model or the reparametrized Arrhenius model. They allow increasing the goodness-to-fit of TTIs for a more accurate assessment of food quality (Gao et al., 2020). TTIs can also be integrated with RFID tags or bar codes. This solution improves their performance, increases functionality, and convenience of use in comparison with standard TTI labels (Ahmed et al., 2018).

3.1.1 *Diffusion-based indicators*

Diffusion-based TTIs usually consist of a porous matrix and a color dye (chemical compound, e.g., fatty acid ester) that melts within a given temperature range.

This temperature-induced melting of the dye chemical causes its diffusion in the porous matrix (e.g., high-quality blotting paper) and results in color band development, which reflects cumulative information of temperature (Taoukis, 2001; Gao et al., 2020). Depending on the dye chemical used (its melting temperature range and concentration at the origin), the threshold temperature and the speed of diffusion can be adjusted to different food storing conditions, to reflect food spoilage most accurately. In indicators of this type, the activation occurs through the adhesion of two materials. They can be stored separately for a relatively long time at ambient temperature before being used. The TTI is activated by adhesion of the two materials that before use can be stored separately for a long period at ambient temperature. Commercially available diffusion-based TTIs are Monitor Mark by 3M (USA) (Taoukis, 2001) (Fig. 8.2A) or Tempix (Tempix AB, Sweden) (Fig. 8.2B,C). Lee and Rahman (2014) described it as a rectangular flat laminate containing layers of paper, film, adhesive, or other components and is used to monitor the storage and transportation conditions of food products. One end of the porous wick track strip is placed over a reservoir pad containing a blue-dyed chemical that has the desired melting point. A removable activation strip acts as a barrier between the indicator and the reservoir before activation. These components are covered by a paperboard layer containing viewing windows and are topped with a clear protective film. Before using

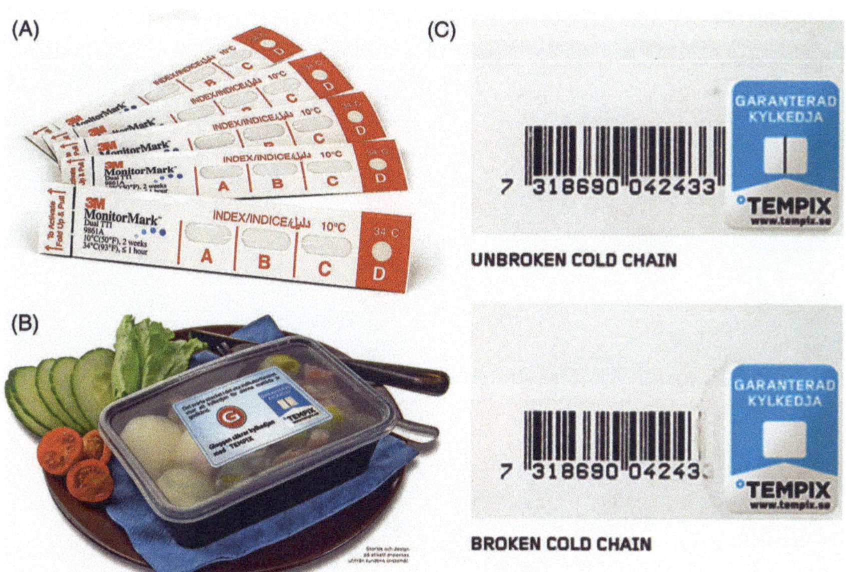

FIGURE 8.2 Examples of diffusion-based time-temperature indicators. MonitorMark by 3M (A), Tempix TTI label (B), Tempix label when the cold chain was and was not broken (C) *(Source: 3M (2020) (A); Tempix TTI (2020) (B); Tempix (2020) (C).)*

the activation strips, they must be conditioned to prevent a premature response. This is usually done by keeping the indicator for a minimum of 2 h at or below a certain temperature (in a fridge or freezer)—indicator's threshold above which diffusion of the dye. This action ensures that the response chemical is in a solid-state and is ready for use. The reservoir pad and the end of the porous wick indicator track come in contact upon removal of the activation strip. When exposed to temperatures above the critical response temperature, the chemical in the reservoir starts to melt (expressed as a blue color) and moves along the graduated device in a temperature-dependent manner.

3.1.2 Enzymatic indicators

An example of an enzymatic TTI is CheckPoint by Vitsab International AB (Sweden) (Fig. 8.3A). It is in the form of an adhesive label that can be attached to the packaging. This TTI is based on a pH-dependent color change, which results from controlled enzymatic hydrolysis of a lipid substrate (US Patents 4,043,871 and 4,284,719). CheckPoint consists of two distinct sealed compartments (pouches) containing a mixture of an aqueous solution of lipolytic enzyme and a pH-indicating dye in one pouch. The second pouch contains the lipid substrate absorbed in a pulverized PVC carrier and suspended in an aqueous medium. Compounds like glycerin, tricapronate (tricaproin), tripelargolin, tributyrin, and mixed esters of polyvalent alcohols and organic acids, can be used as

FIGURE 8.3 CheckPoint L5-8 Smart TTI Seafood Label before activation *(white)* and activated *(green)* (A). CheckPoint label L5-8 Smart TTI Seafood Label showing thermal exposure over the safe levels recommended by the FDA (B). *(Source: TTI Label, 2020.)*

substrates. In order to vary the response based on different temperature dependencies, different types of enzyme-substrate combinations and concentrations can be used (Lee and Rahman, 2014; Meng et al., 2018). To activate the label, pressure needs to be applied to the plastic mechanically (manually or by online automation), which breaks the barrier between the pouches causing the mixing of enzyme and substrate. The next step is the subsequent reaction that promotes the degradation of the substrate and liberates fatty acids. This chemical reaction results in pH decrease and the whole process can be observed on the label in the form of a color change. The round field of the indicator changes its color from deep green to bright yellow or orange-red (Fu and Labuza, 1992; Ellouze and Augustin, 2010; Taoukis, 2010; Eden et al., 2011; Lee and Rahman, 2014) (Fig. 8.3B). CheckPoint labels maintain long shelf life if kept chilled before activation (Koutsoumanis and Gougouli, 2015).

3.1.3 Polymer-based TTIs

TTIs can also be based on a solid-state polymerization reaction, for example, Fresh-Check (Temptime Corporation, USA) (Fig. 8.4). In this indicator, the polymerization ability of disubstituted diacetylene crystals is utilized. This reaction occurs during the reaction of the controlled crystal lattice in the solid-state, resulting in a colored polymer. As a result of temperature changes over time, this polymer changes color and becomes darker. So reading this TTI is by comparing the color of the active center of the label with the color of the surrounding ring, which is the reference color. These indicators are active from the time of their production and in order to slow down their response prior application significantly, they need to be stored at low freezing temperatures (Fu and Labuza, 1992; Ellouze and Augustin, 2010; Taoukis, 2010; Eden et al., 2011; Lee and Rahman, 2014; Koutsoumanis and Gougouli, 2015).

FIGURE 8.4 Fresh-Check TTI showing when the product is safe and not safe for consumption. *(Source: Fresh-Check, 2020.)*

3.2 Examples of commercial TTIs application

However, there has been a significant development in the cold chain logistics over the past decades; temperature control during distribution and storage is still a significant problem of perishable foods. TTIs can be used as temperature-dependent monitoring tools in these chains. By reflecting the history of food temperature in real-time, they can contribute to predicting the remaining shelf life. Thus, TTIs have more advantages than those commonly used so far printed "sell-by-date" labels. This fact is, among other things, due to a mismatch between the printed best-before date and the real quality of the product (Albrecht et al., 2019). Gao and co-authors distinguished several factors responsible for the successful use of TTIs on a large scale. They pointed to their simplicity, low cost, reliability, accuracy, but also on relevant laws and policies (of enterprises, states). Another critical factor in this aspect is consumer acceptance of this type of smart labels. At the root of the vast potential for broad commercialization of TTIs also lies a reduction in costs and increased accuracy (Gao et al., 2020). Although the application of TTIs in food products has increased due to their many advantages, there are still some critical points that need to be kept in mind by manufacturers related to their use. If they are to be used as indicators for assessing the quality (suitability for consumption) of foods, they must be appropriately calibrated, so that their response coincides with the deterioration of food products. It is also necessary to include, among other things, the chemical, physical, microbiological, and sensory properties of products to which they are applied (e.g., pH, color, sensory evaluation, total microbial count, VBN, TBA, or water holding capacity) which are subject to change depending on storage period and processing method of particular food products (Chun et al., 2013).

A system monitoring temperature conditions based on TTIs would go in line with already existing Hazard Analysis Critical Control Point method (HACCP) and Good Manufacturing and Hygiene Practices as monitoring of critical control points may be achieved by using TTI labels/tags. The TTI tag should then be selected or pre-adjusted so that the activation energy required to produce a color change is similar to that required to initiate significant deterioration in the food being monitored. Such a system would also help in facilitating the scheduling of product distribution employing least shelf life first-out (LSFO) or shortest-remaining shelf life (SRSL) policies. At the heart of those policies is the idea that the products approaching the end of their shelf life are moved first at the warehouse and in the retail system (Fu and Labuza, 1992; Taoukis, 2010; Eden et al., 2011). A decrease in food waste and better use of resources goes in line with the EU's Sustainable Consumption and Production and Sustainable Industrial Policy (SCP/SIP) Action Plan (Commission 2008). As a result of many studies, TTIs are a useful tool for monitoring the temperature history of food products in cold supply chains. Their widespread use can contribute to reducing food waste, enabling more accurate monitoring of the cold chain and cold storage, and ensuring increased food safety (Gao et al., 2020). Different from the optimal temperature and prolonged duration of adverse conditions have a significant impact on the generation of food waste in the supply chain, in particular in case of products sensitive to temperature changes (e.g., frozen food and fresh vegetables or fruits) (Mohebi and Marquez, 2015; Fang et al., 2017). Unfavorable temperature conditions affect the monitored product (or its environment) as well as the TTI used and accelerate the reaction (chemical, enzymatic, electrochemical, mechanical, or microbiological), which is the basis of its activity and provide recognizable outcome such as color change (Sohail et al., 2018).

3.2.1 Dairy products

Dairy products are an example of perishable foods—those likely to spoil, decay, or become unsafe to consume if not kept refrigerated or frozen. When it comes to their quality, it strongly depends on several factors, including conditions of processing, storage, and transportation. Due to temperature fluctuations, microbial growth, protein denaturation, fat acid decay, and carbohydrate fermentation occurs, which inevitably affect the deterioration of these products. For this reason, crucial in reducing losses along the distribution chain and during production is monitoring and controlling temperature (Gao et al., 2020). Claeys and co-authors reviewed different intrinsic TTIs for thermally processed milk. They took into account the kinetics of TTIs responses, and microbial growth of milk pathogens, different heat treatment processes of milk (e.g., pasteurization, sterilization, UHT) (Claeys et al., 2002). Another study by de Fazio Aguiar and co-authors shows the development of an enzymatic TTI for high-temperature short-time (HSTS) pasteurization process of milk. The authors pointed out that an enzymatic TTI based on enzyme alkaline phosphatase from bovine intestinal mucosa (ALP) had the potential for the assessment of continuous thermal processes in the range 70°C–80°C (pasteurization conditions) (De Fazio Aguiar et al., 2012).

3.2.2 Meat products

Meat products are susceptible to microbial contamination during processing, storage, and transportation that can cause economic loss and pose food poisoning threats (Gao et al., 2020). They are rich in nutrients and are a source of essential exogenous amino acids (Ahmed et al., 2018). Some commercially available TTIs, for example, OnVu by Freshpoint (Switzerland) (Fig. 8.5) or an enzymatic TTI by Vitsab International AB (Sweden) can be used to monitor time-temperature fluctuations within the supply chain (Albrecht et al., 2019; Chun et al., 2013; Kreyenschmidt et al., 2010;). OnVu is used on all packages of Kneuss fresh chicken produced and distributed by Ernst Kneuss Geflügel A.G. (Switzerland), and the (eO) TTI is applied by Monoprix on packed fresh pork (Taoukis 2010). The Keep-it TTI is used by a Norwegian retail chain Rema 1000, an e-shop Kolonial.no, but also by Oslo Universitetssykehus (hospital), Gilde, Nortura, and Stange (meat and meat products manufacturers) (Biegańska, 2017). Moreover, TTIs can also be used for evaluating the sterilization and pasteurization process (Ahmed et al., 2018).

FIGURE 8.5 Examples of OnVu labels. *(Source: SEDEF, 2020.)*

3.2.3 Fish and fish products

TTIs can also be applied in the estimation of quality and remaining shelf life of frozen seafood during the whole cold chain (Gao et al., 2020). The CheckPoint (Vitsab International AB) is, for example, applied to vacuum- or modified atmosphere (MA) packed fresh seafood imported to the United States. The FDA's Import Alert #16-125 covers the import of these products. They are placed on detention by FDA unless their importers are on the Green List. One has to provide information to the FDA that controls are in place to prevent *Clostridium botulinum* toxin formation or need to provide a visual indication of a potential problem in order to qualify for the list. The Green list includes manufacturers, shippers, and importers. The L5-8 CheckPoint TTI response conforms to these requirements (Taoukis, 2010; Biegańska, 2017). Another Vitsab TTI (Fresh Check) was applied to turbot quality monitoring; in the study, five kinds of bacteria were investigated (ones that are related to turbot freshness), and among them, *Pseudomonas* spp. was confirmed to show the fastest growth. Whereas, the reliability of a photochromic TTI OnVu by Freshpoint (Switzerland) was evaluated to ensure the marketing of Cobia (*Rachycentron canadum*)(Gao et al., 2020). Godehav and Salmar (Norwegian fish and fish products manufacturers) use Keep-it TTI for monitoring the remaining shelf life of their products (Biegańska, 2017). Lidl Finland uses Tempix TTIs for securing the cold chain of their fresh fish brand ISOKARI (Lidl Finland, 2020).

3.2.4 Vegetables

Fresh or ready-to-eat (RTE) vegetables that require different storing temperatures are stored in walk-in-coolers, cooled food displays, or mixed produce freight containers that are adjusted to a specific temperature (too low or too high for some of the stored products) sometimes greater or equal to 9°C (Kou et al., 2014; Koutsoumanis and Gougouli, 2015; Badia-Melis et al., 2016). One of the factors which can significantly influence the sensory attributes and nutritional values of chilled or frozen vegetables during the cold chain is temperature fluctuation (Gao et al., 2020). Commercialized enzymatic TTI (Vitsab

International AB) was applied to check mushroom quality (Agaricus bisporus Pilát). The tests were done at constant, and variable temperature conditions and responses of both mushrooms and the selected TTI were measured (Bobelyn et al., 2006).

It is noteworthy that the majority of TTI applications are within the meat and fishing industry. Presumably, this is the effect of some balance between accuracy, cost, and consumer acceptance. On the other hand, higher prices and added values of meat and fishery products than that of dairy, fruit, and vegetables, can offset the cost of TTIs, which in turn can result in high consumer acceptance (Gao et al., 2020). According to Choi et al. (2020), more than US$50 billion annual cost society foodborne illnesses, which result in more than 120, 000 hospitalizations and 3000 deaths only in the United States. According to the World Health Organization, almost 1 in 10 people fall ill after eating contaminated food worldwide. The World Bank also noted that there is a cost of US$ 110 billion because of lost productivity and medical expenses annually in low- and middle-income economies as a result of unsafe food (Jafee, 2019).

3.3 Overview of TTIs strengths and weaknesses

The requirements of the food industry limit TTIs capability to fulfill them because TTIs still meet some challenges. One of them is focusing on TTIs modeling as a single quality indicator, and not considering other relevant quality attributes which change during food deterioration. This approach can be observed in existing researches. This multi-quality-indicator modeling should be comprehensively considered and integrated. Such manner of studies should then result in a single TTI, which could precisely reflect multiple quality indicators, resulting in the integration of multiple quality parameters, and providing versatile information of the food shelf life.

Most of the known TTIs, require many chemicals and relevant, unique materials for their preparation. The complexity of the manufacturing process can increase the cost, and cause difficulty during manufacturing. In order to avoid the food safety issue, TTIs should be developed/improved to decrease their complexity and reduce cost, as well as to obtain simplification. Only when those requirements are met, TTIs could benefit the food industry, and the broader availability of safer and cheaper TTIs could motivate the industry to widely adopt this smart technology (Gao et al., 2020).

In addition, diffusion-based TTIs are not as sensitive to temperature as the factors responsible for food spoilage. Moreover, they require rigid packaging, which protects the porous structure from external impact because they are not flexible. What is more, they need to be modularized in a durable box to separate the colored materials from the porous medium, which in turn makes them inefficient and expensive. Another drawback is the need to physically separate the thermal sensor and display element because they are sensitive to external shocks that may prevent their proper operation. Besides, the TTI sensor alone is sensitive to temperature changes, which translates into its effectiveness because the temperature does not affect the operation of the display. Not only that, but the very structure and operation of the indicator raise some reservations. Consumers, in turn, are afraid of potential exposure to the chemical compounds indicator is made of. Their concerns are related to the potential leakage of these substances into food. Not without significance is the fact that this type of TTI requires a precise application of heat to its chemical element to make it work properly. This fact is quite cumbersome and can be an obstacle to large-scale implementation. Partial heating or a temperature gradient can provide an incorrect result; it also requires a thick hard case (Choi et al., 2020).

TTIs are widespread for fresh food products, perishables, meat and meat products, fish, and seafood, dairy products, and mushroom stored under chilled or frozen. The indicators for real-time visual monitoring of industrial thermal processes (e.g., pasteurization) are still limited. The use of TTIs in food labels during pasteurization of meat products can offer the client, a safer, more interactive food with nominal shelf life ensured. On the other hand, the manufacturer is looking for the possibility of identifying inefficiencies during heat treatment, which allows reducing costs and errors, improving the performance in obtaining the product, and thereby increasing the profits (Brizio and Prentice, 2015). Although deviations from the desired temperature or meeting a specific temperature range in food processing, and within the cold chain can be monitored utilizing TTIs, there are challenges that they need to face. Some perishable food products (e.g., fruit and vegetables) are also sensitive to temperatures below their optimum range. Storage in lower temperatures than those required can lead to damage. Despite many advantages, TTIs are not developed yet, to detect such exposure (Sharrock, 2012). Deterioration of perishable food products may yield waste that has to be disposed of. Accordingly, the cost of disposal plays a vital role in the industry, therefore having an impact on their operations and giving a reason to consider it for modeling inventory (Suárez Díaz et al., 2020). The application of IP can help facilitate this process and help reduce product deterioration and losses.

4 Food freshness indicators

The assurance of food quality and safety is the desire of every participant in the food supply chain (Fuertes et al., 2016). The direct information about the current state of packaged food (its ripening or spoilage) can be provided by freshness indicator, which has this advantage over correlative markers like TTI or integrity indicators (Sharrock, 2012). The freshness indicators are designed to be sensitive to metabolites released during the food ripening or spoilage (Tichoniuk et al., 2017a). The selection of target metabolites depends on the type of monitored food products, their chemical content, and the mechanisms of chemical changes of the product (Fang et al., 2017; Poyatos-Racionero et al., 2018). The increase of the concentration of such chemical compounds like different organic acids (e.g., acetic acid or lactic acid), ethanol, glucose, carbon dioxide, volatile nitrogen compounds, or sulfur compounds indicates the food spoilage mainly associated with the growth of microorganisms (Sohail et al., 2018). The freshness indicators are designed to be sensitive to changes in the chemical composition of the atmosphere in the packaging (mainly via variation of pH of its recognition element); therefore, it must be placed inside the packaging to be exposed to the target metabolites (Kalpana et al., 2019). They are usually developed as simple colorimetric labels changing their colors (or another element of their graphical layout) when the food product ripening or deterioration (and releasing the target metabolites) (Tichoniuk et al., 2017a).

Unfortunately, matching and validating freshness indicators is very difficult to perform (Shaefer and Cheung, 2018. It is impossible to develop a universal freshness indicator because the way how particular food products ripen or spoil can be different. Fresh fruit and vegetables keep their enzymatic mechanisms still active after harvest. Therefore, this type of food products requires more attention to their freshness and quality. Most dairy products in commercial circulation undergo sterilization process due to which they are essentially semi-finished or finished products, and spoilage of those products could be avoided. However, for fresh meat and fishery products, the only process for retaining their quality is keeping them at low temperature straight after slaughter or fishing. This fact is the reason why meat and fishery products are more sensitive to microbial contamination. The so-called "danger zone" for perishable foods is temperatures in the range of 5°C–60°C. Their exposure within this range stimulates the growth of pathogens and spoilage microorganisms, like *Escherichia coli, Staphylococcus aureus, Campylobacter* spp., *Listeria monocytogenes, Salmonella* spp., and *norovirus*. Entirely spoiled foods are easily distinguished through sensory changes (e.g., tactility, appearance, smell, and flavor), but foods in the process of deterioration are challenging to detect using only one sense and can be the cause of acute food poisoning to a large number of individuals (Choi et al., 2020). Once optimized, freshness indicator also could react differently to factors appearing in the distribution chain like changes in temperature altering the volatility of metabolites or affecting the way the products mature/deteriorate (Morsy et al., 2016; Prabhakar et al., 2020). The development of freshness indicators to a defined group of food products generates additional costs, therefore, it could be incurred only in case of application of the IP for the distribution of products that are particularly susceptible to spoilage (e.g., raw fish) or of products of a certain degree of maturity desired by consumers (e.g., fresh fruits). These products should be valuable for consumers to be able to bear the cost of additional security in the form of a freshness indicator (Wang, 2019).

The high susceptibility to deterioration due to the chemical composition and a high risk of microbial contamination makes fish and seafood complicated food products for processing and transportation in the distribution chain (FAO, 2014). Maintaining their safety and quality is a big challenge for producers and distributors. The quality of fishery products can be assessed using various methods (sensory, microbiological, chemical, and physicochemical), but usually, these tools cannot be used in fast recognition of the current state of the product (Oehlenschläger, 2014; Parlapani et al., 2015). The evaluation of fish freshness is based mainly on the detection of trimethylamine (TMA) or assessment of total volatile basic nitrogen (TVB-N) content in the food packaging atmosphere (Prabhakar et al., 2020). A conventional method for the analysis of TVB-N level engages a long-lasting extraction of volatile nitrogenous bases from food samples by a solution of perchloric acid, followed by alkalization of the extract and subsequent steam distillation in which an acid receiver adsorbs volatile base compounds. Finally, the concentration of TVB-N is measured by titration of the adsorbed bases (Commission Decision 95/149/EC of 8 March, 1995). For this reason, many freshness indicators are being developed, focusing on quickly identifying the level of deterioration in fish or seafood (Morsy et al., 2016; Cierpiszewski et al., 2017; Kuswandi, 2017a; Aghaei et al., 2018; Gokoglu, 2020; Jia et al., 2019). Usually, the recognition of target metabolites is based on the application of pH-sensitive dyes, which changes the indicator color proportionally to the severity of food deterioration and the higher concentration of its degradation products. Artificial chemical dyes provide the colorimetric response (e.g., methyl red, bromocresol green, cresol red), natural dyes (e.g., curcumin, grape peel extract) or other polymeric compounds (e.g., polyaniline) (Kuswandi, 2017a; Tichoniuk et al., 2017b; Zhai et al., 2017). The response of freshness indicators is comparable with the increasing level of TVB-N, changes in microbial condition, and sensory properties of monitored fish and seafood products. The freshness sensors provide fast and reliable information about the current state of food products. Despite a large number of reports on the development of freshness indicators for fishery products, most of them are still at the experimental stage.

Food freshness indicators are developed not only for fish and seafood. Much research is devoted for performing similar colorimetric arrays for other perishable foods like red meat, poultry, dairy products, fermented or fresh fruits and vegetables (Kuswandi et al., 2014a,b, 2017b; Shukla et al., 2019; Chen et al., 2018; Poyatos-Racionero et al., 2018; Kurek et al., 2019; Tirtashi et al., 2019; Goodarzi et al., 2020). Their construction and mechanism of action are similar to colorimetric pH-sensitive indicators sensitive to the volatile products of food deterioration and metabolites formed in the developing microbiological contamination of the food. Many of them require further development to enhance the stability of applied pH-sensitive dyes regarding the conditions in the food supply chain, usability in the industrial application on packaging surface on a large scale, and legibility for consumers reaching for products in intelligent packaging (Shaefer and Cheung, 2018). Most of these aspects have been resolved in ripeness indicators for fresh fruits and vegetables that are already available in the market for some time (Kalpana et al., 2019). Ripeness indicators are sensitive to volatile chemical compounds released by the fruit when it acquires consumption maturity. A well-known example of ripeness

FIGURE 8.6 ripeSense freshness indicator. *(Source: ripeSense, 2020.)*

indicators is ripeSense released onto the market in 2004 in the United States of America and in New Zealand. The indicator is sensitive to the aroma compounds released from ripening fruits and gradually change the color of its label from red to orange during ripening (ripeSense, 2020). The overripe fruits cause a yellow coloring of the indicator label (Fig. 8.6). ripeSense indicator is applied to packed pears, but it is developed to be applicable for other fruits like kiwi, mango, avocado, or melon.

Both commercially available ripeness indicators and still being developed other freshness sensors could reduce the amount of food waste in the supply chain by precise monitoring the condition of packaged food. Nevertheless, the indicators require precise recognition of the positioning in food packaging and conditions of their final application to provide reliable information about the state of monitored food (Tichoniuk, 2019). Usually, wholesaler or a retailer is not in favor of precise freshness indicators, because apart from the risk of incorrect indication of the intelligent packaging, it limits commercial usefulness of the product. Much more desired by all participants of the food supply chain are integrity indicators.

5 Other intelligent packaging systems

5.1 Integrity indicators

The chemical composition of the packaging atmosphere is susceptible to many factors present in the production stage and the distribution chain. The composition of the gases inside the package can be distorted as a result of changes in the packaged product or the loss of packaging tightness. Integrity (gas) indicators are often used as support for other packaging systems such as aseptic or modified atmosphere packaging (MAP), to maintain its full functionality (Müller and Schmid, 2019; Tichoniuk, 2019). The integrity indicator generally

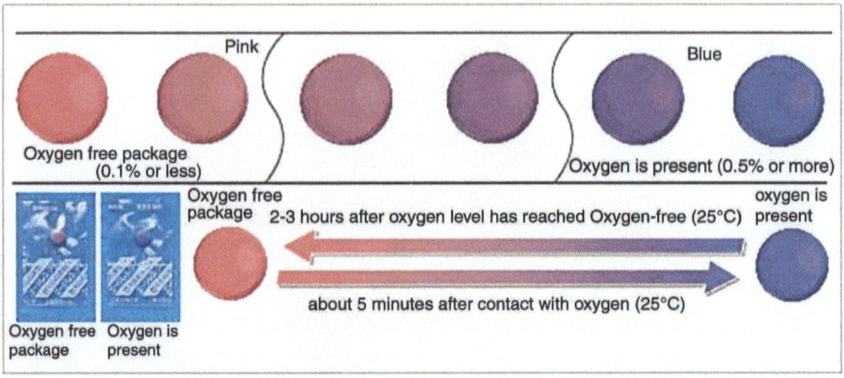

FIGURE 8.7 Ageless Eye oxygen indicator. *(Source: MGC, 2020.)*

has the form of a label, tablet, or printing on the inner wall of the packaging. It can recognize an undesirable gaseous component of the atmosphere inside the packaging caused by leakage or released from the packaged product during its spoilage (Yam et al., 2005). These indicators must be in direct contact with the atmosphere inside the package as they monitor the gaseous environment in the packaging. The indicator changes its color in contact with target gases such as oxygen, carbon dioxide, water vapor, ethanol, or hydrogen sulfide and informs that the quality and safety of the product may be affected (Fang et al., 2017). An increased amount of oxygen in the packaging may indicate improper closing of the packaging, its leakage or release of oxygen from the packed products, which may be residual oxygen dissolved in the product or the result of biochemical changes of the product. Shillingford with co-authors presents an unusual non-toxic surface coating for the inner sides of the packaging, which in the presence of oxygen stains irreversibly (Shillingford et al., 2016). The coating is based on selectively permeable rubber containing colorless oxygen-sensitive organic dyes precursors (darkly pigmented synthetic melanin). The rubber layer controls the rate of oxygen diffusion and thus facilitates the regulation of the indicator sensitivity. One of the most recognizable and commercially available integrity indicators is Ageless Eye developed by Mitsubishi Gas Chemical Co. (MGC, 2020). In the presence of oxygen in the packaging atmosphere, the indicator changes its color from pink to blue (Fig. 8.7). The reaction is reversible, but the dynamics of the color change depending on the ambient temperature. Ageless Eye is often applied to confirm the effectiveness of oxygen absorbers and packaging tightness.

The exceptional utility of intelligent packaging offers Insignia After Opening Freshness Timer developed by Insignia Technologies Ltd. from Scotland (Insignia Smart Labels, 2020). The smart label indicates (by the change of color of its center dot) how long a food package has been opened. It is especially important for products packed in MAP technology. The indicators called "after opening timer" are usually applied on the inside of the packaging lid. After filling the packaging with carbon dioxide, the indicator label turns yellow and remains in this color until the packaging is opened and carbon dioxide is released. Prolonged and frequent opening of the packaging causes the indicator to discolor first to red (corresponding to information: use soon) and then to purple (indicating loss of freshness of the packaged food). The discoloration rate is also affected by the ambient temperature (the higher one speeds it up), but both the indicator activation temperature and the discoloration rate can be adapted to the packaging of the selected product. The indicator can be applied both in refrigerated and ambient temperatures. Another version of the indicator offered by Insignia Technologies Ltd., is the so-called Pull-Timer, which can be used on the outside of the packaging (e.g., glass jars with pasteurized food). The indicator activation relies on the detachment of the pull tab from its surface and the exposition of the indicator active part to the ambient atmosphere. The company also offers leak detection labels, which, together with after opening timers, seem to be particularly useful for retailers and food consumers, but also in catering and by restaurants. All participants of the food distribution chain can take advantage of other indicators provided by Insignia Technologies Ltd., such as FRESHTAG timer for total shelf life monitoring of packed products or CO_2 MONITOR for the visual indication of carbon dioxide content inside load units prepared in MAP technology.

5.2 Data carriers

Data carriers are considered IP systems that facilitate the flow of information in the supply chain. Data carriers do not monitor the quality of packaged products, but they are intended to guarantee the possibility of their unequivocal identification, automation, protection against theft, or counterfeits (Müller and Schmid, 2019). This group includes barcodes-different variations of one-dimensional barcodes and two-dimensional (2D) codes (e.g., Quick Response (QR) codes), RFID systems and their combinations with various sensors (e.g., temperature or mechanical shock sensors) (Ghaani et al., 2016; Fang et al., 2017).

Barcodes are symbols that can be scanned electronically. Barcodes are used to encode information such as product numbers and serial numbers. They enable information to be coded for later quick and precise reading throughout the entire supply chain. Barcodes are currently the most popular type of product identification. There are several types of bar codes, each designed to be used in a different application in the supply chain (GS1, 2020). A one-dimensional bar code consisting of a group of printed bars and spaces and sometimes numerals and is used for storing information and data. Two-dimensional bar codes, often called 2D bar codes, have a square or rectangular shape and contain many small dots arranged in a unique pattern. Two-dimensional barcodes are more secure because the information stored in these codes is easily encrypted. There are several types of 2D barcode symbols, for example, QR codes.

RFID technology gains much attention because it is indicated as the successor of optical communication in the supply chain, providing contactless and non-line-of-sight to collect real-time data (Müller and Schmid, 2019). RFID systems are products that use radio signals to exchange data between the label and the reader wirelessly. At the same time, no direct contact between these two elements is needed. The necessary components of RFID technology are:

1. tag consisting of a microchip, antenna and sometimes battery,
2. readers consisting of an antenna and electronic control modules,
3. computers with databases and control systems (Hunt et al., 2007).

The use of RFID technology allows each product to be assigned a unique code.

RFID tags are divided into the groups of active, passive, and semi-active (or semi-passive) devices concerning their construction (mainly related to the energy source), functionalities, and amount of transferred/stored data. The main feature of active tags is their independent power source, supplying the RFID chip. Thanks to the antenna powered by the internal battery, the signals sent to the reader are more reliable than in passive tags. However, the need to use the energy contained in the battery limits the lifetime of this indicator. In this case, the data transmission speed is high, and the number of tags that can be read simultaneously is high. The disadvantages of active RFID are high costs and tag size, which is quite large. Passive RFID tags do not have an independent power source. In such a system, the energy needed to power the RFID chip is taken from the electromagnetic field emitted by the reader (Bibi et al., 2017). The data transmission speed for passive RFID depends on the operating frequency. These tags are characterized by low cost, long operational life, small size, and lightweight. Semi-passive tags are combination between passive and active tags. In such systems, the battery only powers the microprocessor, while the receiving antenna is powered as in the case of passive tags with the energy electromagnetic field emitted by the reader antenna. RFID systems can be classified according to the frequency range used. Four frequency standards are used in RFID technology: low frequency—LF (125–134 kHz), high-frequency HF (13.56 MHz), ultra-high frequency UHF (860–960 MHz), and microwave: (2.45–5.8 GHz).

RFID technologies are among the systems called Auto-ID (Bibi et al., 2017; Kumari et al., 2015). Collecting data via the electromagnetic field is the basic functionality of RFID tags and related electronic accessories. The use of RFID in the supply chain can improve communication for more efficient product traceability and inventory management (Fang et al., 2017), which concerning food product storage and transportation of food products, could reduce food waste in the whole supply chain. Data obtained with the use of RFID allow for making better business decisions, facilitate decision making, and help in making changes throughout the process. RFID information can differentiate a company from competitors and change the entire supply/value chain (Delen et al., 2007). The next step in using RFID technology in packaging is to combine a microchip with a sensor, for example, temperature, humidity, biogenic, and amines. Eom and coworkers proposed a meat freshness monitoring system based on an intelligent RFID tag. The RFID tag was connected to a gas sensor, temperature sensor, and humidity sensor, which allowed them to obtain a relationship between the freshness of the meat and the released gases. The proposed RFID tag measured temperature, humidity, and the presence of ammonia during meat distribution and storage. The results of the experiments showed that the system could quickly and accurately determine the freshness of meat (Eom et al., 2014). RFID tag, which contains temperature and relative humidity detection functions, has also been used to monitor the cold chain in the intercontinental logistics chain of fresh fish (Abad et al., 2009). RFID tags integrated into the conductive composite film were also investigated, allowing the detection of biogenic amines associated with food spoilage (Fiddes et al., 2014).

5.3 Sensors

Sensors in packaging systems are the next step of the evolution of intelligent food packaging, but due to the development of electronic technologies and new packaging materials, their predictions are more accurate compared to indicators. Sensors are the group of analytical items capable of detecting, locating, or quantify target substance or phenomenon (so-called analyte), and to describe its physical or chemical characteristics (Sohail et al., 2018). Usually, the sensor consists of two components cooperating:

- receptor—detecting the presence, activity, composition, or concentration of the analyte;
- transducer—converting measured characteristic into a useful analytical signal (electrical, optical, chemical, or thermal one).

In addition to the two elements described earlier, signal processing electronics and a signal display unit are also needed (Neethirajan et al., 2009). An optimal sensor should provide required specificity and selectivity of reaction toward target analyte or monitored physicochemical parameter, quick response time, adequate lifetime, suitable size and construction, and low cost of production (Ghaani et al., 2016). The developed sensors (usable in IP systems) are, for example, time and temperature recorders, gas sensors (mainly for carbon dioxide or oxygen), or biosensors for the detection of microbiological contamination (Müller and Schmid, 2019). The sensing devices are designed to work independently or in combination with more advanced data carriers (e.g., active RFID tags) or complex monitoring systems. The electronic temperature sensors are the most popular sensing element applicable to packaging in the food supply chain. Various temperature monitoring devices are available, of which the most popular are thermocouples, thermistors, resistive temperature sensors (RTDs), and infrared devices (WatElectrical, 2019). Thermocouples are the most adaptable, inexpensive, and extensive range devices. Thermistors are semiconductor devices whose resistance alters as temperature changes.

They work well for susceptible measurements in a limited temperature range, but the relationship between temperature and resistance is non-linear. RTD sensors use the phenomenon of metal resistance changes with temperature. They operate linearly over a wide temperature range, are very stable, and are characterized by higher precision and resolution than thermocouples. Resistive temperature sensors use a precision wire, usually made of platinum, as a detection element. Infrared sensors use radiant heat to detect temperature from a distance (Farnell, 2020).

Sensors for detecting gaseous substances or microbiological contaminants are more technically advanced and sophisticated in their structure. These devices usually have nanoscale components to enhance their functionality in the detection of target chemical compounds and to reduce the detection time from days to hours or even minutes (Fuertes et al., 2016). The most applicable gas sensors are based on both electrochemical and optical transducer that are connected with piezo-electric crystal microbalance, amperometric or potentiometric sensing arrays, organic conducting polymers, luminescence, or infrared recognition systems (Ghaani et al., 2016; Müller and Schmid, 2019). Gas sensors are designed mainly to detect oxygen (Baleizao et al., 2008; Pankaj et al., 2016) or carbon dioxide (Borisov et al., 2007; Aydogdu et al., 2011; Borchert et al., 2013) in the packaging headspace and directly monitor the freshness of packed food products.

A particular group of sensing devices is biosensors using in its recognition layer biological materials such as enzymes, immune system components (antigens and antibodies), nucleic acids, or whole cells (Tichoniuk, 2017c). Biosensors are widely developed as analytical devices for food quality monitoring. However, examples of their use in packaging are somewhat limited (Dincer et al., 2019). Immune-sensor based on an antibody of a specific pathogen immobilized on a membrane attached to the barcode was developed by SIRA Technologies company (USA) and offered as the Food Sentinel Systems. The barcode label system is capable of detecting pathogens continuously in fish, poultry, meat, and some liquid-product packages (Lee and Rahman, 2014). The growth of contaminating microflora results in the color change of barcode, and it makes it impossible to be read by an optical scanner—a flexible biosensor for the detection of aflatoxins and toxic metabolites of *E. coli* 0157, *Listeria* spp. Alternatively, *Salmonella* spp. were offered by Flex Alert Company Ltd. (Canada). The system employs RFID communication to provide data to real-time access to information in the field, during storage, and in packaged food products (Flex-Alert, 2020). Another commercially available system is Toxin Guard, provided by Toxin Alert Inc. (USA). The immunoassay employs antigen-antibody reactions conducted in a thin layer of flexible polymer film and results in a color change of the sensor in case of bacterial toxin presence. The Toxin Guard is capable of detecting pathogens or other microorganisms that spoil food products, for example, *Campylobacter* spp., *Listeria* spp., or *Salmonella* spp. (Han, 2014).

Sensor technologies are still being developed thanks to advances in nanotechnology and the development of new detection systems. Nevertheless, the application of physicochemical sensors and biosensors in various parts of the food supply chain could improve its controllability and reduce the scale of food waste more efficiently (Shaefer and Cheung, 2018).

6 Intelligent packaging in food processing and distribution

Packaging plays a crucial role in the whole supply chain in protecting and portioning food until it reaches the consumer. It is also often used for food storage before the final consumption (Poyatos-Racionero et al., 2018). Properly selected packaging reduces food losses in transport and storage, but the innovations in packaging materials, design, and labeling create the opportunity to improve efficiency in this area (Verghese et al., 2013). The critical reasons for the occurrence of wasted food in the supply chain at the stages of food processing, packaging, and distribution are listed on the left side of Fig. 8.8.

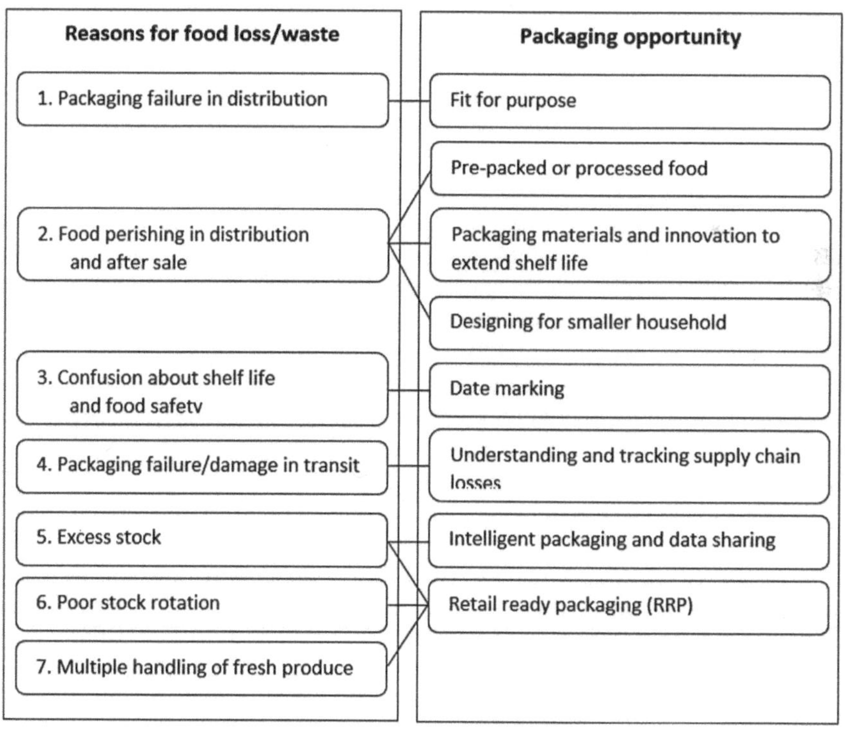

FIGURE 8.8 Reasons for food losses in food processing and packaging (1–3) and distribution (4–7) with the opportunities to reduce the losses with correct packaging. *(Own work on the basis of Verghese et al., 2013.)*

Many sources of food waste in the supply chain originate in inappropriate stock management, packaging mismatches, and inadequate communication about the properties of packaged products (Verghese et al., 2013; Müller and Schmid, 2019). Perishable food with short shelf life (e.g., fruit and vegetables, meat, seafood), as well as fresh ready-made meals such as salads, soups, and ready-to-eat food, can quickly become waste if they are not stored properly, or their "best-before" dates are missed. The reduction of food waste can be possible through the widespread introduction of IP technologies. The IP systems can indicate real-time expiration data, trace the product in the supply chain, or monitor its surrounding (Shaefer and Cheung, 2018). Retailers would benefit by gaining confidence that they do not offer deteriorated products to consumers. On the other hand, consumers could take advantage of being satisfied that the purchased products have not been exposed to inappropriate temperature conditions. However, regardless of the undeniable advantages, intelligent indicator technologies have not conquered the food packaging market, and only a few solutions are widely available (Pennanen et al., 2014).

The IP indicators could be applied in the food supply chain for more precise product quality monitoring and dynamic adjustment of their shelf life (DSL attitude) (Buisman et al., 2019). Highly perishable products with the shelf life shorter than 2 weeks have conservatively established expiration dates with the FSL to ensure general food safety. On the other hand, the FSL attitude could lead to food waste in the supply chain when the products are still of good quality but have passed their expiration dates and must be disposed of. The food waste risk could also increase when consumers selectively choose products with a more extended expiration date from the group of the same products available on the store shelf. The other benefit of the application of TTI labels and freshness indicators is the possibility of the indication of food spoilage before the expiration date (printed on the packaging surface) in case of improper food storage or transportation in the supply chain. Pennanen et al. (2014) research demonstrated that French, Greek, and German participants considered TTIs to increase food safety and security. This study also showed consumer concerns and maybe even limited confidence in the market activities of food producers and retailers. The consumers from all countries engaged in the research (Finland, France, Germany, and Greece) opted for the statement that TTI would allow greater confidence in individual participants in the food distribution chain. The respondents expressed TTIs usefulness for many aspects of food consumption, including selection criteria of food in a store, controlling food waste at home. On the other hand, they were suspicious of retailers trying to manipulate with labels, the technology itself and showed concerns about a possible rise in product prices (Verghese et al., 2015). These factors should be taken into account by a retailer when making a decision about the use of intelligent packaging for their products. As it can be beneficial for both the parties, but need to be implemented with care and thorough knowledge of the product, the intelligent solutions applied, and consumer demand.

The economic effect of the implementation of TTIs (tags) in the inventory system of perishable food (fish products) on the effectiveness of price differentiation policies models was evaluated by Avi Herbon research group (2012). Three types of food packaging were compared: without TTI tag, with TTI tag informing about exceeding the critical temperature, and with TTI indicator enabling the prediction of the real expiration date of the product. The study shows that simple TTIs could increase the profit by the reduction of food waste, but the more sophisticated TTI tags were not so favorable because of the cost of the introduction and maintenance of the intelligent packaging system.

7 Summary

IP systems have the potential to help in the reduction of food waste on a significant scale (Mataragas et al., 2019). Nevertheless, their applicability depends strongly on the reliability of presented information, implementation costs (both economic and organizational), and real interest of the members of the food supply chain. The IP systems offer a wide variety of tools such as indicators, efficient data carriers, and developed sensors that could enhance the efficiency of all elements of food supply chain. This effort will, in turn, reduce the waste of food at each stage through more effective monitoring of product quality, precise control of storage and transportation conditions, better communication and data transfer, and other possibilities that are offered by IP systems. IP systems are intended for use on an industrial scale should also be able to withstand the loads associated with packaging production operations, for example, elevated temperature and pressure, mechanical stress, and other physical and chemical processing factors. Additionally, the IP systems must be included in the production processes in a cost-effective way, which makes labels (indicators) printed or directly attached to packaging materials more effective solutions on an industrial scale (Ahmed et al., 2018).

The novel food packaging technologies such like IP systems, in addition to their primary function of product protection increase the impact on food quality and safety. IP can have a more significant impact on effective supply chain management, because it enables us to acquire detailed knowledge of products throughout the entire chain and helps maintain food quality, for example, by finding critical control points (Majid et al., 2018). They also make possible a better management and delivery scheduling. The benefits of new packaging technologies and their adaptation to the needs of specific food supply chains should be considered along with the understanding and perception of such technologies by consumers and other participants of the supply chain. Intelligent food packaging can provide support for cooperation and data transfer by the introduction of realistic expiration dates, efficient product traceability, and control of transportation/storage conditions to give better "on-demand" feedback to various supply chain stakeholders. The application of those technologies in packaging can provide suppliers information on product condition and logistics usability, which further can be applied to lower product flow time in the supply chain, prolong its shelf life, diminish the probability of product spoilage, and increase the possibility to cut down food waste in the supply chain (Verghese et al., 2015). Besides, IP (as part of smart packaging technology) allows for a new market development related to digitization, which could be a part of the concept of Industry 4.0 (Shaefer and Cheung, 2018).

References

Abad, E., Palacio, F., Nuin, M., González de Zárate, A., Juarros, A., Gómez, J. M., et al. (2009). RFID smart tag for traceability and cold chain monitoring of foods: Demonstration in an intercontinental fresh fish logistic chain. *Journal of Food Engineering*, *93*, 394–399. https://doi.org/10.1016/j.jfoodeng.2009.02.004.

ActInPack (2018). Food waste—Active & intelligent packaging in relation to food waste. Available from: <http://www.actinpak.eu/food-waste/> Accessed: 01.07.2019.

Adekomaya, O., Jamiru, T., Sadiku, R., & Huan, Z. (2016). Sustaining the shelf life of fresh food in cold chain—A burden on the environment. *Alexandria Engineering Journal*, *55*(2), 1359–1365. https://doi.org/10.1016/j.aej.2016.03.024.

Aghaei, Z., Emadzadeh, B., Ghorani, B., & Kadkhodaee, R. (2018). Cellulose acetate nanofibres containing alizarin as a halochromic sensor for the qualitative assessment of rainbow trout fish spoilage. *Food and Bioprocess Technology*, *11*, 1087–1095. https://doi.org/10.1007/s11947-017-2046-5.

Ahmed, I., Lin, H., Zou, L., Li, Z., Brody, A. L., Qazi, I. M., et al. (2018). An overview of smart packaging technologies for monitoring safety and quality of meat and meat products. *Packaging Technology and Science*, *31*, 449–471. https://doi.org/10.1002/pts.2380.

Akkerman, R., Farahani, P., & Grunow, M. (2010). Quality, safety and sustainability in food distribution: A review of quantitative operations management approaches and challenges. *OR Spectrum*, *32*(4), 863–904. https://doi.org/10.1007/s00291-010-0223-2.

Albrecht, A., Ibald, R., Raab, V., Reichstein, W., Haarer, D., & Kreyenschmidt, J. (2019). Implementation of time temperature indicators to improve temperature monitoring and support dynamic shelf life in meat supply chains. *Journal of Packaging Technology and Research*, *4*, 22–32. https://doi.org/10.1007/s41783-019-00080-x.

Aydogdu, S., Ertekin, K., Suslu, A., Ozdemir, M., Celik, E., & Cocen, U. (2011). Optical CO_2 sensing with ionic liquid doped electrospun nanofibers. *Journal of Fluorescence*, *21*(2), 607–613. https://doi.org/10.1007/s10895-010-0748-4.

Badia-Melis, R., Mc Carthy, U., & Uysal, I. (2016). Data estimation methods for predicting temperatures of fruit in refrigerated containers. *Biosystems Engineering*, *151*, 261–272. https://doi.org/10.1016/j.biosystemseng.2016.09.009.

Baleizao, C., Nagl, S., Schaferling, M., Berberan-Santos, M. N., & Wolfbeis, O. S. (2008). Dual fluorescence sensor for trace oxygen and temperature with unmatched range and sensitivity. *Analytical Chemistry*, *80*(16), 6449–6457. https://doi.org/10.1021/ac801034p.

Bibi, F., Guillaume, C., Gontard, N., & Sorli, B. (2017). A review: RFID technology having sensing aptitudes for food industry and their contribution to tracking and monitoring of food products. *Trends in Food Science & Technology*, *62*, 91–103. http://dx.doi.org/10.1016/j.tifs.2017.01.013.

Biegańska, M. (2017). Shelf-life monitoring of food using time-temperature indicators (TTI) for application in intelligent packaging. *Polish Journal of Commodity Science (Towaroznawcze Problemy Jakości)*, *2*(51), 75–85. http://www.tpj.uniwersytetradom.pl/images/pelne/2_51_2017.pdf.

Bobelyn, E., Hertog, M. L., & Nicolaï, B. M. (2006). Applicability of an enzymatic time temperature integrator as a quality indicator for mushrooms in the distribution chain. *Postharvest Biology and Technology*, *42*(1), 104–114. https://doi.org/10.1016/j.postharvbio.2006.05.011.

Borchert, N. B., Kerry, J. P., & Papkovsky, D. B. (2013). A CO_2 sensor based on Pt-porphyrin dye and FRET scheme for food packaging applications. *Sensors and Actuators B: Chemical*, *176*, 157–165. https://doi.org/10.1016/j.snb.2012.09.043.

Borisov, S. M., Waldhier, M. C., Klimant, I., & Wolfbeis, O. S. (2007). Optical carbon dioxide sensors based on silicone-encapsulated room-temperature ionic liquids. *Chemistry of Materials*, *19*(25), 6187–6194. https://doi.org/10.1021/cm7019312.

Brizio, A. P. D. R., & Prentice, C. (2015). Development of a new time temperature indicator for enzymatic validation of pasteurization of meat products. *Journal of Food Science*, *80*(6), 1271–1276. https://doi.org/10.1111/1750-3841.12889.

Broekmeulen, R. A. C. M., & van Donelaar, K. H. (2019). Quantifying the potential to improve on food waste, freshness and sales for perishables in supermarkets. *International Journal of Production Economics*, *209*, 265–273. https://doi.org/10.1016/j.ijpe.2017.10.003.

Buisman, M. E., Haijema, R., & Bloemhof-Ruwaard, J. M. (2019). Discounting and dynamic shelf life to reduce fresh food waste at retailers. *International Journal of Production Economics*, *209*, 274–284. https://doi.org/10.1016/j.ijpe.2017.07.016.

Cabot, M. I., Luque, A., de las Heras, A., & Aguayo, F. (2019). Aspects of sustainability and design engineering for the production of interconnected smart food packaging. *PLoS ONE*, *14*(5), 1–15. https://doi.org/10.1371/journal.pone.0216555.

Cerqueira, M.A., Nurmi, M., Gregor-Svetec, D. (2018). Intelligent packaging. Industry leaflet. Available from: <http://www.actinpak.eu/leaflets-intelligent-packaging/> Accessed: 01.04.2020.

Chen, H., Zhang, M., Bhandari, B., & Guo, Z. (2018). Applicability of a colorimetric indicator label for monitoring freshness of fresh-cut green bell pepper. *Postharvest Biology and Technology*, *140*, 85–92. https://doi.org/10.1016/j.postharvbio.2018.02.011.

Choi, S., Eom, Y., Kim, S. -M., Jeong, D. -W., Han, J., Koo, J. M., et al. (2020). A self-healing nanofiber-based self-responsive time-temperature indicator for securing a cold-supply chain. *Advanced Materials*, *32*(11), 1–8. https://doi.org/10.1002/adma.201907064.

Chowdhury, E. U., & Morey, A. (2019). Intelligent packaging for poultry industry. *Journal of Applied Poultry Research*, *28*, 791–800. http://dx.doi.org/10.3382/japr/pfz098.

Chun, J. -Y., Choi, M. -J., Lee, S. J., & Hong, G. -P. (2013). Applications of time-temperature integrator (TTI) as a quality indicator of grounded pork patty. *Korean Journal for Food Science of Animal Resources*, *33*(4), 439–447. https://doi.org/10.5851/kosfa.2013.33.4.439.

Cierpiszewski, R., Tichoniuk, M., & Radomska, N. (2017). Application of freshness indicator for monitoring of fish spoilage in model and commercial packaging. *Polish Journal of Commodity Science*, *4*(53), 103–117. https://doi.org/10.19202/j.cs.2017.04.10.

Claeys, W. L., Van Loey, A. M., & Hendrickx, M. E. (2002). Intrinsic time temperature integrators for heat treatment of milk. *Trends in Food Science & Technology*, *13*, 293–311. https://doi.org/10.1016/S0924-2244(02)00164-4.

Dainelli, D., Gontard, N., Spyropoulos, D., Zondervan-van den Beuken, E., & Tobback, P. (2008). Active and intelligent food packaging: legal aspects and safety concerns. *Trends in Food Science & Technology*, *19*, 103–112. https://doi.org/10.1016/j.tifs.2008.09.011.

De Fazio Aguiar, H., Yamashita, A. S., & Gut, J. A. W. (2012). Development of enzymic time-temperature integrators with rapid detection for evaluation of continuous HTST pasteurization processes. *LWT-Food Science and Technology*, *47*(1), 110–116. https://doi.org/10.1016/j.lwt.2011.12.027.

Delen, D., Hardgrave, B. C., & Sharda, R. (2007). RFID for better supply-chain management through enhanced information visibility. *Production and Operations Management (POMS)*, *16*(5), 613–624. September-October 2007, https://doi.org/10.1111/j.1937-5956.2007.tb00284.x.

Dincer, C., Bruch, R., Costa-Rama, E., Fernández-Abedul, M. T., Merkoçi, A., Manz, A., et al. (2019). Disposable sensors in diagnostics, food, and environmental monitoring. *Advanced Materials*, *31*, 1–28. 1806739, https://doi.org/10.1002/adma.201806739.

Dobon, A., Cordero, P., Kreft, F., Østergaard, S. R., Robertsson, M., Smolander, M., et al. (2011). The sustainability of communicative packaging concepts in the food supply chain. A case study: part 1. Life cycle assessment. *The International Journal of Life Cycle Assessment*, *16*, 168–177. https://doi.org/10.1007/s11367-011-0257-y.

Eden, M., Raab, V., Kreyenschmidt, J., HafliÐason, T., Olafsdóttir, G., & Bogason, S. G. (2011). Continuous temperature monitoring along the chilled food supply chain. In J. Hoorfar, K. Jordan, F. Butler, & R. Prugger (Eds.), *Food chain integrity. A holistic approach to food traceability, safety, quality and authenticity* (pp. 115–129). Woodhead Publishing Ltd.. https://doi.org/10.1533/9780857090621.2.115.

Ellouze, M., & Augustin, J. -C. (2010). Applicability of biological time temperature integrators as quality and safety indicators for meat products. *International Journal of Food Microbiology*, *138*, 119–129. https://doi.org/10.1016/j.ijfoodmicro.2009.12.012.

Eom, K. H., Hyun, K. H., Lin, S., & Kim, J. W. (2014). The meat freshness monitoring system using the smart RFID tag. *International Journal of Distributed Sensor Networks*, *2014*(3), 1–9. Article ID 591812, http://dx.doi.org/10.1155/2014/591812.

European Union, European Commission. Communication from the Commission to the European Parliament, the Council, the European Economic and Social Committee and the Committee of the Regions on the Sustainable Consumption and Production and Sustainable Industrial Policy Action Plan, 16 July 2008 (Accessed 15 May 2020).

Fang, Z., Zhao, Y., Warner, R. D., & Johnson, S. K. (2017). Active and intelligent packaging in meat industry. *Trends in Food Science & Technology*, *61*, 60–71. http://dx.doi.org/10.1016/j.tifs.2017.01.002.

FAO (2014). The State of World Fisheries and Aquaculture. Opportunities and Challenges, 2014. Fish Stock Status Highlights, Food and Agriculture Organization of the United Nations, E-ISBN 978-92-5-108276-8 (PDF), Available from: http://www.fao.org/3/a-i3720e.pdf Accessed: 20.04.2020.

FAO (2015). Global Initiative on Food Loss and Waste Redution, Food and Agriculture Organization of the United Nations. Available from: <http://www.fao.org/3/a-i4068e.pdf> Accessed 02.05.2020.

Farnell (2020). Czujniki temperatury (ang. Temperature sensors). Available from: <https://pl.farnell.com/sensor-temperature-sensor-technology#> Accessed 02.05.2020.

Fiddes, L. K., Chang, J., & Yan, N. (2014). Electrochemical detection of biogenic amines during food spoilage using an integrated sensing RFID tag. *Sensors and Actuators B*, *202*, 1298–1304. http://dx.doi.org/10.1016/j.snb.2014.05.106.

Flex-Alert (2020). Flex-Alert RFID & Biosensor Technology Solutions. Available from: <https://flex-alert.com/applications/> Accessed 02.05.2020.

Fresh-Check® Indicator (2020). Fresh-Check. Available from: <http://fresh-check.com/> Accessed 10.04.2020.

Fu, B., & Labuza, T. P. (1992). Considerations for the application of time-temperature integrators in food distribution. *Journal of Food Distribution Research, Food Distribution Research Society*, *23*(1), 9–17. https://doi.org/10.22004/ag.econ.27193.

Fuertes, G., Soto, I., Carrasco, R., Vargas, M., Sabattin, J., & Lagos, C. (2016). Intelligent packaging systems: Sensors and nanosensors to monitor food quality and safety. *Journal of Sensors*, *2016*, 1–8. https://doi.org/10.1155/2016/4046061. article ID 4046061.

Gao, T., Tian, Y., Zhu, Z., & Sun, D. -W. (2020). Modelling, responses and applications of time-temperature indicators (TTIs) in monitoring fresh food quality. *Trends in Food Science & Technology*, *99*, 311–322. https://doi.org/10.1016/j.tifs.2020.02.019.

Ghaani, M., Cozzolino, C. A., Castelli, G., & Farris, S. (2016). An overview of the intelligent packaging technologies in the food sector. *Trends in Food Science & Technology*, *51*, 1–11. https://doi.org/10.1016/j.tifs.2016.02.008.

Gokoglu, N. (2020). Innovations in seafood packaging technologies: A review. *Food Reviews International*, *36*(4), 340–366. https://doi.org/10.1080/87559129.2019.1649689.

Goodarzi, M. M., Moradi, M., Tajik, H., Forough, M., Ezati, P., & Kuswandi, B. (2020). Development of an easy-to-use colorimetric pH label with starch and carrot anthocyanins for milk shelf life assessment. *International Journal of Biological Macromolecules*, *153*, 240–247. https://doi.org/10.1016/j.ijbiomac.2020.03.014.

GS1 (2020). GS1 barcodes. Available from: <www.gs1.org/standards/barcodes> Accessed 20.03.2020.

Han, J. H. (2014). A review of food packaging technologies and innovations. In J. H. Han (Ed.), *Innovations in food packaging* (pp. 3–12). Amsterdam: Elsevier Academic Press. https://doi.org/10.1016/B978-0-12-394601-0.00001-1.

Heising, J. K., Claassen, G. D. H., & Dekker, M. (2017). Options for reducing food waste by quality controlled logistics using intelligent packaging along the supply chain. *Food Additives & Contaminants: Part A*, *34*(10), 1672–1680. https://doi.org/10.1080/19440049.2017.1315776.

Herbon, A., Levner, E., & Cheng, E. (2012). Perishable inventory management and dynamic pricing using TTI technologies. *International Journal of Innovation, Management and Technology*, *3*(3), 262–266. http://www.ijimt.org/papers/235-D0111.pdf.

Holden, N. M., White, E. P., Lange, M. C., & Oldfield, T. L. (2018). Review of the sustainability of food systems and transition using the Internet of Food. *NPJ Science for Food*, *2*(18), 1–7. https://doi.org/10.1038/s41538-018-0027-3.

Hunt, D., Puglia, A., & Puglia, M. (2007). In *RFID: A guide to radio frequency identification*. New Jersey: Wiley Interscience. ISBN: 978-0-470-10764-5.

Insignia Smart Labels (2020). Available from: <https://www.insigniatechnologies.com/about.php> Accessed 20.03.2020.

Jafee, S., et al. (2019). *The Safe Food Imperative. Accelerating Progress in Low- and Middle-Income Countries*. NW, Washington DC: International Bank for Reconstruction and Development/The World Bank Available at: <https://www.worldbank.org/en/topic/agriculture/publication/the-safe-food-imperative-accelerating-progress-in-low-and-middle-income-countries?cid=EXT_WBEmailShare_EXT>. (Accessed 1 May 2020).

Jia, R., Tian, W., Bai, H., Zhang, J., Wang, S., & Zhang, J. (2019). Amine-responsive cellulose-based ratiometric fluorescent materials for real-time and visual detection of shrimp and crab freshness. *Nature Communications*, *10*(795), 1–8. https://doi.org/10.1038/s41467-019-08675-3.

Kalpana, S., Priyadarshini, S. R., Maria Leena, M., Moses, J. A., & Anandharamakrishnan, C. (2019). Intelligent packaging: Trends and applications in food systems. *Trends in Food Science & Technology*, *93*, 145–157. https://doi.org/10.1016/j.tifs.2019.09.008.

Kou, L., Luo, Y., Park, E., Turner, E. R., Barczak, A., & Jurick, W. M., II (2014). Temperature abuse timing affects the rate of quality deterioration of commercially packaged ready-to-eat baby spinach. Part I: Sensory analysis and selected quality attributes. *Postharvest Biology and Technology*, *91*, 96–103. https://doi.org/10.1016/j.postharvbio.2013.12.025.

Koutsoumanis, K. P., & Gougouli, M. (2015). Use of Time temperature integrators in food safety management. *Trends in Food Science & Technology*, *43*, 236–244. https://doi.org/10.1016/j.tifs.2015.02.008.

Kreyenschmidt, J., Christiansen, H., Hübner, A., Raab, V., & Petersen, B. (2010). A novel photochromic time–temperature indicator to support cold chain management. *International Journal of Food Science and Technology*, *45*, 208–215. https://doi.org/10.1111/j.1365-2621.2009.02123.x.

Kumari, L., Narsaiah, K., Grewal, M. K., & Anurag, R. K. (2015). Application of RFID in agri-food sector. *Trends in Food Science & Technology*, *43*, 144–161. http://dx.doi.org/10.1016/j.tifs.2015.02.005.

Kurek, M., Hlupić, L., Ščetar, M., Bosiljkov, T., & Galić, K. (2019). Comparison of two pH responsive color changing bio-based films containing wasted fruit pomace as a source of colorants. *Journal of Food Science*, *84*(9), 2490–2498. https://doi.org/10.1111/1750-3841.14716.

Kuswandi, B., Irmawati, T., Hidayat, M. A., Ahmad, J., & Ahmad, M. (2014). A simple visual ethanol biosensor based on alcohol oxidase immobilized onto polyaniline film for halal verification of fermented beverage samples. *Sensors*, *14*(2), 2135–2149. https://doi.org/10.3390/s140202135.

Kuswandi, B., Jayus, Oktaviana R., Abdullah, A., & Heng, L. Y. (2014). Novel on-package sticker sensor based on methyl red for real-time monitoring of broiler chicken cut freshness. *Packaging Technology and Science*, *27*, 69–81. https://doi.org/10.1002/pts.2016.

Kuswandi, B., Mahawan, A. G., & Nuriman (2017). On-package color indicator for ripeness monitoring of *tape ketan* packaging. *Food Science and Engineering*, *1*(1), 25–36. http://ojs.wiserpub.com/index.php/FSE/article/view/181/112.

Kuswandi, B. (2017). Freshness sensors for food packaging. In *Reference module in food science*. Elsevier. https://doi.org/10.1016/B978-0-08-100596-5.21876-3.

Lee, S. J., & Rahman, A. T. M. M. (2014). Intelligent packaging for food products. In J. H. Han, & S. L. Taylor (Eds.), *Innovations in food packaging* (2nd ed., pp. 171–209). London: Food Science and Technology, International Series, Elsevier. http://dx.doi.org/10.1016/B978-0-12-394601-0.00008-4.

Lee, S. Y., Lee, S. J., Choi, D. S., & Hur, S. J. (2015). Current topics in active and intelligent food packaging for preservation of fresh foods. *Journal of the Science of Food and Agriculture*, *95*(14), 2799–2810. https://doi.org/10.1002/jsfa.7218.

Lidl Finland (2020). Lidl Finland is the first retailer to secure their cold chains. Available from: <https://www.tempix.com/news> Accessed 22.04.2020.

Lin, H., Shavezipur, M., Yousef, A., & Maleky, F. (2016). Prediction of growth of *Pseudomonas fluorescens* in milk during storage under fluctuating temperature. *Journal of Dairy Science*, *99*(3), 1822–1830. https://doi.org/10.3168/jds.2015-10179.

Lu, L., Zheng, W., Lv, Z., & Tang, Y. (2013). Development and application of time–temperature indicators used on food during the cold chain logistics. *Packaging Technology and Science*, *26*, 80–90. https://doi.org/10.1002/pts.2009.

Lydekaityte, J., Tambo, T. (2020). Smart packaging: definitions, models and packaging as an intermediator between digital and physical product management. *The International Review of Retail, Distribution and Consumer Research* 30, published on-line, https://doi.org/10.1080/09593969.2020.1724555.

Müller, P., & Schmid, M. (2019). Intelligent packaging in the food sector: A brief overview. *Foods*, *8*(1), 16. https://doi.org/10.3390/foods8010016.

Majid, I., Nayik, G. A., Dar, S. M., & Nanda, V. (2018). Novel food packaging technologies: Innovations and future prospective. *Journal of the Saudi Society of Agricultural Sciences*, *17*, 454–462. http://dx.doi.org/10.1016/j.jssas.2016.11.003.

Mataragas, M., Bikouli, V. C., Korre, M., Sterioti, A., & Skandamis, P. N. (2019). Development of a microbial time temperature indicator for monitoring the shelf life of meat. *Innovative Food Science and Emerging Technologies*, *52*, 89–99. https://doi.org/10.1016/j.ifset.2018.11.003.

Meng, J., Qian, J., & Tang, Y. (2018). A solid-state time-temperature indicator used in chilled fresh pork monitoring. *Packaging Technology and Science*, *31*, 353–360. https://doi.org/10.1002/pts.2328.

AGELESS EYE, Oxygen Indicator (2020). MSG. Available from: <https://www.mgc.co.jp/eng/products/sc/ageless-eye.html> Accessed 10.04.2020.

Mohebi, E., & Marquez, L. (2015). Intelligent packaging in meat industry: An overview of existing solutions. *Journal of Food Science and Technology*, *52*(7), 3947–3964. https://doi.org/10.1007/s13197-014-1588-z.

Morsy, M. K., Zór, K., Kostesha, N., Alstrøm, T. S., Heiskanen, A., El-Tanahi, H., et al. (2016). Development and validation of a colorimetric sensor array for fish spoilage monitoring. *Food Control*, *60*, 346–352. https://doi.org/10.1016/j.foodcont.2015.07.038.

3M™ MonitorMark™ Time Temperature Indicators (2020). 3 M United States. Available from: <https://www.3m.com/3M/en_US/company-us/all-3m-products/~/MONMARK-3M-MonitorMark-Time-Temperature-Indicators/?N=5002385+3293785721&preselect=8710779+8711414&rt=rud> Accessed 10.04.2020.

Neethirajan, S., Jayas, D. S., & Sadistap, S. (2009). Carbon dioxide (CO2) sensors for the agri-food industry. A review. *Food and Bioprocess Technology*, *2*, 115–121. https://doi.org/10.1007/s11947-008-0154-y.

O'Grady, M. N., & Kerry, J. P. (2008). Smart packaging technology. In F. Toldrá (Ed.), *Meat biotechnology* (pp. 425–451). New York: Springer. https://doi.org/10.1007/978-0-387-79382-5_19.

Oehlenschläger, J. (2014). Seafood quality assessment. In I. S. Boziaris (Ed.), *Seafood processing, IFST advances in food science series, technology, quality & safety* (pp. 359–386). West-Sussex UK: John Willey & Sons Ltd.. https://doi.org/10.1002/9781118346174.ch14.

Pankaj, S. K., Kelly, C. A., Bueno-Ferrer, C., Kerry, J. P., Papkovsky, D. B., Bourke, P., et al. (2016). Application of phosphorescent oxygen sensors in in-package dielectric barrier discharge plasma environment. *Innovative Food Science & Emerging Technologies*, *33*, 234–239. https://doi.org/10.1016/j.ifset.2015.11.005.

Parlapani, F. F., Harautounian, S. A., Nychas, G-J. E., & Boziaris, I. S. (2015). Microbiological spoilage of volatile production of gutted european sea bass stored under air and commercial modified atmosphere package at 2°C. *Food Microbiology*, *50*, 44–53. https://doi.org/10.1016/j.fm.2015.03.006.

Pennanen, K., Focas, C., Kumpusalo-Sanna, V., Keskitalo-Vuokko, K., Matullat, I., Ellouze, M., et al. (2014). European consumers' perceptions of time–temperature indicators in food packaging. *Packaging Technology and Science*, *28*, 303–323. https://doi.org/10.1002/pts.2105.

Pereira, V. A., Jr., de Arruda, I. Z. Q., & Stefani, R. (2015). Active chitosan/PVA films with anthocyanins from *Brassica oleraceae* (red cabbage) as time-temperature indicators for application in intelligent food packaging. *Food Hydrocolloids*, *43*, 180–188. https://doi.org/10.1016/j.foodhyd.2014.05.014.

Poyatos-Racionero, E., Ros-Lis, J. V., Vivancos, J. -L., & Martínez-Máñez, R. (2018). Recent advances on intelligent packaging as tools to reduce food waste. *Journal of Cleaner Production*, *172*, 3398–3409. https://doi.org/10.1016/j.jclepro.2017.11.075.

Prabhakar, P. K., Vatsa, S., Srivastav, P. P., & Pathak, S. S. (2020). A comprehensive review on freshness of fish and assessment: Analytical methods and recent innovations. *Food Research International*, *133*(109157), 1–17. https://doi.org/10.1016/j.foodres.2020.109157.

ripeSense (2020). ripeSense—the next evolution in fresh produce marketing, 2020. Available from: <http://www.ripesense.co.nz/> Accessed 01.04.2020.

Schmidt Rivera, X. C., Leadley, C., Potter, L., & Azapagic, A. (2019). Aiding the design of innovative and sustainable food packaging: Integrating techno-environmental and circular economy criteria. *Energy Procedia*, *161*, 190–197. https://doi.org/10.1016/j.egypro.2019.02.081.

Examples of OnVu labels, 2020. SEDEF. Available from: <http://www.sedef.com/izler_sicaklik_zaman_etiket_de.php> Accessed 20.04.2020.

Shaefer, D., & Cheung, W. M. (2018). Smart packaging: Opportunities and challenges. 51st CIRP conference on manufacturing systems. *Procedia of CIRP*, *72*, 1022–1027. https://doi.org/10.1016/j.procir.2018.03.240.

Sharrock, K. R. (2012). Advances in freshness and safety indicators in food and beverage packaging. In K. L. Yam, & D. S. Lee (Eds.), *Emerging food packaging technologies. Principles and practice* (pp. 175–197). Cambridge: Woodhead Publishing Limited. https://doi.org/10.1533/9780857095664.2.175.

Shillingford, C., Russell, C. W., Burgess, I. B., & Aizenberg, J. (2016). Bioinspired artificial melanosomes as colorimetric indicators of oxygen exposure. *ACS Applied Materials & Interfaces*, *8*(7), 4314–4317. https://doi.org/10.1021/acsami.5b11933.

Shukla, V., Kandeepan, G., & Vishnuraj, M. R. (2019). Development of on package indicator sensor for real-time monitoring of meat quality. *Veterinary World*, *8*(3), 393–397. https://doi.org/10.14202/vetworld.2015.393-397.

Sohail, M., Sun, D. -W., & Zhu, Z. (2018). Recent developments in intelligent packaging for enhancing food quality and safety. *Critical Reviews in Food Science and Nutrition*, *58*(15), 2650–2662. https://doi.org/10.1080/10408398.2018.1449731.

Suárez Díaz, R. D., Paternina-Arboleda, C. D., Martínez-Flores, J. L., & Jimenez-Barros, M. A. (2020). Economic order quantity for perishables with decreasing willingness to purchase during their life cycle. *Operations Research Perspectives*, *7*, 1–6. https://doi.org/10.1016/j.orp.2020.100146.

Taoukis, P. S., & Labuza, T. P. (2003). Time-temperature indicators (TTIs). In R. Ahvenainen (Ed.), *Novel food packaging techniques* (pp. 103–126). Cambridge: Woodhead Publishing Ltd.

Taoukis, P. S. (2001). Modelling the use of time-temperature indicators in distribution and stock rotation. In L. M. M. Tijskens, M. L. A. T. M. Hertog, & B. M. Nicolaï (Eds.), *Food process modelling* (pp. 402–431). Cambdridge: Woodhead Publishing Ltd.

Taoukis, P. S. (2010). Commercialization of time-temperature integrators for foods. In C. J. Doona, K. Kustin, & F. E. Feeherry (Eds.), *Case studies in novel food processing technologies. Innovations in Processing, Packaging, and Predictive Modelling* (pp. 351–366). Woodhead Publishing Ltd.. https://doi.org/10.1533/9780857090713.3.351.

Tempix TTI indicator (2020). AP Food Industry. Available from: <https://www.apfoodonline.com/images/Features/1607VoyagerVenture/Tags%20Comparison.png> Accessed 20.04.2020.

Tempix (2020). Tempix label when cold chain was and wasn't broken, 2020. Available from: <https://www.tempix.com/pictures/gluggen.jpg?v20190821134806> Accessed 20.04.2020.

Tichoniuk, M., Radomska, N., & Cierpiszewski, R. (2017). Application of food freshness indicators in intelligent packaging. In H. Śmigielska (Ed.), *Current trends in commodity science: food safety and analysis of bioactive substances* (pp. 189–202). Poznań: PUEB.

Tichoniuk, M., Radomska, N., & Cierpiszewski, R. (2017). The application of natural dyes in food freshness indicators designed for intelligent packaging. *Studia Eoconomica Posnaniensia*, *5*(7), 19–34. https://doi.org/10.18559/SOEP. 2017.7.2.

Tichoniuk, M. (2017). Electrochemical sensors for fast detection of foodborne pathogens. *Polish Journal of Commodity Science.*, *4*(53), 42–53. https://doi.org/10.19202/j.cs.2017.04.04.

Tichoniuk, M. (2019). Innovative packaging improving food quality and extending its shelf life. *Polish Journal of Commodity Science*, *1*(58), 21–35. https://doi.org/10.192020/j.cs.2019.01.02.

Tirtashi, F. E., Moradi, M., Tajik, H., Forough, M., Ezati, P., & Kuswandi, B. (2019). Cellulose/chitosan pH-responsive indicator incorporated with carrot anthocyanins for intelligent food packaging. *International Journal of Biological Macromolecules*, *136*, 920–926. https://doi.org/10.1016/j.ijbiomac.2019.06.148.

TTI Label (2020). Vitsab. Available from: <http://vitsab.com/en/tti-label/> Accessed 10.04.2020.

Verghese, K., Lewis, H., Lockrey, S., & Williams, H. (2013). In *The role of packaging in minimising food waste in the supply chain of the future*. Sydney: CHEP Australia and RMIT University.

Verghese, K., Lewis, H., Lockrey, S., & Williams, H. (2015). Packaging's role in minimizing food loss and waste across the supply chain. *Packaging Technology and Science*, *28*, 603–620. https://doi.org/10.1002/pts.2127.

Wang, S., Liu, X., Yang, M., Zhang, Y., Xiang, K., & Tang, R. (2015). Review of time temperature indicators as quality monitors in food packaging. *Packaging Technology and Science*, *28*, 839–867. https://doi.org/10.1002/pts.2148.

Wang, L., Wu, Z. & Cao, C., (2019). Technologies and Fabrication of Intelligent Packaging for Perishable Products. *Appl. Sci.*, *9*(4858), 1–17. https://doi.org/10.3390/app9224858.

WatElectrical.com (2019). Know about Various Types of Temperature Sensors. Available from: <https://www.watelectrical.com/6-different-types-of-temperature-sensors-with-their-specifications/> Accessed: 01.05.2020.

Woodhouse, A., Davis, J., Pénicaud, C., & Östergren, K. (2018). Sustainability checklist in support of the design of food processing. *Sustainable Production and Consumption*, *16*, 110–120. https://doi.org/10.1016/j.spc.2018.06.008.

Yam, K. L., Takhistov, P. T., & Miltz, J. (2005). Intelligent packaging: Concepts and applications. *Journal of Food Science*, *70*(1), R1–R10. https://doi.org/10.1111/j.1365-2621.2005.tb09052.x.

Zhai, X., Shi, J., Zou, X., Wang, S., Jiang, C., Zhang, J., et al. (2017). Novel colorimetric films based on starch/polyvinyl alcohol incorporated with roselle anthocyanins for fish freshness monitoring. *Food Hydrocolloids*, *69*, 308–317. https://doi.org/10.1016/j.foodhyd.2017.02.014.

Chapter 9

Active Packaging

João P.F. Carvalho, Carmen S.R. Freire and Carla Vilela*
CICECO–Aveiro Institute of Materials, Department of Chemistry, University of Aveiro, Aveiro, Portugal
**Corresponding author*

1 Introduction

The quest to reduce food spoilage, waste and recalls, and foodborne illness outbreaks, associated with the increasing food demand by the ever-growing population, has led to the inception of the concepts of active and intelligent packaging (Janjarasskul and Suppakul, 2018; Dobrucka and Przekop, 2019). In addition to an inert barrier to external conditions, these strategies involve packaging solutions containing additives that preserve or extend product quality or shelf-life in the first case (i.e., active packaging) and monitor the condition of packaged food by giving information regarding the quality of the packaged food during transportation and storage in the second case (i.e., intelligent packaging), as illustrated in Fig. 9.1. The global market size of active and intelligent packaging was valued at ca. USD 17.5 billion in 2019, and is expected to grow at a rate of 6.78% per year to reach ca. USD 25 billion by 2025 (Mordor Intelligence, 2019).

The concept of active packaging has emerged as a tool that might exhibit multiple active functions that can be divided into active releasing systems and active scavenging systems depending on their properties. In the former systems, compounds are added into the headspace or to the packaged food, and include antimicrobials, carbon dioxide, antioxidants, preservatives, flavors, and ethylene, as shown in Fig. 9.2. In the latter systems, undesired compounds, such as oxygen, carbon dioxide, ethylene, moisture, and odors, are removed from the food or its environment (Fig. 9.2). The judicious selection of the active agents depends on the characteristics of the foodstuff with, for example, antioxidant agents being very important for food products with high fat content (Ganiari et al., 2017; Tian et al., 2013; Domínguez et al., 2018), and ethylene scavengers for fruits and vegetables (Álvarez-Hernández et al., 2018; Vilela et al., 2018; Gaikwad et al., 2020). These active packaging systems can be assembled via incorporation, coating, immobilization, or surface modification into the packaging materials (Bastarrachea et al., 2015), and the selection of active additives is

Sustainable food processing and engineering challenges. http://dx.doi.org/10.1016/B978-0-12-822714-5.00009-7

FIGURE 9.1 The functions of advanced packaging systems. (*Yam et al., 2005; Mihindukulasuriya and Lim, 2014*).

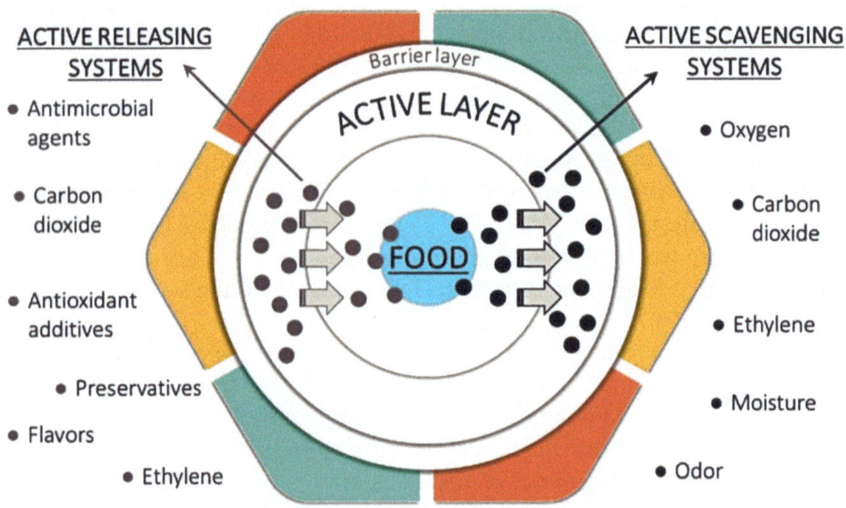

FIGURE 9.2 Active packaging: active releasing and scavenging systems. (*Ahmed et al., 2017; Yildirim and Röcker, 2018*).

following a cognate tendency as that of packaging materials, leaning toward natural-based and environmentally-friendly alternatives (Silva-Weiss et al., 2013; Valdés et al., 2014).

The scientific activity in the realm of active packaging has grown exponentially in the last couple of years. Still, most of the published literature dealing with active agents is only centered on the active properties of the supplemented packaging materials, instead of thoroughly examining their behavior in the presence

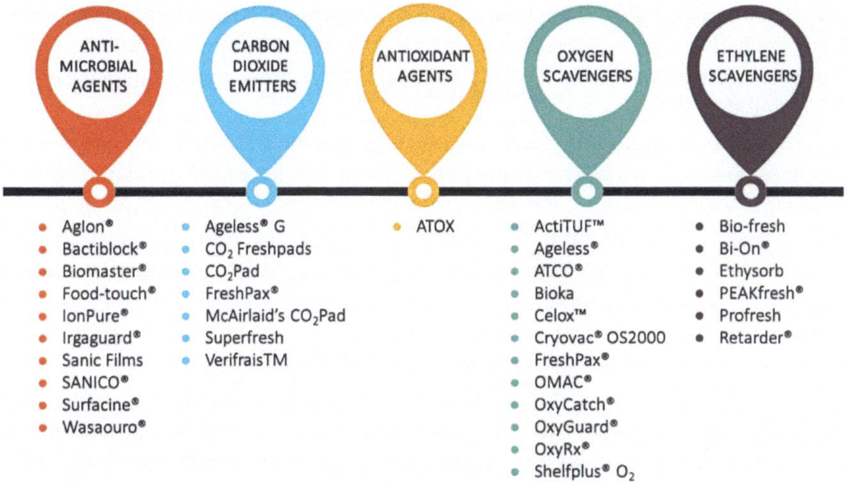

FIGURE 9.3 Examples of commercial active agent systems for active food packaging. (*Vilela et al., 2018*).

of real food products, and therefore does not validate the real efficiency of the active packaging system on concrete applications (Vilela et al., 2018). Nonetheless, the studies that perform proof-of-concept tests are primarily dedicated to foods like meat (Domínguez et al., 2018; Ahmed et al., 2017; Fang et al., 2017; Islam et al., 2017; Schumann and Schmid, 2018) and dairy (Haghighi-Manesh and Azizi, 2017) products, as well as fruits and vegetables (Álvarez-Hernández et al., 2018; Gaikwad et al., 2020; Aziz and Karboune, 2018).

Various exceptional appraisals on food packaging with active properties have been recently published, including the reviews on innovative active, intelligent and bioactive food packaging technologies (Majid et al., 2018), active agents for active food packaging (Vilela et al., 2018; Yildirim et al., 2018), active packaging coatings (Bastarrachea et al., 2015), electrospun nanofibers for active packaging (Zhang et al., 2020), natural additives for active food packaging (Valdés et al., 2015), essential oils for active food packaging (Ataei et al., 2020), active packaging systems for muscle foods (Ahmed et al., 2017; Schumann and Schmid, 2018) and dairy products (Haghighi-Manesh and Azizi, 2017).

The rising scientific activity in this realm is being complemented by the thriving commercialization stories (Fig. 9.3) of agents in the form of sachets and pads, or films and coatings, including (1) antimicrobial agents, such as AgIon, Bactiblock, Biomaster, Food-touch, IonPure, Irgaguard, Sanic Films, SANICO, Surfacine, and Wasaouro; (2) CO_2 emitters like Ageless G, CO_2Pad, FreshPax, McAirlaid's CO_2Pad, Superfresh, and Verifrais,; (3) antioxidant agents such as ATOX; (4) O_2 scavengers like ActiTUF, Ageless, ATCO, Bioka, Celox, Cryovac OS2000, FreshPax, OMAC, OxyCatch, OxyGuard, OxyRx, and Shelfplus O_2; and (5) ethylene scavengers, such as Bio-fresh, Bi-On, Ethysorb, PEAKfresh,

Profresh and Retarder (Vilela et al., 2018). Furthermore, the confluence of more than one active agent in a single system is also being applied to foodstuff that require more than one active function to boost their safety, quality, and shelf-life (Vilela et al., 2018).

Given this evolution of ideas, it seems appropriate to review, as comprehensively as possible, the work carried out on the past years in the domain of active agents for food packaging. Herein, emphasis will be set on antimicrobial agents (e.g., metals, nisin, essential oils, lactoferrin, lysozyme, and chitosan), carbon dioxide emitters (e.g., sodium bicarbonate, ferrous carbonate, and citric acid), antioxidant agents (e.g., essential oils, plant extracts, phenolic compounds, and lignin), oxygen scavengers (e.g., iron, palladium, ascorbic acid, gallic acid, pyrogallol, glucose oxidase, and laccase), and ethylene scavengers (e.g., potassium permanganate, metal oxides, activated carbon, titanium dioxide, and metal organic frameworks (MOFs)), as summarized in Fig. 9.4. An endeavor was made to choose illustrative studies in each of the relevant categories that are taking tangible steps on the road to viable and sustainable active packaging systems.

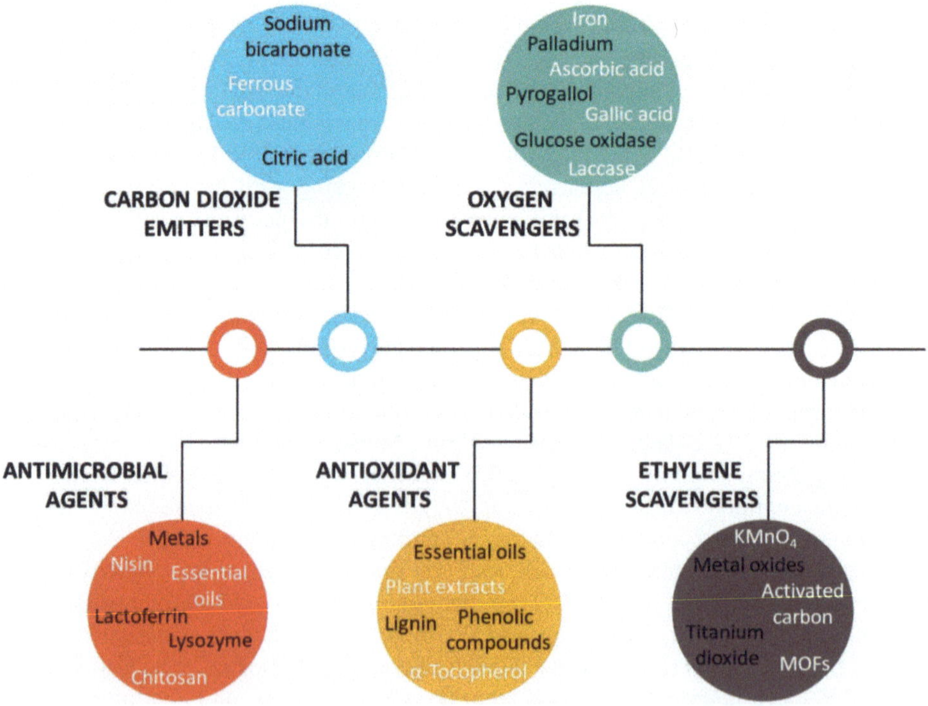

FIGURE 9.4 Active agents for active food packaging. (*Vilela et al., 2018*).

2 Antimicrobial agents

The undesired growth of microorganisms is the major culprit of food spoilage, making antimicrobial agents for food packaging a critical component to avoid the deterioration of food products (Ahmed et al., 2017; Otoni et al., 2016; Al-Tayyar et al., 2020). Antimicrobial agents are able to prevent the spreading of multiple microorganisms, including pathogenic bacteria (e.g., *Salmonella spp.*, *Staphylococcus aureus*, *Listeria monocytogenes*, *Bacillus cereus*, and *Escherichia coli*), spoilage bacteria (e.g., *Pseudomonas*, *Klebsiella*, and *Lactobacillus spp.*), yeasts (e.g., *Candida and Torulopsis*) and molds (e.g., *Rhizopus* and *Aspergillus*) (Vilela et al., 2018).

Numerous commercial forms of antimicrobial agents are currently available, given the relevance of microbial-spoilage on food deterioration (including AgIon, Bactiblock, Biomaster, Food-touch, IonPure, Irgaguard, Sanic Films, SANICO, Surfacine, and Wasaouro) mostly in the form of emitting sachets and absorbent pads, and based on silver, silver zeolite, glucose oxidase, triclosan, chlorine dioxide, ethanol vapor emitting, natamycin, sulfur dioxide, and allyl isothiocyanate, as active additives (Fang et al., 2017). These commercial systems are frequently used to protect packaged fresh and processed meat, fresh and smoked fish, fresh seafood, dairy products, fresh and processed fruits and vegetables, grains, cereals and bakery products, and ready-to-eat meals (Yildirim et al., 2018). The absorbent pads are particularly important when dealing with packaged meat products, which exude liquids that need to be removed to avert microbial growth (Otoni et al., 2016).

Research and scientific publications on this topic are very abundant and explore the antimicrobial potential of numerous types of components: metals (in the form of ions or metal oxides), essential oils, plant extracts, polysaccharides (e.g., chitosan), pure bioactive components (e.g., thymol and carvacrol), peptides, and enzymes or even synthetic antimicrobial agents (e.g., propionic, benzoic and sorbic acids). This section will focus on the latest and most relevant antimicrobial systems based in metals and metal oxides, essential oils and biomacromolecules, individually or in combination, as presented in Table 9.1.

The antimicrobial potential of metals (e.g., silver, copper, gold, and platinum) and metal oxides (e.g., TiO_2 and ZnO) has been thoroughly described and exploited for multiple applications. Silver and silver compounds, like metallic silver (Ag^0), silver ions (most common Ag^+) or silver nanoparticles (Ag NPs), have shown to effectively prevent microorganism proliferation, even in small concentrations, presenting great antimicrobial effect, and very little toxicity toward humans (Dakal et al., 2016). In a recent work, Ag NPs were successfully synthetized through eco-friendly procedures using four phenolic compounds, namely gallic and pyrogallic acid, rutin, and quercetin (Marrez et al., 2019). The Ag NPs in cellulose acetate films revealed antimicrobial capacity against multiple bacteria *(S. aureus, B. cereus, Salmonella typhi, E. coli, Klebsiella pneumoniae*) at very low concentrations (between 16.7 and 46.7 $\mu g.mL^{-1}$), with no toxic effect identified (Marrez et al., 2019).

TABLE 9.1 Examples of antimicrobial agents included into polymer matrices.

Antimicrobial agent	Film forming polymer	Food product	Microorganism	References
ZnO NPs	PLA	Fresh-cut apple	Bacteria, yeast, and fungi	(Li et al., 2017)
ZnO/chitosan	PLA	—	*E. coli*	(Olmo et al., 2020)
TiO_2 NPs	Chitosan	—	*S. aureus, E. coli, Salmonella typhimurium, Pseudomonas aeruginosa, Aspergillus, Penicillium*	(Siripatrawan and Kaewklin, 2018)
Clove essential oil and ZnO nanorods	Gelatin	Shrimp	*L. monocytogenes* *S. typhimurium*	(Ejaz et al., 2018)
Clove essential oil	Soy protein isolate	Muscle fillets of bluefin tuna (*Thunnus thynnus*)	*Pseudomonas* spp. Lactic acid bacteria and enterobacteria	(Echeverría et al., 2018)
Oregano essential oil	PHBV	—	*S. aureus* *E. coli*	(Figueroa-Lopez et al., 2019)
Chestnut spiny burs and roasted hazelnut skins extracts	Pullulan	—	*S. aureus*	(Esposito et al., 2020)
Nisin	PHB/PCL with organo-clays	Ham	*Lactobacillus plantarum* CRL691	(Correa et al., 2017)
Lactoferrin	Bacterial cellulose	Fresh sausages	*S. aureus, E. coli*	(Padrão et al., 2016)
Lactoferrin and lysozyme	PET	Fresh salmon	H_2S-producing bacteria	(Rollini et al., 2016)
Lysozyme nanofibers	Pullulan	—	*S. aureus*	(Silva et al., 2018)
Chitosan and essential oils of cinnamon and ginger	Chitosan	Pork slices	Meat bacteria	(Wang et al., 2017)
Poly(sulfobetaine methacrylate)	Bacterial cellulose	—	*S. aureus, E. coli*	(Vilela et al., 2019)
Feruloylated compounds	NFC and arabinoxylans	—	*S. aureus, E. coli, Candida albicans*	(Moreirinha et al., 2020)

Abbreviations: NFC (nanofibrillated cellulose), NPs (nanoparticles), PCL (poly(ε-caprolactone)), PET (poly(ethylene terephthalate)), PHB (poly(hydroxybutyrate)), PHBV (poly(3-hydroxybutyrate-co-3-hydroxyvalerate)), PLA (poly(lactic acid)).

As previously mentioned, other metals and metal-containing compounds, namely gold, copper, zinc oxide, and titanium dioxide, have also displayed antimicrobial activity (Hoseinnejad et al., 2017; Akhidime et al., 2019). Recent works using zinc oxide in a poly(lactic acid) (PLA) matrix (Li et al., 2017) or titanium dioxide nanoparticles in chitosan films (Siripatrawan and Kaewklin, 2018) proved the potential of both metal oxides for antimicrobial purposes for food packaging applications.

Essential oils have also been described to compromise microbial growth, making them an alternative for food packaging applications. These essential oils, namely aromatic mixtures of low molecular weight compounds (mostly phenolic compounds, such as flavonoids and phenolic acids), are usually produced by plants and have been applied for the safe food-packaging of fruits, vegetables, cheese, and fish/meat (Maisanaba et al., 2017). However, high concentrations of essential oils are required to achieve a desirable antimicrobial effect, which might affect the intrinsic organoleptic characteristics of the food products. A wide array of essential oils have shown antimicrobial potential (e.g., thyme, oregano, pimento, clove, citron, lemon verbena, lemon balm, and cypress leaf), but the mechanism behind such potential is still not clearly identified (Aziz and Karboune, 2018). As an example, oregano essential oil, rosemary extract, and green tea extract were incorporated in a mat of electrospun fibers of poly(3-hydroxybutyrate-co-3-hydroxyvalerate) (PHBV), with oregano essential oil revealing the most notorious effect against *S. aureus* and *E. coli* (Figueroa-Lopez et al., 2019). Another very relevant example is the use of clove essential oil together with soy protein isolate and montmorillonite for the preservation of bluefin tuna under refrigerated storage (Echeverría et al., 2018). Extensive reviews on the use of essential oils for antimicrobial food-packaging can be easily found, providing in-depth description of this specific class of compounds (Maisanaba et al., 2017; Ribeiro-Santos et al., 2017; Atarés and Chiralt, 2016).

Antimicrobial potential has also been found in biomacromolecules like certain peptides (e.g., nisin and lactoferrin), enzymes (e.g., lysozyme), and polysaccharides (chitosan). Nisin is currently the only peptide approved by the European Union (as E234), the World Health Organization (WHO), and the United Stated Food and Drug Administration (FDA) for food-related antimicrobial applications, and is frequently used as a natural food preservative in fruits, dairy products, and meat. Furthermore, it is currently classified as a GRAS (generally recognized as safe) peptide (Preciado et al., 2016). An in-depth review on the production, mechanism of action and applications of nisin and other antimicrobials is available (Santos et al., 2018; Bahrami et al., 2019). As a representative example, nanocomposite films composed of nisin and poly(hydroxybutyrate)/poly(ε-caprolactone) (PHB/PCL) were effective against *Lactobacillus plantarum*, which is a meat spoilage bacteria (Correa et al., 2017).

Lactoferrin is an iron-binding glycoprotein that has similarly shown antimicrobial activity against bacteria, fungi, and parasites. Bacterial cellulose edible films loaded with lactoferrin have proven to be effective against *S. aureus* and

E. coli (Padrão et al., 2016). Moreover, carboxymethylcellulose films loaded with both lactoferrin and lysozyme revealed a potentially synergic antimicrobial effect against *Listeria innocua* (Barbiroli et al., 2012).

Lysozyme's antimicrobial potential is well described against Gram-positive bacteria, given this enzyme's ability to hydrolyze the β-1,4-glucosidic linkages between N-acetylmuramic acid and N-acetylglucosamine in the peptidoglycans present on the cell walls of these bacteria, compromising the membrane integrity (Aziz and Karboune, 2018). On the other hand, Gram-negative bacteria can resist this effect due to the lipopolysaccharide protective layer on their outer membrane (Aziz and Karboune, 2018). The application of this enzyme in poly(vinyl alcohol)/β-cyclodextrin films together with cinnamon essential oil significantly improved their antimicrobial effect against *L. monocytogenes* and *Salmonella enteritidis* (Feng et al., 2017). On a similar approach, lysozyme nanofibers have been incorporated into pullulan films, granting them with antimicrobial activity against *S. aureus* (Silva et al., 2018). Worth mentioning here is the fact that all pathogenic *Staphylococcal* strains, including *S. aureus*, are completely resistant to lysozyme, namely a cationic cell wall-lytic enzyme, but the nanofibrillation of lysozyme bestows antimicrobial activity to lysozyme nanofibers, presumably because of the exposed peptides (Silva et al., 2018).

Chitosan, a polysaccharide obtained from the N-deacetylation of chitin, with remarkable and thoroughly studied antimicrobial effect (against multiple bacteria and fungi) is often combined with other agents for food-packaging applications (Wang et al., 2018). For instance, combinations with metal oxides (Olmo et al., 2020), essential oils (Wang et al., 2017; Hafsa et al., 2016) and ellagic acid (Vilela et al., 2017) have been described, always with significant inhibition of microbial proliferation. A very comprehensive and in-depth review focused on chitosan as an active compound for active packaging (alone or in combination with other molecules) can be found on a very recent work (Kumar et al., 2020).

More instructive literature about the current domain of antimicrobial agents for active packaging can be found elsewhere (Vilela et al., 2018; Ahmed et al., 2017; Yildirim et al., 2018; Otoni et al., 2016; Al-Tayyar et al., 2020; Sofi et al., 2018; Benbettaïeb et al., 2019; Azeredo et al., 2019; Topuz and Uyar, 2020).

3 Carbon dioxide emitters

Carbon dioxide (CO_2) emitters, just like antimicrobial agents, can reduce the growth of microorganisms and, thereby, are extensively used in the food industry to preserve quality and extend self-life of food products. This gaseous molecule is soluble in the foods' aqueous and fat phases, which leads to the formation of carbonic acid and consequent acidification of the foodstuff (Vilela et al., 2018). The antimicrobial power of CO_2 is assumedly driven by an interchange between, for example, the alteration of the bacterial cell membrane, inhibition of bacterial enzymes, and cytoplasmic pH changes, whose combination outcomes in the extension of the lag-phase and, hence, growth retardation of several aerobic spoilage bacteria (Vilela et al., 2018; Yildirim et al., 2018; Sivertsvik et al., 2002). According to Fig. 9.3, various CO_2 emitter systems are commercially available, including Superfresh, CO_2Pad, CO_2 Freshpads, Verifrais, Ageless G, FreshPax M and McAirlaid's CO_2Pad (Yildirim et al., 2018; Kerry, 2014; Coma, 2008). CO_2 emitters can also be applied together with other active agents like oxygen scavengers. Herein, Ageless G and FreshPax M (Fig. 9.3) are commercial examples of dual function systems of CO_2 emitters and O_2 scavengers based on either ferrous carbonate or a mixture of ascorbic acid and sodium bicarbonate (Yildirim et al., 2018; Coma, 2008). Moreover, CO_2 emitters are mostly employed on fresh meat and fish (Yildirim et al., 2018).

In terms of scientific activity, the number of published studies in the domain of CO_2 emitters is limited, as portrayed in Table 9.2, and the majority involves the combination between two compounds, namely sodium bicarbonate ($NaHCO_3$) and an organic acid that is usually citric or ascorbic acid. Furthermore, and since the total amount of carbon dioxide present in the headspace of the package is vital for the antimicrobial effect, CO_2 emitters are frequently used in association with the modified atmosphere packaging (MAP) technology (Yildirim et al., 2018), as indeed exemplified by the studies listed in Table 9.2. The CO_2 emitter systems are typically applied in the form of a sachet or pad, and often combined with a liquid absorber, which enables the reaction of liquids exuded from the food product with the active ingredients, decreasing the pH and triggering the production of carbon dioxide (Vilela et al., 2018; Yildirim et al., 2018).

The partnership between sodium bicarbonate and citric acid dry powder, placed into a liquid absorber pad, has the notorious flexibility of allowing the adjustment of the ratio between active ingredients, thereby controlling the resulting pH to be compatible with the normal pH range of a given food product (Hansen, Moen, Rødbotten, Berget, & Pettersen, 2016; Holck, Pettersen, Moen,

TABLE 9.2 Examples of carbon dioxide emitters included into different matrices.

Active agent	Matrix	Food product	References
$NaHCO_3$ and citric acid	Absorbent pad Dri-Loc + MAP (60% CO_2, 40% N_2)	Cod (*Gadus morhua*)	(Hansen, Moen, Rødbotten, Berget, & Pettersen, 2016)
$NaHCO_3$ and citric acid	Sachet + MAP (100% CO_2)	Chicken	(Holck et al., 2014)
$NaHCO_3$ and citric acid	MAP (60% CO_2, 40% N_2)	Reindeer meat	(Pettersen et al., 2014)

Abbreviations: MAP (modified atmosphere packaging).

& Sørheim, 2014). This fine adjustment is the main benefit of this CO_2 emitter system. In fact, Hansen and coworkers established a model for the calculation of the necessary amounts of active agents to preserve salmon fillets, by taking into account variables like the weight and surface area of the product, tray size and gas volume to product volume ratio (g/p ratio) (Hansen, Høy, & Pettersen, 2009). Furthermore, the effect of the CO_2 releasing system on the quality, shelf life, and MAP package sizes has been documented in food systems (Table 9.2), including cod (Hansen, Moen, Rødbotten, Berget, & Pettersen, 2016), reindeer meat (Pettersen et al., 2014), and chicken (Holck et al., 2014).

The combination between sodium bicarbonate and ascorbic acid is another example of an equivalent CO_2 releasing system where the organic acid acts simultaneously as the acidifier promoting CO_2 formation, and as a reducing agent (i.e., O_2 scavenger (Cruz et al., 2012)). The antioxidant activity of ascorbic acid may well be advantageous in packaging of high fat food products, by decelerating lipid oxidation and the appearance of rancid taste and odors, which will give rise to quality deterioration (Yildirim et al., 2018).

Comprehensive discussions regarding carbon dioxide emitters are accessible in the relevant literature (Vilela et al., 2018; Ahmed et al., 2017; Fang et al., 2017; Yildirim et al., 2018; Coma, 2008).

4 Antioxidant releasers

Oxidative degradation of food products is one of the most relevant causes of food spoilage, making antioxidant releasers a domain of enormous interest for food-packaging applications (Vilela et al., 2018; Gómez-Estaca et al., 2014). The oxidative reactions that affect oxidation-sensitive products result in significant modifications of the characteristics of the food products, namely appearance of off-flavors and odors, aspect changes (e.g., color alterations due to pigments degradation) and decrease of nutritional value, which is triggered by the degradation of fatty acids, proteins, and other macromolecules (Bastarrachea et al., 2015; Sanches-Silva et al., 2014).

Antioxidant releasers are vital to protect oils, nuts, seeds, and meat/fishery products with high fat contents (Ganiari et al., 2017). These additives are commonly included in the packaging material (and not directly in the food formulations) to circumvent any impact on the organoleptic features of the food product. Actually, most of the currently available systems are enclosed in the packaging layer, or included in the form of sachets, pads, or labels (Vilela et al., 2018; Gómez-Estaca et al., 2014; Sanches-Silva et al., 2014). As specified in Fig. 9.3, ATOX film is an example of a commercial antioxidant releaser fabricated by Artibal, S.A. (Spain), whose antioxidant activity is credited to oregano essential oils (Domínguez et al., 2018; Fang et al., 2017).

The antioxidant releasers can prevent the oxidative modification of food products by a multitude of different mechanisms. For categorizing purposes, antioxidant agents can be grouped as either primary (i.e., chain-breaking) or

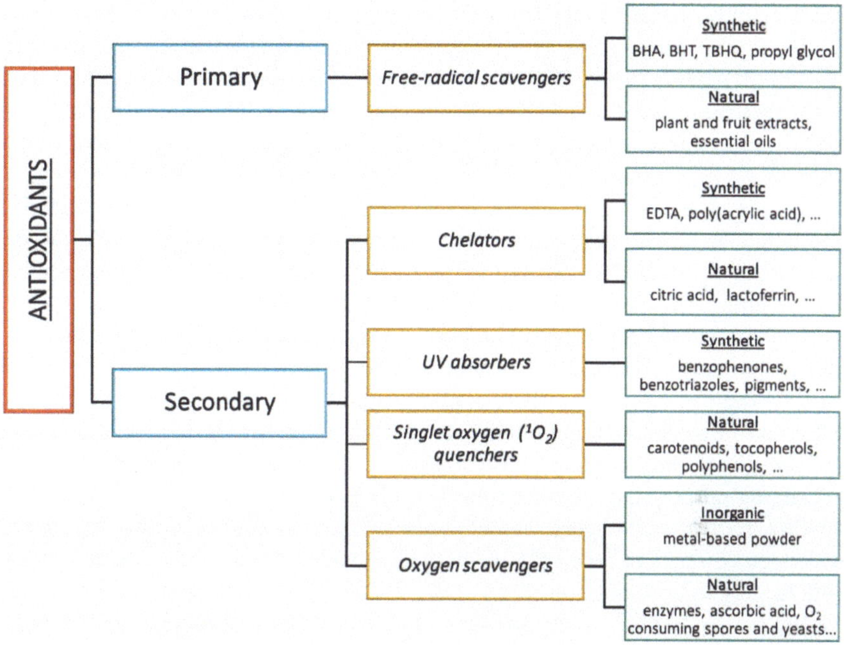

FIGURE 9.5 Classification of antioxidant compounds based on their mechanism of action. (*Tian et al., 2013; Vilela et al., 2018*).

secondary (i.e., preventive) antioxidants (Tian et al., 2013; Vilela et al., 2018), as shown in Fig. 9.5. The action of primary antioxidants, also known as free-radical scavengers, relays on their reaction with free radicals to originate relatively stable compounds that will not participate in additional initiation or propagation reactions. On the other hand, secondary antioxidants effectively reduce or prevent the phenomenon of undesired oxidative reactions, and comprise metal chelators, UV absorbers, singlet oxygen ($^1O_2$) quenchers, and oxygen scavengers (Islam et al., 2017). Certain antioxidant agents might, however, present more than one mechanism of action, and thus belong to more than one group. Moreover, many of these agents (like the $^1O_2$ quenchers, metal chelators, and UV absorbers) exhibit potential for other active food packaging applications (e.g., antimicrobial properties) (Tian et al., 2013).

Current research on this topic has been deviating from the use of synthetic antioxidants like butylated hydroxytoluene (BHT), butylated hydroxyanisole (BHA), and tert-butylhydroquinone (TBHQ) and moving toward natural alternatives (Fig. 9.5), including essential oils and plant or food extracts (Ganiari et al., 2017; Domínguez et al., 2018; Aziz and Karboune, 2018; Sanches-Silva et al., 2014; Pokorný, 2007; Kumar et al., 2015; Amorati et al., 2013). This change of direction is mostly attributed to the suspected systemic effects of the synthetic antioxidants on human health (Nieva-Echevarría et al., 2015). The use of natural antioxidants is clearly an option to outsmart such limitations, but the use of natural extracts and essential oils is actually limited by a relevant constraint, namely the need of a substantial concentration of the natural antioxidants to achieve the desired effect (Tian et al., 2013; Vilela et al., 2018). Plants and fruits extracts (e.g., cashew, mango, rosemary, oregano, green tea, grape seed, green apple, mint, murta, and pomegranate peel), essential oils (e.g., apricot kernel, cinnamon, lemongrass, clove bud, thyme, ginger, oregano, pimento, and bergamot), and other natural pure compounds (e.g., tocopherol, carvacrol, quercetin, catechin, thymol, caffeic acid, and ascorbic acid) have emerged as natural antioxidant alternatives to the synthetic counterparts.

Given so, this section will comprise a concise description of the latest advances on each of the classes of antioxidants described earlier, except for the most widely explored active packaging technology, the oxygen scavengers, which will be covered individually on the following section. Special attention will be given here to natural and non-toxic alternatives for antioxidant food packaging purposes, as summarized in Table 9.3.

Free-radical scavengers, namely primary antioxidants, are able to donate hydrogen to reactive free radicals and form stable compounds which are unable to perpetrate oxidation (Tian et al., 2013). The synthetic BHA, BHT, and TBHQ belong to this group, but some plant extracts, essential oils, and tocopherols are also known to trigger this kind of antioxidant mechanism (Sanches-Silva et al., 2014). Notable reviews on this matter are available in the literature (Islam et al., 2017; Maisanaba et al., 2017), and representative examples of works using natural free-radical scavengers include, for instance, the use of hydroxypropyl-methylcellulose-based films loaded with green tea extract (Wrona et al., 2017), pullulan films containing protein nanofibers (Silva et al., 2018) or chestnut spiny burs and roasted hazelnut skins extracts (Esposito et al., 2020), and the widespread use of chitosan films with a vast array of antioxidant compounds, such as ellagic acid (Vilela et al., 2017), kombucha tea extract (Ashrafi and Jokar, 2018), mango leaf (Rambabu et al., 2019) or apricot kernel essential oil (Priyadarshi et al., 2018) have been described.

Alternatively, secondary antioxidant additives reduce or avoid the oxidative reactions. Metallic chelating agents act by compromising the ability of metallic pro-oxidants to potentiate lipid-oxidation reactions in food products. These metallic pro-oxidants (mostly iron or copper derivatives), form stable products when bound to synthetic metal chelators like ethylenediaminetetraacetic acid (EDTA) and poly(acrylic acid) (PAA), or natural antioxidants like citric acid or certain iron-binding proteins (e.g., lactoferrin) (Tian et al., 2013). Relevant publications on this topic describe, for instance, the grafting of acrylic acid from a polypropylene (PP) surface by photo-grafting technique (Lin et al., 2016) or the use of polyphenols (catechol and catechin) on PP films for the chelation of Fe^{3+} (Roman et al., 2016). Another recent publication reported the successful synthesis of a chelating copolymer based on poly(*n*-butyl acrylate) and iminodiacetic acid through emulsion polymerization, and its application on PP films (Lin and

TABLE 9.3 Examples of antioxidant releasers included into different matrices.

Active agent	Matrix	Food product	References
Chestnut spiny burs and roasted hazelnut skins extracts	Pullulan	—	(Esposito et al., 2020)
Rosemary extract	Cassava starch	Aqueous and fatty foods	(Piñeros-Hernandez et al., 2017)
Green tea extract	HPMC with PLA nanoparticles	—	(Wrona et al., 2017)
Naringin (and citrus extract)	Chitosan	—	(Iturriaga et al., 2014)
Buriti oil	Chitosan	—	(Silva et al., 2016)
Ellagic acid	Chitosan	—	(Vilela et al., 2017)
Catechin or catechol	PP	—	(Roman et al., 2016)
Ferulic acid and feruloylated-AX oligosaccharides	NFC and AX	—	(Moreirinha et al., 2020)
Bixin	Cassava starch	Sunflower oil	(Pagno et al., 2016)
Lysozyme nano-fibrils	Pullulan	—	(Silva et al., 2018)
AA	PP	—	(Lin et al., 2016)
HA	PP	—	(Lin et al., 2016)
HA	PP	Liquid and semi-liquid foods	(Roman et al., 2015)
HA	PP	—	(Tian et al., 2013)
HA	PET	—	(Johnson et al., 2015)
IDA	PP	—	(Lin et al., 2016)
IDA	PP-g-PAA	—	(Lin and Goddard, 2018)
Lignin	Alginate	—	(Aadil et al., 2016)
Lignin nanoparticles	PVA and chitosan	—	(Yang et al., 2016)
Boric acid	Chitosan and gelatin	—	(Ahmed and Ikram, 2016)

Abbreviations: AA (acrylic acid), AX (arabinoxylans), HA (hydroxamic acid), HPMC (hydroxypropyl methylcellulose), IDA (iminodiacetate), NFC (nanofibrillated cellulose), PAA (poly(acrylic acid)), PET (poly(ethylene terephthalate)), PLA (poly(lactic acid)), PP (polypropylene), PVA (poly(vinyl alcohol)).

Goddard, 2018). Older works by the same team had already described the use of iminodiacetate and other derivatives for the same purposes (Lin et al., 2016; Lin and Goddard, 2018).

UV absorbers (like benzotriazoles, benzophenones, and some pigments like phtalocyanine or TiO_2) can absorb the UV radiation that could, otherwise, cause the photo-oxidation of lipids in sensitive food products. These absorbers are frequently added into food-packaging materials (like plastics) with insufficient light-barrier properties, to block UV light and circumvent eventual oxidative reactions of oily products caused by radiation. Works on UV absorbers include the use of many different compounds: feruloylated arabinoxylo-oligosaccharides (Moreirinha et al., 2020), boric acid (Ahmed and Ikram, 2016), ellagic acid (Vilela et al., 2017), lignin (Aadil et al., 2016), naringin (Iturriaga et al., 2014), and others (Kwon et al., 2018).

$^1O_2$ is normally generated in foodstuffs by the presence of light and certain photosensitizers (like myoglobin, riboflavin, or chlorophyll). This highly unstable molecule potentiates the development of oxidative reactions and generation of lipid hydroperoxides. $^1O_2$ quenchers have the ability to deplete the excess energy of $^1O_2$ and consequently prevent the photo-oxidation of lipids (Bradley and Min, 1992). They are probably the most underexplored class of antioxidant compounds for food packaging, and the few publications on the topic focus on the use of natural agents (like carotenoids, tocopherols, and polyphenols) (Tian et al., 2013). An interesting work by Colín-Chávez and colleagues described the combined use of TiO_2 and carotenoids (from marigold extract) to modify high-density polyethylene films for the active packaging of soybean oil. According to the authors, the simultaneous use of TiO_2 and carotenoids protected the latter from UV degradation and was able to significantly delay the oxidation of soybean oil, tripling the time before undesired oxidation (Colín-Chávez et al., 2014).

Further insights into the field of antioxidant agents can be explored in the appropriate literature (Ganiari et al., 2017; Domínguez et al., 2018; Islam et al., 2017; Aziz and Karboune, 2018; Maisanaba et al., 2017; Benbettaïeb et al., 2019; Topuz and Uyar, 2020; Gómez-Estaca et al., 2014; Sanches-Silva et al., 2014; Pokorný, 2007; Kumar et al., 2015; Amorati et al., 2013).

5 Oxygen scavengers

Multiple food products are susceptible to oxygen and the existence of residual headspace oxygen on packaged foods can impair their quality and shelf-life, by enabling the growth of aerobic microorganisms or promoting the oxidation of the food product (Dey and Neogi, 2019). Consequently, sensorial, color, and nutritional changes take place and food spoilage occurs (Yildirim et al., 2018). Given so, oxygen must often be excluded from the food packaging of certain products. The food industry uses mainly gas flushing or modified atmosphere packaging (MAP) processes to eliminate oxygen from packaged foods or, more effectively, oxygen scavengers that control residual oxygen (Vilela et al., 2018). As outlined in Fig. 9.3, examples of commercial products with oxygen scavenger technology include ActiTUF, Ageless, ATCO, Bioka, Celox, Cryovac OS2000, Enzyme-based FreshMax, FreshPax, OMAC, OxyGuard, OxyCatch, OxyRx, and Shelfplus O_2 (Wyrwa and Barska, 2017; Realini and Marcos, 2014). Regarding food products, oxygen scavenger systems are mostly applied to (sliced) cooked meat products, grated cheese, (par-baked) bakery products, fruit and vegetable juices, seeds, nuts, oils, fat-containing instant powders, fried snacks, and dried meat products (Yildirim et al., 2018).

The mechanism of oxidation is contingent on the used active compounds, and can encompass iron, platinum group metals, unsaturated hydrocarbon, α-tocopherol, ascorbic acid, enzyme, and microorganisms based systems, as recently reviewed (Dey and Neogi, 2019). Most commercial oxygen scavengers available include iron-based sachets containing iron powder, whose mechanism of action is moisture activated and based on iron oxidation (Arvanitoyannis and Oikonomou, 2012; Cooksey, 2010). Still, sachet-based applications have some limitations mostly associated with the risk of accidental rupture and thus automatic content ingestion, as well as the prerequisite of a supplementary packaging operation stage, the unsuitability for beverages, and the inserts aesthetics that vary from country to country (Vilela et al., 2018). So, engineering alternatives based on polymeric packaging films or containers as matrices for oxygen scavenger inclusion are detours to thwart those limitations (Yildirim et al., 2018). Illustrative examples of such oxygen scavenger technologies applied to food products are outlined in Table 9.4 and comprise iron nanoparticles, palladium, titanium oxide nanotubes, ascorbic acid, phenolic compounds (e.g., pyrogallol, gallic acid, and α-tocopherol) and enzymes as active oxygen scavenging agents in polymeric matrices (e.g., silicon, PET, low-density polyethylene (LDPE), PLA).

An interesting publication concerning alternative oxygen scavenger systems includes the study of a non-metallic oxygen scavenger composed of activated carbon and sodium L-ascorbate (Lee et al., 2018). According to Lee and coworkers, this oxygen scavenger system exhibited an oxygen-scavenging volume *ca.* 2.2 times higher than those of commercial iron powder-containing oxygen scavengers. Furthermore, the utilization of these ascorbic acid-based oxygen scavengers during storage of raw meatloaves validated their potential for lipid oxidation inhibition and reduction of microbial growth of lactic acid bacteria, yeasts, and molds (Lee et al., 2018).

Another original development was reported by Gaikwad and coworkers who fabricated films of LDPE containing a non-metallic oxygen scavenger based on pyrogallol (i.e., a phenolic compound with high oxygen scavenging ability) (Gaikwad et al., 2017a,b). These films have high oxygen scavenging capacity (0.816 mL O_2 per cm^2 after 8 days) under storage conditions of elevated humidity (75%) and temperature (60°C) (Gaikwad et al., 2017a). Moreover, soybean oil samples packaged with LDPE/pyrogallol films kept their stability during the

TABLE 9.4 Examples of oxygen scavengers loaded into different matrices.

Active agent	Matrix	Food product	References
Zero valent iron nanoparticles	Silicon	—	(Foltynowicz et al., 2017)
Palladium	PET/SiO_x	Ham	(Yildirim et al., 2015)
Palladium	PET/SiO_x	Cooked cured ham	(Hutter et al., 2016)
Palladium	PET/SiO_x	Linseed oil	(Faas et al., 2020)
Titanium oxide nanotubes	—	—	(Tulsyan et al., 2017)
Ascorbic acid	Activated carbon	Meatloaves	(Lee et al., 2018)
Gallic acid	LDPE	—	(Ahn et al., 2016)
Gallic acid	Multilayered bio-based film	—	(Pant et al., 2017)
Pyrogallic acid	LDPE	Peeled garlic	(Singh et al., 2019)
Pyrogallol	LDPE	Soybean oil	(Gaikwad et al., 2017a,b)
α-Tocopherol	PLA microparticles	—	(Scarfato et al., 2017)
Glucose oxidase	Poly(ethylene-vinyl acetate)	—	(Wong et al., 2017)
$FeO(OH) \times H_2O$, Fe_2O_3 and ascorbic acid	LDPE/PET laminated films	—	(Wołosiak-Hnat et al., 2019)
CNCs and Pd nanoparticles	Poly(ethylene-co-vinyl alcohol)	—	(Cherpinski et al., 2019)

Abbreviations: CNCs (cellulose nanocrystals), LDPE (Low-density polyethylene), PET (poly(ethylene terephthalate)), PLA (poly(lactic acid)).

30 days storage period (Gaikwad et al., 2017b). As an extension to these studies, Singh and coworkers modified a LDPE film via blending with sodium carbonate and coating with pyrogallic acid (Singh et al., 2019). According to this study, the oxygen scavenging film was used to store peeled garlic at 5°C for 30 days, and during this period the quality of the fresh product was retained.

Oxygen scavenger systems can also be used in tandem with other active agents like antimicrobial additives. For instance, the commercial Ageless oxygen scavenger iron-based sachet (one of the first developed sachets (Otoni et al., 2016)) was combined with ginger essential oil as an antimicrobial agent in a multilayer film plastic pouch of ethylene-vinyl alcohol, aiming at extending the shelf-life of fresh cobia fish steaks stored at 2°C (Remya, Mohan,

Venkateshwarlu, Sivaraman, & Ravishankar, 2017). This packaging system was able to decrease and maintain the oxygen concentration to less than 0.01% during storage, and concurrently reduce or inhibit the growth of *Pseudomonas* spp., lactic acid bacteria, and *Brochothrix thermosphact* (Remya, Mohan, Venkateshwarlu, Sivaraman, & Ravishankar, 2017).

Additional reading about the state of affairs in oxygen scavenging systems can be reached in the appropriate literature (Yildirim and Röcker, 2018; Cruz et al., 2012; Dey and Neogi, 2019; Gaikwad et al., 2018).

6 Ethylene absorbers

The ripening and senescence of fruits and vegetables is accelerated by phytohormones, such as ethylene (C_2H_4), which is a volatile unsaturated hydrocarbon molecule (Álvarez-Hernández et al., 2018; Gaikwad et al., 2020). Hence, it is important from the standpoint of food quality and shelf-life that the post-harvesting ethylene levels in the atmosphere surrounding fresh food products during shipping, storage, and handling are monitored and controlled. So, ethylene absorbers have emerged as a strategy to enhance the quality and extend shelf-life of several fresh food products (Gaikwad et al., 2020), namely climacteric fruits and vegetables. As depicted in Fig. 9.3, several ethylene absorber systems are commercially available, including Bio-fresh, Ethylene Control Power Pellet, Ethysorb, EvertFresh Green Bags, Retarder, PEAKfresh, Profresh, and Bi-On (Álvarez-Hernández et al., 2018; Yildirim et al., 2018; Wyrwa and Barska, 2017).

Ethylene absorbers based on potassium permanganate ($KMnO_4$) supported on inert matrices, like silica gel or alumina, are the most widely used systems (Álvarez-Hernández et al., 2019). Herein, $KMnO_4$ oxidizes C_2H_4, which is associated with a color change from purple to brown in this process, demonstrating the C_2H_4-scavenging capacity. As an illustrative example, silica (SiO_2) and alumina (Al_2O_3) nanoparticles impregnated with $KMnO_4$ were used as a nanostructured platform to oxidize ethylene by simply following the color change from purple to brown, indicating ethylene removal (Spricigo, Foschini, Ribeiro, Corrêa, & Ferreira, 2017). Although $KMnO_4$ has potential as ethylene absorber in the form of sachets during storage of fresh food products (e.g., tomatoes (Köstekli et al., 2016)), this inorganic compound is toxic, presents restricted long-term efficacy in high-moisture environments and, thus, cannot be applied in direct contact with food products (Álvarez-Hernández et al., 2018; Yildirim et al., 2018).

In this context, metal oxides (e.g., silica gel and activated alumina), layer silicates and zeolites (e.g., clays, vermiculite, zeolite), nanoparticles, and activated carbon (Álvarez-Hernández et al., 2018; Yildirim et al., 2018) are examples of surrogate systems for ethylene removal that can be included into the package material or supplied in sachets for insertion into packages or storage environments. Some studies have been published about ethylene absorbers in the area

TABLE 9.5 Examples of ethylene absorbers supported on different matrices.

Active agent	Matrix	Food product	References
$KMnO_4$	SiO_2 and Al_2O_3 nanoparticles	Tomato	(Spricigo, Foschini, Ribeiro, Corrêa, & Ferreira, 2017)
$KMnO_4$	Sepiolite	Apricot	(Álvarez-Hernández et al., 2020)
Palladium and $KMnO_4$ promoted nano-zeolite	—	Tomato	(Mansourbahmani et al., 2018)
$PdCl_2$ and $CuSO_4$	Activated carbon powder	Broccoli	(Cao, Li, Wu, Jiang, & Qu, 2015)
TiO_2 nanoparticles	Chitosan	—	(Siripatrawan and Kaewklin, 2018)
TiO_2 nanoparticles	Chitosan	Cherry tomatoes	(Kaewklin et al., 2018)
Ce-doped TiO_2 with dispersed Pt atoms	Cellulose scaffold	—	(Guo et al., 2019)
Cu- and Al-based MOF	—	Banana	(Chopra et al., 2017)
HNTs	PE	Banana, tomato, and strawberry	(Tas et al., 2017)
Porous clay heterostructures	PP	—	(Srithammaraj et al., 2018)

Abbreviations: HNTs (halloysite nanotubes), MOF (metal organic framework), PE (polyethylene), PP (polypropylene).

of active food packaging and are summarized in Table 9.5. A newly published research includes the study of Kaewklin and coworkers who developed chitosan films containing nanosized titanium dioxide (TiO_2) (Siripatrawan and Kaewklin, 2018) to preserve quality and prolong storage life of cherry tomatoes (Kaewklin et al., 2018). Moreover, the authors purport their study as the first dealing with the use of chitosan and TiO_2 nanocomposite films as an ethylene scavenger for postharvest handling of a climacteric fruit. These films present ethylene photodegradation that can hinder the ripening process and extend the storage life of this vegetable crop (Kaewklin et al., 2018).

In a distinct study, metal organic frameworks (MOFs) were tested as ethylene absorbers with banana fruits, where copper-based MOF (Basolite C300) and aluminum-based MOF (Basolite A520) potentially sorb, store, and release gaseous compounds that influence plant physiology, namely ethylene and 1-methylcyclopropene (1-MCP), an ethylene action inhibiter (Chopra et al., 2017). In another contribution, palladium-promoted nano-zeolite was used as a tool for the simultaneous extension of self-life and preservation of quality characteristics of tomato fruits during storage (Mansourbahmani et al., 2018). Furthermore, the authors demonstrated that this ethylene absorber system is more efficient than $KMnO_4$-promoted nano-zeolite, 1-MCP, $CaCl_2$, salicylic acid, and UV-C radiation, on the maintenance of postharvest quality of tomatoes (Mansourbahmani et al., 2018).

Once more, the combination of two or more active agents in the same packaging system is also reported. As mentioned earlier, Siripatrawan and Kaewklin fabricated nanocomposite films composed of chitosan and TiO_2 nanoparticles for the simultaneous application as ethylene scavenging and antimicrobial active food packaging (Siripatrawan and Kaewklin, 2018). According to this study, the films exhibited an increasing ethylene photodegradation with the increase of the TiO_2 concentration and antimicrobial activity against Gram-positive (*S. aureus*) and Gram-negative (*E. coli*, *S. typhimurium*, and *P. aeruginosa*) bacteria and fungi (*Aspergillus* and *Penicillium*) (Siripatrawan and Kaewklin, 2018).

Additional aspects about the current situation in ethylene absorber systems to expand the postharvest life of fruits and vegetables are accessible in numerous appraisals (Álvarez-Hernández et al., 2018, 2019; Gaikwad et al., 2020; Zhang et al., 2017; Sadeghi et al., 2020).

7 Conclusion

Innovations in food packaging originated the concept of active packaging, well-defined as packaging systems that "deliberately incorporate components that would release or absorb substances into or from the packaged food or the environment surrounding the food" (Yildirim and Röcker, 2018). Thus, the active packaging technology exhibits multiple active functions, alongside from acting as a physical barrier between the food product and the external conditions. The present chapter ventured into the advances in the realm of active agents for active food packaging. Emphasis was set on active functions, such as antimicrobial activity, carbon dioxide emitting, antioxidant activity, oxygen and ethylene scavenging (Fig. 9.4), as well as on packaging materials derived from renewable resources, namely polysaccharides, proteins and PLA, as part of the concept of sustainable development.

The design of active packaging systems should take into consideration the characteristics of the food, like pH, water activity, and nutritional components, and the activity, stability, migration, and toxicity of the active agents, to yield systems capable of improving the shelf-life, safety, and quality of food products. The major obstacle in developing active packaging systems is ineluctably the simultaneous preservation of the mechanical and barrier properties of the original materials, and the activity of the active agents during the whole activity of shipping, storage, and handling the food packaging materials.

Notwithstanding the magnitude and variety of the research activities performed in the last years, where the active agents used for active packaging are following the same tendency as the packaging materials toward natural-based eco-friendly alternatives, not all systems will reach commercial translation. In fact, most literature is centered on the active properties of the supplemented packaging materials, instead of examining their behavior in the presence of real food products. Nonetheless, the pertinency of this subject will persist to attract interest from both academia and industry.

Acknowledgments

This work was developed within the scope of the project CICECO—Aveiro Institute of Materials, UIDB/50011/2020 & UIDP/50011/2020, financed by national funds through the Portuguese Foundation for Science and Technology (FCT)/MCTES. The AgroForWealth project (CENTRO-01-0145-FEDER-000001) is acknowledged for the research grant to J.P.F. Carvalho. FCT is also acknowledged for the research contract under Scientific Employment Stimulus to C. Vilela (CEECIND/00263/2018).

References

Aadil, K. R., Prajapati, D., & Jha, H. (2016). Improvement of physcio-chemical and functional properties of alginate film by *Acacia* lignin. *Food Packaging and Shelf Life*, *10*, 25–33.

Ahmed, S., & Ikram, S. (2016). Chitosan and gelatin based biodegradable packaging films with UV-light protection. *Journal of Photochemistry and Photobiology B: Biology*, *163*, 115–124.

Ahmed, I., Lin, H., Zou, L., Brody, A. L., Li, Z., Qazi, I. M., et al. (2017). A comprehensive review on the application of active packaging technologies to muscle foods. *Food Control*, *82*, 163–178.

Ahn, B. J., Gaikwad, K. K., & Lee, Y. S. (2016). Characterization and properties of LDPE film with gallic-acid-based oxygen scavenging system useful as a functional packaging material. *Journal of Applied Polymer Science*, *133*, 44138.

Akhidime, I. D., Saubade, F., Benson, P. S., Butler, J. A., Olivier, S., Kelly, P., et al. (2019). The antimicrobial effect of metal substrates on food pathogens. *Food and Bioproducts Processing*, *113*, 68–76.

Al-Tayyar, N. A., Youssef, A. M., & Al-hindi, R. (2020). Antimicrobial food packaging based on sustainable bio-based materials for reducing foodborne pathogens: A review. *Food Chemistry*, *310*, 125915.

Álvarez-Hernández, M. H., Artés-Hernández, F., Ávalos-Belmontes, F., Castillo-Campohermoso, M. A., Contreras-Esquivel, J. C., Ventura-Sobrevilla, J. M., et al. (2018). Current scenario of adsorbent materials used in ethylene scavenging systems to extend fruit and vegetable postharvest life. *Food Bioprocess Technology*, *11*(3), 511–525.

Álvarez-Hernández, M. H., Martínez-Hernández, G. B., Avalos-Belmontes, F., Castillo-Campohermoso, M. A., Contreras-Esquivel, J. C., & Artés-Hernández, F. (2019). Potassium permanganate-based ethylene scavengers for fresh horticultural produce as an active packaging. *Food Engineering Review*, *11*(3), 159–183.

Álvarez-Hernández, M. H., Martínez-Hernández, G. B., Avalos-Belmontes, F., Miranda-Molina, F. D., & Artés-Hernández, F. (2020). Postharvest quality retention of apricots by using a novel sepiolite–loaded potassium permanganate ethylene scavenger. *Postharvest Biology and Technology, 160,* 111061.

Amorati, R., Foti, M. C., & Valgimigli, L. (2013). Antioxidant activity of essential oils. *Journal of Agricultural and Food Chemistry, 61,* 10835–10847.

Arvanitoyannis, I. S., & Oikonomou, G. (2012). Active and intelligent packaging. In I. S. Arvanitoyannis (Ed.), *Modified atmosphere and active packaging technologies* (pp. 628–654). Boca Raton: CRC Press.

Ashrafi, A., Jokar, M., & Nafchi, A. M. (2018). Preparation and characterization of biocomposite film based on chitosan and kombucha tea as active food packaging. *International J Biological Macromolecules, 108,* 444–454.

Ataei, S., Azari, P., Hassan, A., Pingguan-Murphy, B., Yahya, R., & Muhamad, F. (2020). Essential oils-loaded electrospun biopolymers: A future perspective for active food packaging. *Advances in Polymer Technology, 2020,* 9040535.

Atarés, L., & Chiralt, A. (2016). Essential oils as additives in biodegradable films and coatings for active food packaging. *Trends in Food Science & Technology, 48,* 51–62.

Azeredo, H. M. C., Otoni, C. G., Corrêa, D. S., Assis, O. B. G., Moura, M. R., & Mattoso, L. H. C. (2019). Nanostructured antimicrobials in food packaging—Recent advances. *Biotechnology Journal, 14*(12), 1900068.

Aziz, M., & Karboune, S. (2018). Natural antimicrobial/antioxidant agents in meat and poultry products as well as fruits and vegetables : A review. *Critical Reviews in Food Science and Nutrition, 58*(3), 486–511.

Bahrami, A., Delshadi, R., Jafari, S. M., & Williams, L. (2019). Nanoencapsulated nisin: An engineered natural antimicrobial system for the food industry. *Trends in Food Science & Technology, 94,* 20–31.

Barbiroli, A., Bonomi, F., Capretti, G., Iametti, S., Manzoni, M., Piergiovanni, L., et al. (2012). Antimicrobial activity of lysozyme and lactoferrin incorporated in cellulose-based food packaging. *Food Control, 26*(2), 387–392.

Bastarrachea, L. J., Wong, D. E., Roman, M. J., Lin, Z., & Goddard, J. M. (2015). Active Packaging Coatings. *Coatings, 5*(4), 771–791.

Benbettaïeb, N., Debeaufort, F., & Karbowiak, T. (2019). Bioactive edible films for food applications: mechanisms of antimicrobial and antioxidant activity. *Critical Review in Food Science and Nutrition, 59*(21), 3431–3455.

Bradley, D. G., & Min, D. B. (1992). Singlet oxygen oxidation of foods. *Critical Review in Food Science and Nutrition, 31*(3), 211–236.

Cao, J., Li, X., Wu, K., Jiang, W., & Qu, G. (2015). Preparation of a novel $PdCl_2$-$CuSO_4$-based ethylene scavenger supported by acidified activated carbon powder and its effects on quality and ethylene metabolism of broccoli during shelf-life. *Postharvest Biology and Technology, 99,* 50–57.

Cherpinski, A., Biswas, A., Lagaron, J. M., Dufresne, A., Kim, S., Buttrum, M., et al. (2019). Preparation and evaluation of oxygen scavenging nanocomposite films incorporating cellulose nanocrystals and Pd nanoparticles in poly(ethylene-co-vinyl alcohol). *Cellulose, 26*(12), 7237–7251.

Chopra, S., Dhumal, S., Abeli, P., Beaudry, R., & Almenar, E. (2017). Metal-organic frameworks have utility in adsorption and release of ethylene and 1-methylcyclopropene in fresh produce packaging. *Postharvest Biology and Technology, 130,* 48–55.

Colín-Chávez, C., Vicente-Ramírez, E. B., Soto-Valdez, H., Peralta, E., & Auras, R. (2014). The release of carotenoids from a light-protected antioxidant active packaging designed to improve the stability of soybean oil. *Food and Bioprocess Technology, 7*(12), 3504–3515.

Coma, V. (2008). Bioactive packaging technologies for extended shelf life of meat-based products. *Meat Science*, *78*(1–2), 90–103.

Cooksey, K. (2010). Oxygen scavenging packaging systems. In *Encyclopedia of polymer science and technology*. John Wiley & Sons, Inc.

Correa, J. P., Molina, V., Sanchez, M., Kainz, C., Eisenberg, P., & Massani, M. B. (2017). Improving ham shelf life with a polyhydroxybutyrate/polycaprolactone biodegradable film activated with nisin. *Food Packaging and Shelf Life*, *11*, 31–39.

Cruz, R. S., Camilloto, G. P., Pires, A. C., & dos, S. (2012). Oxygen scavengers: An approach on food preservation. In A. A. Eissa (Ed.), *Structure and function of food engineering* (pp. 21–42). Rijeka, Croatia: InTech.

Dakal, T. C., Kumar, A., Majumdar, R. S., & Yadav, V. (2016). Mechanistic basis of antimicrobial actions of silver nanoparticles. *Front in Microbiology*, *7*, 1831.

Dey, A., & Neogi, S. (2019). Oxygen scavengers for food packaging applications: A review. *Trends in Food Science & Technology*, *90*, 26–34.

Dobrucka, R., & Przekop, R. (2019). New perspectives in active and intelligent food packaging. *Journal of Food Processing and Preservation*, *43*(11), e14194.

Domínguez, R., Barba, F. J., Gómez, B., Putnik, P., Kovačević, D. B., Pateiro, M., et al. (2018). Active packaging films with natural antioxidants to be used in meat industry: A review. *Food Research Internationa*, *113*, 93–101.

Echeverría, I., López-Caballero, M. E., Gómez-Guillén, M. C., Mauri, A. N., & Montero, M. P. (2018). Active nanocomposite films based on soy proteins-montmorillonite- clove essential oil for the preservation of refrigerated bluefin tuna (*Thunnus thynnus)* fillets. *International Journal of Food Microbiology*, *266*, 142–149.

Ejaz, M., Arfat, Y. A., Mulla, M., & Ahmed, J. (2018). Zinc oxide nanorods/clove essential oil incorporated Type B gelatin composite films and its applicability for shrimp packaging. *Food Packaging and Shelf Life*, *15*, 113–121.

Esposito, T., Silva, N. H. C. S., Almeida, A., Silvestre, A. J. D., Piccinelli, A., Aquino, R. P., et al. (2020). Valorisation of chestnut spiny burs and roasted hazelnut skins extracts as bioactive additives for packaging films. *Industrial Crop and Products*, *151*, 112491.

Faas, N., Röcker, B., Smrke, S., Yeretzian, C., & Yildirim, S. (2020). Prevention of lipid oxidation in linseed oil using a palladium-based oxygen scavenging film. *Food Packaging and Shelf Life*, *24*, 100488.

Fang, Z., Zhao, Y., Warner, R. D., & Johnson, S. K. (2017). Active and intelligent packaging in meat industry. *Trends in Food Science & Technology*, *61*(2), 60–71.

Feng, K., Wen, P., Yang, H., Li, N., Lou, W. Y., Zong, M. H., et al. (2017). Enhancement of the antimicrobial activity of cinnamon essential oil-loaded electrospun nanofilm by the incorporation of lysozyme. *RSC Advances*, *7*(3), 1572–1580.

Figueroa-Lopez, K. J., Vicente, A. A., Reis, M. A. M., Torres-Giner, S., & Lagaron, J. M. (2019). Antimicrobial and antioxidant performance of various essential oils and natural extracts and their incorporation into biowaste derived poly(3-hydroxybutyrate-co-3-hydroxyvalerate) layers made from electrospun ultrathin fibers. *Nanomaterials*, *9*(2), 1–22.

Foltynowicz, Z., Bardenshtein, A., Sängerlaub, S., Antvorskov, H., & Kozak, W. (2017). Nanoscale, zero valent iron particles for application as oxygen scavenger in food packaging. *Food Packaging and Shelf Life*, *11*, 74–83.

Gaikwad, K. K., Singh, S., & Lee, Y. S. (2017). A new pyrogallol coated oxygen scavenging film and their effect on oxidative stability of soybean oil under different storage conditions. *Food Science and Biotechnology*, *26*(6), 1535–1543.

Gaikwad, K. K., Singh, S., & Lee, Y. S. (2017). A pyrogallol-coated modified LDPE film as an oxygen scavenging film for active packaging materials. *Progress in Organic Coatings*, *111*, 186–195.

Gaikwad, K. K., Singh, S., & Lee, Y. S. (2018). Oxygen scavenging films in food packaging. *Environmental Chemistry Letters*, *16*, 523–538.

Gaikwad, K. K., Singh, S., & Negi, Y. S. (2020). Ethylene scavengers for active packaging of fresh food produce. *Environmental Chemistry Letters*, *18*, 269–284.

Ganiari, S., Choulitoudi, E., & Oreopoulou, V. (2017). Edible and active films and coatings as carriers of natural antioxidants for lipid food. *Trends in Food Science & Technology*, *68*, 70–82.

Gómez-Estaca, J., López-de-Dicastillo, C., Hernández-Munoz, P., Catalá, R., & Gavara, R. (2014). Advances in antioxidant active food packaging. *Trends in Food Science & Technology*, *35*(1), 42–51.

Guo, H., Warnicke, P., Griffa, M., Müller, U., Chen, Z., Schaeublin, R., et al. (2019). Hierarchical porous wood cellulose scaffold with atomically dispersed Pt catalysts for low-temperature ethylene decomposition. *ACS Nano*, *13*(12), 14337–14347.

Hafsa, J., Smach, M. ali., Ben Khedher, M. R., Charfeddine, B., Limem, K., Majdoub, H., et al. (2016). Physical, antioxidant and antimicrobial properties of chitosan films containing *Eucalyptus globulus* essential oil. *LWT—Food Science and Technology*, *68*, 356–364.

Haghighi-Manesh, S., & Azizi, M. H. (2017). Active packaging systems with emphasis on its applications in dairy products. *Journal of Food Processing Engineering*, *40*, e12542.

Hansen, AÅ, Høy, M., & Pettersen, M. K. (2009). Prediction of optimal CO_2 emitter capacity developed for modified atmosphere packaging of fresh salmon fillets (*Salmo salar* L.). *Packaging Technology and Science*, *22*(4), 199–208.

Hansen, AÅ, Moen, B., Rødbotten, M., Berget, I., & Pettersen, M. K. (2016). Effect of vacuum or modified atmosphere packaging (MAP) in combination with a CO_2 emitter on quality parameters of cod loins (*Gadus morhua*). *Food Packaging and Shelf Life*, *9*, 29–37.

Holck, A. L., Pettersen, M. K., Moen, M. H., & Sørheim, O. (2014). Prolonged shelf life and reduced drip loss of chicken filets by the use of carbon dioxide emitters and modified atmosphere packaging. *Journal of Food Protection*, *77*(7), 1133–1141.

Hoseinnejad, M., Jafari, S. M., & Katouzian, I. (2017). Inorganic and metal nanoparticles and their antimicrobial activity in food packaging applications. *Critical Reviews in Microbiology*, 1–21.

Hutter, S., Rüegg, N., & Yildirim, S. (2016). Use of palladium based oxygen scavenger to prevent discoloration of ham. *Food Packaging and Shelf Life*, *8*, 56–62.

Islam, R. U., Khan, M. A., & Islam, S. U. (2017). Plant derivatives as promising materials for processing and packaging of meat-based products—Focus on antioxidant and antimicrobial effects. *J of Food Processing and Preservation*, *41*(2), e12862.

Iturriaga, L., Olabarrieta, I., Castellan, A., Gardrat, C., & Coma, V. (2014). Active naringin-chitosan films: Impact of UV irradiation. *Carbohydrate Polymers*, *110*, 374–381.

Janjarasskul, T., & Suppakul, P. (2018). Active and intelligent packaging: The indication of quality and safety. *Critical Reviews in Food Science and Nutrition*, *58*(5), 808–831.

Johnson, D. R., Tian, F., Roman, M. J., Decker, E. A., & Goddard, J. M. (2015). Development of iron-chelating poly(ethylene terephthalate) packaging for inhibiting lipid oxidation in oil-in-water emulsions. *Journal of Agricultural and Food Chemistry*, *63*(20), 5055–5060.

Kaewklin, P., Siripatrawan, U., Suwanagul, A., & Lee, Y. S. (2018). Active packaging from chitosan-titanium dioxide nanocomposite film for prolonging storage life of tomato fruit. *International Journal of Biological Macromolecules*, *112*, 523–529.

Kerry, J. P. (2014). New packaging technologies, materials and formats for fast-moving consumer products. In J. H. Han (Ed.), *Innovations in Food Packaging* (pp. 549–584). San Diego: Elsevier.

Köstekli, M., Özdzikicierlev, O., Cortés, C., Zulueta, A., Esteve, M. J., & Frígola, A. (2016). Role of potassium permanganate ethylene on physicochemical properties, during storage of five different tomato cultivars. *MOJ Food Processing & Technology*, *3*(2), 00069.

Kumar, Y., Yadav, D. N., Ahmad, T., & Narsaiah, K. (2015). Recent trends in the use of natural antioxidants for meat and meat products. *Comprehensive Reviews in Food Science and Food Safety*, *14*(6), 796–812.

Kumar, S., Mukherjee, A., & Dutta, J. (2020). Chitosan based nanocomposite films and coatings: Emerging antimicrobial food packaging alternatives. *Trends in Food Science & Technology*, *97*, 196–209.

Kwon, S., Orsuwan, A., Bumbudsanpharoke, N., Yoon, C., Choi, J., & Ko, S. (2018). A short review of light barrier materials for food and beverage packaging. *Korean Journal of Packaging Science and Technology*, *24*(3), 141–148.

Lee, J. -S., Chang, Y., Lee, E. -S., Song, H. -G., Chang, P. -S., & Han, J. (2018). Ascorbic acid-based oxygen scavenger in active food packaging system for raw meatloaf. *Journal of Food Science*, *83*(3), 682–688.

Li, W., Li, L., Cao, Y., Lan, T., Chen, H., & Qin, Y. (2017). Effects of PLA film incorporated with ZnO nanoparticle on the quality attributes of fresh-cut apple. *Nanomaterials*, *7*(8), 207.

Lin, Z., & Goddard, J. M. (2018). Photocurable coatings prepared by emulsion polymerization present chelating properties. *Colloids Surfaces B Biointerfaces*, *172*, 143–151.

Lin, Z., Decker, E. A., & Goddard, J. M. (2016). Preparation of metal chelating active packaging materials by laminated photografting. *Journal of Coatings Technology and Research*, *13*(2), 395–404.

Maisanaba, S., Llana-Ruiz-Cabello, M., Gutiérrez-Praena, D., Pichardo, S., Puerto, M., Prieto, A. I., et al. (2017). New advances in active packaging incorporated with essential oils or their main components for food preservation. *Food Reviews International*, *33*(5), 447–515.

Majid, I., Nayik, G. A., Dar, S. M., & Nanda, V. (2018). Novel food packaging technologies: Innovations and future prospective. *Journal of Saudi Society of Agricultural Sciences*, *17*(4), 454–462.

Mansourbahmani, S., Ghareyazie, B., Zarinnia, V., Kalatejari, S., & Mohammadi, R. S. (2018). Study on the efficiency of ethylene scavengers on the maintenance of postharvest quality of tomato fruit. *Journal of Food Measurement and Characterization*, *12*, 691–701.

Marrez, D. A., Abdelhamid, A. E., & Darwesh, O. M. (2019). Eco-friendly cellulose acetate green synthesized silver nano-composite as antibacterial packaging system for food safety. *Food Packaging and Shelf Life*, *20*, 100302.

Mihindukulasuriya, S. D. F., & Lim, L. T. (2014). Nanotechnology development in food packaging: A review. *Trends in Food Science and Technology*, *40*(2), 149–167.

Mordor Intelligence (2019). Active and intelligent packaging market—Growth, trends, and forecast (2020-2025). Available from: https://www.mordorintelligence.com/industry-reports/active-and-intelligent-packaging-market-industry. Accessed 2020 Apr 8.

Moreirinha, C., Vilela, C., Silva, N. H. C. S., Pinto, R. R. J., Almeida, A., Rocha, M. A. M., et al. (2020). Antioxidant and antimicrobial films based on brewers spent grain arabinoxylans, nanocellulose and feruloylated compounds for active packaging. *Food Hydrocolloids*, *108*, 105836.

Nieva-Echevarría, B., Manzanos, M. J., Goicoechea, E., & G Guillén, M. D. (2015). 2,6-Di-tert-butyl-hydroxytoluene and its metabolites in foods. *Comprehensive Reviews in Food Science and Food Safety*, *14*(1), 67–80.

Olmo, J. A. -D., Pérez-Álvarez, L., Ruiz-Rubio, L., & Vilas-Vilela, J. L. (2020). Antibacterial chitosan electrostatic/covalent coating onto biodegradable poly(L-lactic acid). *Food Hydrocolloids*, *105*, 105835.

Otoni, C. G., Espitia, P. J. P., Avena-Bustillos, R. J., & McHugh, T. H. (2016). Trends in antimicrobial food packaging systems: Emitting sachets and absorbent pads. *Food Research International, 83*, 60–73.

Padrão, J., Gonçalves, S., Silva, J. P., Sencadas, V., Lanceros-Méndez, S., Pinheiro, A. C., et al. (2016). Bacterial cellulose-lactoferrin as an antimicrobial edible packaging. *Food Hydrocolloids, 58*, 126–140.

Pagno, C. H., de Farias, Y. B., Costa, T. M. H., Rios, A., de, O., & Flôres, S. H. (2016). Synthesis of biodegradable films with antioxidant properties based on cassava starch containing bixin nanocapsules. *Journal of Food Science and Technology, 53*(8), 3197–3205.

Pant, A. F., Sangerlaub, S., & Muller, K. (2017). Gallic acid as an oxygen scavenger in bio-based multilayer packaging films. *Materials, 10*, 489.

Pettersen, M. K., Hansen, AÅ, & Mielnik, M. (2014). Effect of different packaging methods on quality and shelf life of fresh reindeer meat. *Packaging Technology Science, 27*(12), 987–997.

Piñeros-Hernandez, D., Medina-Jaramillo, C., López-Córdoba, A., & Goyanes, S. (2017). Edible cassava starch films carrying rosemary antioxidant extracts for potential use as active food packaging. *Food Hydrocolloids, 63*, 488–495.

Pokorný, J. (2007). Are natural antioxidants better- and safer- than synthetic antioxidants? *European Journal of Lipid Science and Technology, 109*(6), 629–642.

Preciado, G. M., Michel, M. M., Villarreal-Morales, S. L., Flores-Gallegos, A. C., Aguirre-Joya, J., Morlett-Chávez, J., et al. (2016). Bacteriocins and its use for multidrug-resistant bacteria control. In *Antibiotic resistance: Mechanisms and new antimicrobial approaches* (pp. 329–349). Academic Press.

Priyadarshi, R., Sauraj, Kumar, B., Deeba, F., Kulshreshtha, A., & Negi, Y. S. (2018). Chitosan films incorporated with Apricot (*Prunus armeniaca*) kernel essential oil as active food packaging material. *Food Hydrocolloids, 85*, 158–166.

Rambabu, K., Bharath, G., Banat, F., Show, P. L., & Cocoletzi, H. H. (2019). Mango leaf extract incorporated chitosan antioxidant film for active food packaging. *International Journal of Biological Macromolecules, 126*, 1234–1243.

Realini, C. E., & Marcos, B. (2014). Active and intelligent packaging systems for a modern society. *Meat Science, 98*(3), 404–419.

Remya, S., Mohan, C. O., Venkateshwarlu, G., Sivaraman, G. K., & Ravishankar, C. N. (2017). Combined effect of O_2 scavenger and antimicrobial film on shelf life of fresh cobia (*Rachycentron canadum*) fish steaks stored at 2°C. *Food Control, 71*, 71–78.

Ribeiro-Santos, R., Andrade, M., Melo, N. R. de, & Sanches-Silva, A. (2017). Use of essential oils in active food packaging: Recent advances and future trends. *Trends in Food Science and Technology, 61*, 132–140.

Rollini, M., Nielsen, T., Musatti, A., Limbo, S., Piergiovanni, L., Hernandez Munoz, P., et al. (2016). Antimicrobial performance of two different packaging materials on the microbiological quality of fresh salmon. *Coatings, 6*(1), 6.

Roman, M. J., Decker, E. A., & Goddard, J. M. (2015). Performance of nonmigratory iron chelating active packaging materials in viscous model food systems. *Journal of Food Sciences, 80*(9), E1965–E1973.

Roman, M. J., Decker, E. A., & Goddard, J. M. (2016). Retaining oxidative stability of emulsified foods by novel nonmigratory polyphenol coated active packaging. *Journal of Agricultural and Food Chemistry, 64*(27), 5574–5582.

Sadeghi, K., Lee, Y., & Seo, J. (2020). Ethylene scavenging systems in packaging of fresh produce: A review. *Food Review International*, 1–22.

Sanches-Silva, A., Costa, D., Albuquerque, T. G., Buonocore, G. G., Ramos, F., Castilho, M. C., et al. (2014). Trends in the use of natural antioxidants in active food packaging: a review. *Food Additives & Contaminants Part A*, *31*(3), 374–395.

Santos, J. C. P., Sousa, R. C. S., Otoni, C. G., Moraes, A. R. F., Souza, V. G. L., Medeiros, E. A. A., et al. (2018). Nisin and other antimicrobial peptides: Production, mechanisms of action, and application in active food packaging. *Innovative Food Science and Emerging Technologies*, *48*, 179–194.

Scarfato, P., Avallone, E., Galdi, M. R., Di Maio, L., & Incarnato, L. (2017). Preparation, characterization, and oxygen scavenging capacity of biodegradable α-tocopherol/PLA microparticles for active food packaging applications. *Polymer Composites*, *38*, 981–986.

Schumann, B., & Schmid, M. (2018). Packaging concepts for fresh and processed meat—Recent progresses. *Innovative Food Science and Emerging Technologies*, *47*, 88–100.

Silva, M. F., Lopes, P. S., Da Silva, C. F., & Yoshida, C. M. P. (2016). Active packaging material based on buriti oil—*Mauritia flexuosa* L.f. (*Arecaceae*) incorporated into chitosan films.. *Journal of Applied Polymer Science*, *133*(12), 43210.

Silva, N. H. C. S., Vilela, C., Almeida, A., Marrucho, I. M., & Freire, C. S. R. (2018). Pullulan-based nanocomposite films for functional food packaging: exploiting lysozyme nanofibers as antibacterial and antioxidant reinforcing additives. *Food Hydrocolloids*, *77*, 921–930.

Silva-Weiss, A., Ihl, M., Sobral, P. J. A., Gómez-Guillén, M. C., & Bifani, V. (2013). Natural additives in bioactive edible films and coatings: Functionality and applications in foods. *Food Engineering Reviews*, *5*(4), 200–216.

Singh, S., Gaikwad, K. K., & Lee, Y. S. (2019). Development and application of a pyrogallic acid-based oxygen scavenging packaging system for shelf life extension of peeled garlic. *Scientia Horticulturae (Amsterdam)*, *256*, 108548.

Siripatrawan, U., & Kaewklin, P. (2018). Fabrication and characterization of chitosan-titanium dioxide nanocomposite film as ethylene scavenging and antimicrobial active food packaging. *Food Hydrocolloids*, *84*, 125–134.

Sivertsvik, M., Jeksrud, W. K., & Rosnes, J. T. (2002). A review of modified atmosphere packaging of fish and fishery products—significance of microbial growth, activities and safety. *International Journal of Food Science and Technology*, *37*(2), 107–127.

Sofi, S. A., Singh, J., Rafiq, S., Ashraf, U., Dar, B. N., & Nayik, G. A. (2018). A comprehensive review on antimicrobial packaging and its use in food packaging. *Current Nutrition & Food Science*, *14*(4), 305–312.

Spricigo, P. C., Foschini, M. M., Ribeiro, C., Corrêa, D. S., & Ferreira, M. D. (2017). Nanoscaled platforms based on SiO_2 and Al_2O_3 impregnated with potassium permanganate use color changes to indicate ethylene removal. *Food and Bioprocess Technology*, *10*(9), 1622–1630.

Srithammaraj, K., Magaraphan, R., & Manuspiya, H. (2018). Influence of thiol groups on the ethylene adsorption and conductivity properties of the modified porous clay heterostructures (PCHS) using as ethylene scavenger in smart packaging. *Polym Bulletin*, *75*(9), 3951–3969.

Tas, C. E., Hendessi, S., Baysal, M., Unal, S., Cebeci, F. C., Menceloglu, Y. Z., et al. (2017). Halloysite nanotubes/polyethylene nanocomposites for active food packaging materials with ethylene scavenging and gas barrier properties. *Food and Bioprocess Technology*, *10*(4), 789–798.

Tian, F., Decker, E. A., & Goddard, J. M. (2013). Controlling lipid oxidation of food by active packaging technologies. *Food & Function*, *4*, 669–680.

Topuz, F., & Uyar, T. (2020). Antioxidant, antibacterial and antifungal electrospun nanofibers for food packaging applications. *Food Research International*, *130*, 108927.

Tulsyan, G., Richter, C., & Diaz, C. A. (2017). Oxygen scavengers based on titanium oxide nanotubes for packaging applications. *Packaging Technology and Science*, *30*, 251–256.

Valdés, A., Mellinas, A. C., Ramos, M., Garrigós, M. C., & Jiménez, A. (2014). Natural additives and agricultural wastes in biopolymer formulations for food packaging. *Frontiers in Chemistry*, *2*, 6.

Valdés, A., Mellinas, A. C., Ramos, M., Burgos, N., Jiménez, A., & Garrigós, M. C. (2015). Use of herbs, spices and their bioactive compounds in active food packaging. *RSC Advances*, *5*, 40324–44035.

Vilela, C., Pinto, R. J. B., Coelho, J., Domingues, M. R. M., Daina, S., Sadocco, P., et al. (2017). Bioactive chitosan/ellagic acid films with UV-light protection for active food packaging. *Food Hydrocolloids*, *73*, 120–128.

Vilela, C., Kurek, M., Hayouka, Z., Röcker, B., Yildirim, S., Antunes, M. D. C., et al. (2018). A concise guide to active agents for active food packaging. *Trends in Food Science and Technology*, *80*, 212–222.

Vilela, C., Moreirinha, C., Domingues, E. M., Figueiredo, F. M. L., Almeida, A., & Freire, C. S. R. (2019). Antimicrobial and conductive nanocellulose-based films for active and intelligent food packaging. *Nanomaterials*, *9*(7), 980.

Wang, Y., Xia, Y., Zhang, P., Ye, L., Wu, L., & He, S. (2017). Physical characterization and pork packaging application of chitosan films incorporated with combined essential oils of cinnamon and ginger. *Food and Bioprocess Technology*, *10*(3), 503–511.

Wang, H., Qian, J., & Ding, F. (2018). Emerging chitosan-based films for food packaging applications. *Journal of Agricultural and Food Chemistry*, *66*(2), 395–413.

Wołosiak-Hnat, A., Zych, K., Mężyńska, M., Kifonidis, A., Dajworski, M., Lisiecki, S., et al. (2019). LDPE/PET laminated films modified with FeO(OH) × H_2O, Fe_2O_3, and ascorbic acid to develop oxygen scavenging system for food packaging. *Packaging Technol and Science*, *32*(9), 457–469.

Wong, D. E., Andler, S. M., Lincoln, C., Goddard, J. M., & Talbert, J. N. (2017). Oxygen scavenging polymer coating prepared by hydrophobic modification of glucose oxidase. *Journal of Coatings Technology and Research*, *14*(2), 489–495.

Wrona, M., Cran, M. J., Nerín, C., & Bigger, S. W. (2017). Development and characterisation of HPMC films containing PLA nanoparticles loaded with green tea extract for food packaging applications. *Carbohydrate Polymers*, *156*, 108–117.

Wyrwa, J., & Barska, A. (2017). Innovations in the food packaging market: active packaging. *European Food Research and Technology*, *243*, 1681–2192.

Yam, K. L., Takhistov, P. T., & Miltz, J. (2005). Intelligent packaging: Concepts and applications. *Journal of Food Science*, *70*(1), R1–R10.

Yang, W., Owczarek, J. S., Fortunati, E., Kozanecki, M., Mazzaglia, A., Balestra, G. M., et al. (2016). Antioxidant and antibacterial lignin nanoparticles in polyvinyl alcohol/chitosan films for active packaging. *Industrial Crops and Products*, *94*, 800–811.

Yildirim, S., & Röcker, B. (2018). Active packaging. In *Nanomaterials for food packaging: materials, processing technologies, and safety issues* (pp. 173–202). Elsevier.

Yildirim, S., Röcker, B., Rüegg, N., & Lohwasser, W. (2015). Development of palladium-based oxygen scavenger: Optimization of substrate and palladium layer thickness. *Packaging Technology and Science*, *28*(8), 710–718.

Yildirim, S., Röcker, B., Pettersen, M. K., Nilsen-Nygaard, J., Ayhan, Z., Rutkaite, R., et al. (2018). Active packaging applications for food. *Comprehensive Reviews of Food Science and Food Safety*, *17*, 165–199.

Zhang, J., Cheng, D., Wang, B., Khan, I., & Ni, Y. (2017). Ethylene control technologies in extending postharvest shelf life of climacteric fruit. *Journal of Agricultural and Food Chemistry*, *65*(34), 7308–7319.

Zhang, C., Li, Y., Wang, P., & Zhang, H. (2020). Electrospinning of nanofibers: Potentials and perspectives for active food packaging. *Comprehensive Reviews of Food Science and Food Safety*, *19*, 479–502.

Chapter 10

Food Digestion Engineering

Kaja Kristensen[a], Noémie David-Rogeat[b], Norah Alshammari[c], Qingsu Liu[d], Molly Muleya[a], Syahrizal Muttakin[e,f], Luca Marciani[c], Serafim Bakalis[g], Tim J. Foster[a] and Ourania Gouseti[g,*]

[a]School of Biosciences, University of Nottingham, Sutton Bonington Campus, United Kingdom; [b]Division of Plant and Crop Sciences, University of Nottingham, NIAB EMR, Crop Science and Production System, United Kingdom; [c]School of Medicine, University of Nottingham, University Park Campus, United Kingdom; [d]Faculty of Engineering, University of Nottingham, University Park Campus, United Kingdom; [e]School of Chemical Engineering, University of Birmingham, Edgbaston Campus, United Kingdom; [f]Indonesian Agency for Agricultural Research and Development, Indonesia; [g]Department of Food Science, University of Copenhagen, Frederiksberg, Copenhagen, Denmark

**Corresponding author*

1 Background

Digestion is a vital process in the body that takes place in the gastrointestinal tract (GIT) and breaks down consumed food to nutrients and waste. Inappropriate, over- and under-consumption of foods have been shown to increase the risk for a spectrum of diseases, such as type II diabetes or cardiovascular diseases, while a balanced diet can potentially reduce these risks (Malik et al., 2010; Abuajah et al., 2015). Improving and maintaining a healthy diet is also thought to help when treating major depression events (Jacka et al., 2017). Nutritional needs may vary for different groups. For example, elderly people need considerable protein intake to limit muscle loss (Paddon-Jones & Rasmussen, 2009); while people with celiac disease must completely remove gluten from their diet (Ludvigsson & Green, 2011). With an increasing need for sustainable healthy foods, understanding digestive processes becomes a key in ensuring high nutritional quality of human diet. The EAT-Lancet diet is an example of a sustainable or planetary diet that has been proposed (Willett et al., 2019).

This chapter will firstly introduce the physiology of the digestive system and its components, before exploring different methods used to engineer it. In vivo (means "in the living"), in vitro (means "in the glass"), and in silico (refers to computer models) methods are all used in food engineering laboratories, more or less commonly, and the advantages and drawbacks of each of them will be presented.

Sustainable food processing and engineering challenges. http://dx.doi.org/10.1016/B978-0-12-822714-5.00010-3

2 Digestive system

The digestive system is composed of the digestive tract and supportive organs comprising mainly of the salivary glands, pancreas, liver, and gallbladder. The digestive tract can be divided into four phases, displayed in Fig. 10.1 Oral processing, where particle sizes are reduced and a bolus is produced; gastric phase, where food is stored before moving to the small intestine, and some chemical and enzymatic breakdown occur; small intestinal digestion, where chyme is further broken down and nutrients are absorbed; and large intestinal digestion, which is the final stage of the process and where fermentation occurs, water is removed, and waste material is excreted (Norton et al., 2014).

In oral processing, teeth apply a weak force ranging from 50–150 N during mastication to degrade food while the tongue places it correctly (Mioche & Peyron, 1995). Saliva is present throughout the process at a neutral pH ranging between 5.6 and 7.6 for healthy individuals and with a flow ranging from 0.77 to 4.15 mL/min when stimulated (Chen, 2009). Saliva is a complex mixture composed of roughly 98% water and 2% organic and inorganic substances such as electrolytes and mucins, allowing lubrification of the ingested food

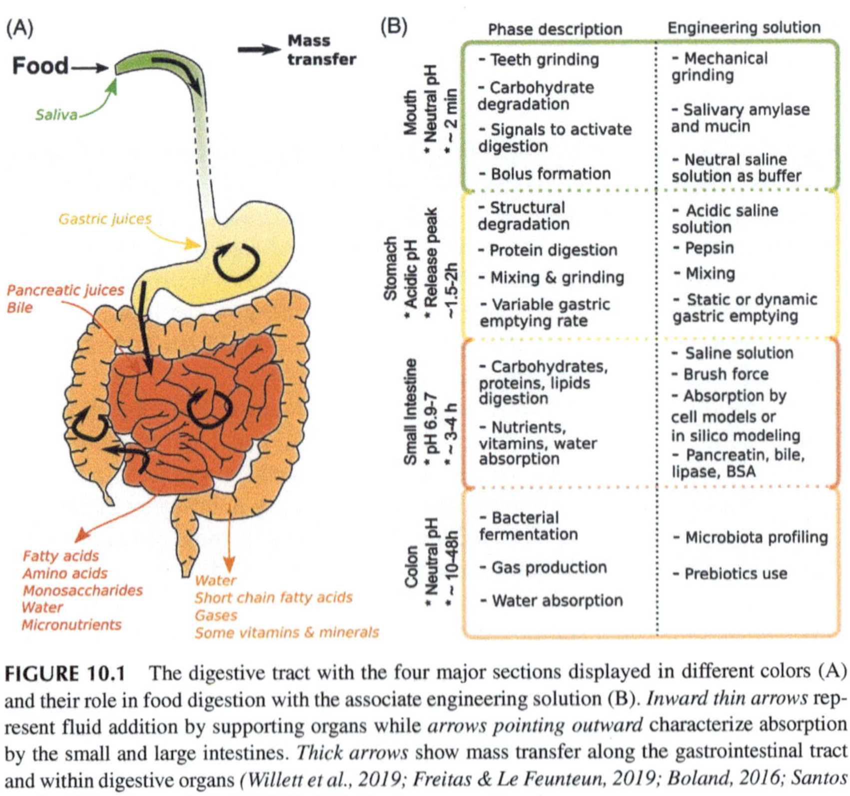

FIGURE 10.1 The digestive tract with the four major sections displayed in different colors (A) and their role in food digestion with the associate engineering solution (B). *Inward thin arrows* represent fluid addition by supporting organs while *arrows pointing outward* characterize absorption by the small and large intestines. *Thick arrows* show mass transfer along the gastrointestinal tract and within digestive organs *(Willett et al., 2019; Freitas & Le Feunteun, 2019; Boland, 2016; Santos et al., 2019; Guyton & Hall, 2010). BSA*, Bovine serum albumin.

(Chen, 2009). Early starch digestion also occurs in the mouth due to the presence of the enzyme α-amylase in the saliva at a concentration around 45 U/mL and which is active at an optimal pH of 7.4 (Gouseti et al., 2019). This enzyme is responsible for the hydrolysis of around 80% of bread starch and 30% of pasta starch (Freitas & Le Feunteun, 2019). After 15–20 chewing strokes, food particles of small size are re-assembled and lubricated into a viscoelastic mass called a bolus (Bornhorst et al., 2016). At the end of oral processing, the bolus is swallowed along the esophagus.

The food passes through the esophagus by sequential contractions to the stomach. This compartment is strongly acidic due to acid secretion that can reach 30 mEq $H^+ h^{-1}$ in the feeding state, an almost threefold increase from the fasted state (Bornhorst & Singh, 2014). The bolus is mixed with digestive juices containing gastric acids and digestive enzymes to form a chyme consisting of a liquid phase and solid particles.

The stomach is greatly heterogeneous in terms of pH distribution and mixing and can be divided into four anatomical sections: fundus, body, antrum, and pylorus. The fundus and proximal part of the body act as a reservoir and can extend greatly during the fed state. Food arriving in the fundus is relatively unmixed with the gastric secretions, and the pH is believed to be relatively high in this upper part, potentially allowing further carbohydrate digestion by salivary amylase, which becomes nonactive at pH of 4.0 or lower. Stronger mixing of food with the gastric secretions and further particle reduction is achieved in the antrum and distal part of the body. The digesta is subjected to strong peristaltic waves at a frequency fluctuating between 2.6 and 3 waves per minute in the antrum (Bornhorst & Singh, 2014).

The forces generated by gastric motility cause mechanical grinding to the food bolus. The stomach acts as a sieve, with liquids located in the antrum and large solid particles mainly placed in the body, allowing small particles and liquid to leave quicker than larger ones. Contractions in the pylorus (the area where chyme passes from the stomach to the small intestine) propel large particles larger than roughly 2 mm back into the main body of the stomach for further size reduction via retropulsion at a force ranging from 0.2 to 1.89 N (Norton et al., 2014). This process leads to an important mixing between the food and the digestive juices and further grinding of particles.

Gastric emptying into the small intestine starts shortly after the start of a meal, and it is usually completed after 3–4 h for solids (Norton et al., 2014). This is a complex process that is still not fully understood, and it depends on multiple factors such as food structure and nutritional density. A highly fibrous meal, for example, can impact the gastric emptying rate due to a range of factors, possibly including a high chyme viscosity (Padayachee et al., 2017).

Small ground particles are exiting the stomach pass to the duodenum, the first 30 cm of the small intestine. The duodenum secretes bicarbonate to neutralize gastric acids and to create an appropriate environment for the intestinal digestive enzymes to work (Boland, 2016). The small intestine is also composed of the jejunum and ileum and represents a total extended area of around 30 m^2 (Helander and Fändriks, 2014). Nutrients in the duodenum and jejunum activate the secretion of cholecystokinin hormone, associated with hunger suppression. In addition, nutrients, particularly lipids, present in the ileum can trigger signals to the stomach that slow stomach emptying or reduce proximal gastrointestinal motility, also called "ileal break" (Maljaars et al., 2008). It is noted that such feedback mechanisms are typically difficult to replicate in a laboratory.

The chyme entering the small intestine is mixed with bile from the gallbladder and pancreatic secretions. These secretions contain a range of hydrolytic enzymes, including α-amylase and proteases. The small intestinal walls are composed of enterocytes bordered by hundreds of microvilli that host digestive enzymes and also greatly increase the surface area of the small intestinal wall to facilitate nutrient absorption. Nutrients released in the small intestine become bioaccessible for absorption through the intestinal wall. The portion of bioaccessible compounds that is absorbed into the blood system and is available for host cells represents its bioavailability (Santos et al., 2019).

Movements in the intestine ensure continuous contact between the chyme and the intestinal wall, as well as the mixing of food particles and intestinal digestive juices. The small intestine shows two main motility patterns. Peristalsis is a propulsive movement responsible for the movement of the food along the tube at a velocity of 0.5–2.0 cm/s, usually triggered by the distention of the gut (Guyton & Hall, 2010). The distention of the intestine also triggers localized concentric contractions called segmentation. Segmentation divides the intestine into segments, promoting progressive mixing of the food and secretions. Both movements contribute to the mixing, propulsion, and absorption of food nutrients through the small intestine.

From the distal part of the small intestine, the chyme is passed to the colon, or large intestine, through the ileocecal junction. The colon is an anaerobic environment with an estimated 10^{13} bacteria covering its mucosa, and the number can vary depending on diet and health, which are key in gut-brain communication (Evrensel & Ceylan, 2015). Bacteria in the proximal part of the colon are performing fermentation of residual food components that have not been absorbed in the proximal part of the digestive tract. These undigested components include dietary fibers that are fermented into short-chain fatty acids, which can affect brain functions, and gases depending on the health and diversity of the microbiota (Gill et al., 2018). At the same time, the digesta passing through the colon via segmentation and peristaltic movements become less liquid due to water absorption through the large intestinal walls (Guyton & Hall, 2010). The second half of the colon is used as a storage of feces until defecation.

Multiple control points are present through the GIT, from the level of mastication to the rate of transit. Comprehensive and complete food engineering systems are thus critical to understanding the fate of food during digestion within the human body.

3 In vivo digestion

The development of in vivo models, where the term "in vivo" comes from Latin and refers to the studies carried out in a living organism, is crucial to help understand the physiological responses to food and its digestion. In vivo studies provide a powerful tool for studying the link between food and health, and can ultimately inform the design of in vitro models (Bohn et al., 2018). In vivo studies can be carried out in cells, animals, and humans using different techniques and measurements. Here animal studies related to food digestion are briefly considered before describing a range of human study techniques and parameters of interest.

3.1 Animal studies

Digestion models can be developed in small (rodent) and large animals, with the choice of animal model dependent on what type of prediction one aims to make (Baker, 2008) as well as cost and type of experimental procedures. Some larger animals may offer advantages such as increased ease of access due to their size (e.g., for intubation or surgical procedures). Small piglets are often used in studies related to neonatal nutrition (Le Huërou-Luron et al., 2018). Pigs are considered to be one of the animal models closest to humans for digestion (Miller & Ullrey, 1987; Ziegler et al., 2016; Roura et al., 2016). An example of the use of pig models is the study of amino acid metabolism (Riedijk et al., 2007). Animal models can include invasive surgical procedures to access different organs of the gastrointestinal tract, allow infusion of specific nutrients at given locations through the digestive tract, or collection of samples at specific time intervals using catheters. Some animal feeding studies involve sacrificing the animals at predetermined times to retrieve and examine the contents of the gut using analytical techniques.

There are, however, differences between animal and human eating behavior, digestive physiology, and anatomical structure. Therefore, the interpretation of the results of food digestion from animal models can be a real challenge, and, in some cases, it may be difficult to extrapolate the results to humans.

3.2 Human studies

Human nutritional studies examine the link between a health outcome, for example, a blood biomarker such as cholesterol (Barbé et al., 2013), and a specific dietary intervention (exposure). The design of nutritional studies can be experimental or observational, and it follows well-established categories (Fig. 10.2).

The various in vivo study designs have limitations to be considered. Some may neglect information about the temporal relationship between exposure and outcome (Solem, 2015). Others do not control the diet of the participants, limiting specificity (Hartung & Touchette, 2009). Large scale human experimental studies can be expensive to run and challenging because of their complexity

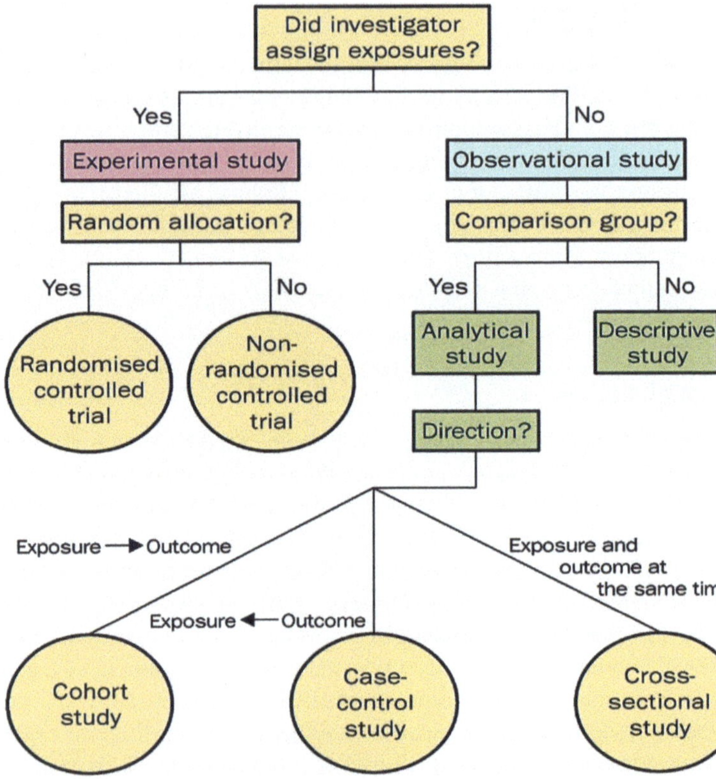

FIGURE 10.2 Flow diagram for the classification of clinical research studies. *(Reproduced with permission from Grimes & Schulz 2002.)*

(Thiese, 2014). Another main problem is how to ensure good compliance with the protocol in a large number of subjects. Well-controlled feeding studies often provide entire diets for prolonged periods to assess the effect of specific dietary interventions (Most et al., 2003).

3.2.1 Parameters of interest in human studies

Digestion and nutrition studies in humans can assess a wide range of parameters of interest.

Questionnaires are widely used to assess satiety and eating behavior. Appetite responses such as fullness and satiety are assessed using a visual analog scale (VAS) as a standard tool to assess hunger, satiety, fullness, desire to eat, and prospective food consumption (PFC). The VAS scale is a measurement tool for subjective characteristics, usually on a scale ranging between 0 and 100 mm (Flint et al., 2000). Eating behavior can be measured using tools such as 24 h recall and 3-days recall capturing diet and intake history. Food frequency questionnaires (FFQ) are dietary tools commonly used in nutritional studies to provide information on the participant's typical diet intake over a period. Questionnaires are not expensive and often easy to complete though the quality of the data may depend on the participants' accuracy. For instance, under-reporting has been observed in some obesity studies (Rodríguez et al., 2019).

Blood and urine samplings offer the possibility to measure objective markers of the body's response to feeding. Examples include urine nitrogen as a marker of protein intake using blood sampling (Picó et al., 2019). Glucose, insulin, and other gut hormone peptides can be measured. Several gastrointestinal hormone peptides are released by the enteroendocrine cells, including glucagon-like peptide 1, peptide YY, and cholecystokinin. These peptides transit through the gut, stimulate secretions, and affect insulin and appetite (Bruen et al., 2012; Huda et al., 2006; Batterham et al., 2003; Little et al., 2006).

Breath test detects hydrogen (and sometimes also methane) to monitor carbohydrate malabsorption and intolerance to lactose, fructose, and sucrose. This is possible as poorly digested carbohydrate is broken down by the bacteria in the large intestine, which can result in gas production (Major et al., 2017). Some of the luminal gastrointestinal gas is absorbed in the blood and released by the lungs. Breath tests are not very expensive and reasonably easy to perform (Braden et al., 2007).

It is of value to understand human digestive processes. This can be achieved, for example, by using intubation techniques or exploiting existing ileostomy procedures in certain types of patients. Naso-gastric or naso-duodenal intubation with catheters are used to assess gut motility, infuse selected nutrients to selected parts of the bowel bypassing the oral cavity and/or also the stomach (Maljaars et al., 2012) and aspirate samples for bench analysis (Sullivan et al., 2014; Boutrou et al., 2013). Intraluminal manometry is often used to study gut motility by measuring pressure changes in the gut (Smout & Mundt, 2009).

To evaluate gastrointestinal pH, pressure, and motility, a less invasive method involves the use of a wireless capsule in the place of the common catheter to provide information on pressure, temperature, and pH of the examined gastric part. Moreover, a wireless telemetric motility capsule was used to determine the gastric emptying time (Koziolek et al., 2015).

The main disadvantage of intubation methods is the invasiveness of the procedure, mainly causing discomfort to the participants. The aspirates may represent only a small part of the digestive tract, and some thicker or more solid components of digestion cannot be aspirated.

Stoma or Ileostomy patients have been used to study digestion to obtain effluent samples from lower parts of the bowel during digestion. This is an effective method, though it is from patients with a range of conditions and may not represent healthy physiology (Edwards et al., 2015).

Considerable attention has been given to the study of the human gut microbiota. The term encompasses a very large number of microorganisms performing a range of functions, including the expression of enzymes that allow fermentation of complex carbohydrates into metabolites such as short-chain fatty acids (Thursby & Juge, 2017). Some of the metabolites produced may have physiological effects, including controlling gut hormones and appetite (Valdes et al., 2018). The increasing number of studies of the gut microbiota and its role in digestion has been facilitated by cost reduction and an increase in throughput of sequencing methods.

3.3 Medical imaging techniques

Medical imaging methods can provide unique insights into the behavior of food in the human digestive system. These methods have been increasingly used in nutrition and food digestion studies, with the stomach being the particular focus of attention. Parameters of interest include gastric volumes and emptying. Examples of medical imaging techniques include X-rays, CT, ultrasound, and magnetic resonance imaging (MRI).

An ultrasound scan also called a sonogram, is a process that uses high-frequency sound waves to form an image of a specific part of the body. It can provide a clear image of the gastric antrum and has been used to assess gastric emptying. It was used in a digestion study, estimating the gastric emptying time of a pasta meal that contained inulin prebiotic (Russo et al., 2011). Ultrasound is not very expensive, and it is safe as it uses non-ionizing radiation. However, ultrasound has limitations. The presence of air in the abdomen can reduce the view of the fundus; gastric emptying is often inferred from the antral views, and its performance highly dependent on the skills of the operator.

X-ray and computed tomography are well-established and available imaging methods. However, they use ionizing radiation and have not been used extensively to study digestion.

MRI is a non-invasive, modern medical imaging technique that is frequently used to create detailed images of the organs and tissues within the body. MRI uses radiofrequency to transmit and receive signals from the nuclei of the hydrogen atoms in the body. The technique has been used to investigate gastrointestinal function and foods and drinks in the gastrointestinal tract (Alyami et al., 2015), as shown in Fig. 10.3.

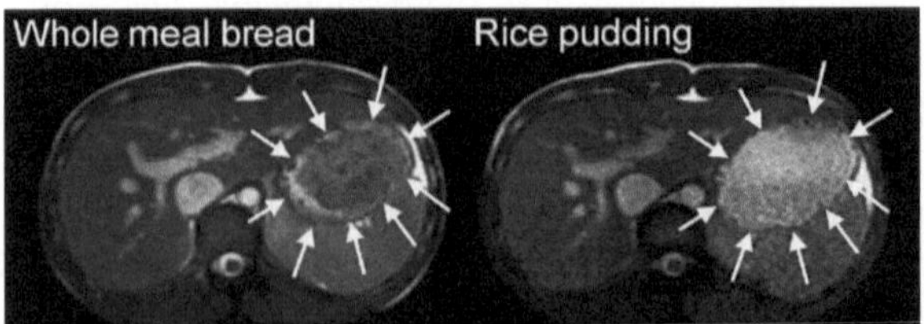

FIGURE 10.3 Axial MRI images of whole wheat bread *(left)* and rice pudding *(right)* meals inside the stomach of healthy human volunteers. The *white arrows* indicate the boundaries of the stomach. *(Reproduced under Creative Commons (Attribution-Noncommercial-ShareAlike) license from Marciani et al., 2013.)*

The advantages of MRI, in addition to the use of non-ionizing radiation, are the good image contrast for foods and fluids in the GIT and good resolution (in terms of millimetric macroscopic scale). It is well-suited to provide an assessment for different parameters and functions such as gastric motility and emptying. Some limitations include being relatively expensive and the possible effects of postural differences as the common MRI scanners have a horizontal bore where the subjects lie on the scanner bed. When food components layer in the stomach, posture may make a marked difference to gastric emptying (Boulby et al., 1997).

Feeding studies have also benefitted greatly from the ability of imaging to assess brain responses to stimuli. Signals from sensory (gustatory, olfactory, and somatosensory receptors) and hormonal responses to food reach primary and secondary areas of the brain. In response to these stimuli, changes in brain blood flow, volume, and oxygenation occur, and these can be imaged using functional MRI (fMRI). There are numerous examples of such studies of brain function, including the oral perception of fat taste (Eldeghaidy et al., 2011), thermal tasting (Skinner et al., 2018), and brain responses to food portion size and energy density (English et al., 2017).

3.4 Outlook

Allowing for the difficulties and costs involved, the more desirable future nutrition and digestion human studies will be well-controlled and randomized controlled trial studies comparing multiple parameters across satiety, food intake, and physiological responses to selected dietary interventions.

4 In vitro digestion

In vitro digestive models are used to study digestion outside of the body, and their popularization has substantially boosted research in the area of digestion. However, mimicking the complexity of digestive processes in the lab is challenging. In one approach, digestive processes have been paralleled to unit operations typically used in chemical engineering plants (Bornhorst et al., 2016).

4.1 Types of in vitro digestion models, applications, and limitations

In vitro models are generally divided into two categories: Static and dynamic. The difference is the absence or presence of temporal changes during simulated digestion, respectively. The changes can be based on mechanical movements (such as grinding, mixing, mass transfer, peristalsis) and/or physiological changes (such as dynamic pH changes). In vitro models can also be mono- or multicompartmental, depending on whether they simulate a single or multiple digestive regions (compartment), respectively.

4.1.1 Static digestion models

Static digestion models have gained popularity as a fast and flexible way to study digestion. They may involve one, two, three, or four stages relating to oral, gastric, small intestinal, and large intestinal digestion, respectively. Typically, the temperature, electrolytes, and enzyme type and concentrations are controlled. Mixing is typically fast (e.g., by using rotators, shaking incubators, magnetic stirrers, etc.), and the pH is adjusted at the beginning of each stage and remains constant throughout the experiment. The common measurements in static models include the concentration of intermediate products or that of endpoint products before absorption, such as testing of the sugar concentration released from starchy foods. There have been more than 36 protocols of in vitro static digestion processes reported for carbohydrate digestion (Woolnough et al., 2008). Likewise, methods to study protein digestion have also been widely varied. In an effort to standardize and harmonize the digestion parameters, the COST Action INFOGEST has proposed an international protocol published in 2014 (Minekus et al., 2014) and further refined in 2019 (Brodkorb et al., 2019). This consensus aims to increase the reproducibility of static in vitro methods, which make nearly 89% of the total in vitro digestion experiments (Lucas-González et al., 2018). The advantages of using static in vitro digestion models include their simplicity and overall ease of use. They are typically cheaper to run than dynamic models, and they can provide good lab-to-lab reproducibility. Comparisons of standardized static models have shown promising correlations with in vivo methods in carbohydrate, protein, and lipid degradation (Brandon et al., 2006; Egger et al., 2017; Goñi et al., 1997).

These models are mainly reported simulating physiological conditions of the upper GIT and study the deformation, delivery, or release of functional ingredients. For lower GIT and absorption, bioaccessibility and bioavailability of active molecules can be further studied by fermentation, membrane-based, and cell-culture models (Deferme et al., 2007; Mosele et al., 2015).

However, static models may only reproduce limited aspects of the complex and dynamic digestion processes. They fail to mimic dynamic changes as conditions such as pH or enzyme activities are kept constant. Static models also lack the inclusion of mixing and mass transfer. The motility of the digestive tract and fluid flow are excluded.

4.1.2 Dynamic digestion model

Dynamic digestion models are designed to simulate the dynamic nature of digestion, such as mechanical motion or real-time changes of the biochemical conditions. In some dynamic digestion models, the shape of the model realistically represents the shape of the relevant in vivo organ (e.g., "J" shaped stomach and tube-like small intestine), and the capacity is such as to handle realistic food or digesta portions. There are also dynamic models in which the biochemical environment that determines the pH and/or enzyme or bile concentration, as well as the flow of digesta, changes dynamically to mimic the in vivo process. In addition, there are dynamic digestion models that realistically replicate the motility of the digestive system and these are used to study the effect of mixing and mass transfer on digestion (Bohn et al., 2018; Lucas-González et al., 2018; Marciani et al., 2001; Moxon et al., 2016).

Like static, dynamic digestion models may be mono- or polycompartmental, and they study different aspects of digestion. There are, for example, mouth models that replicate mastication and bolus formation, others that focus on studying the breakdown of food matrices, or the release of volatile compounds (Salles et al., 2007; Benjamin et al., 2012). Gastric systems such as human gastric simulator, dynamic gastric model, and dynamic in vitro rat stomach series can mimic the flow of food, gastric secretions, and emptying (Wu et al., 2017). Intestinal systems such as the dynamic duodenum model and the human duodenum model can mimic the segmentation movements of the small intestine and have been used to measure food breakdown and nutrient absorption rates (Tharakan et al., 2010; Hussain et al., 2015). Semi-dynamic models also exist with both static and dynamic elements.

Table 10.1, summarizes some aspects of existing dynamic in vitro models, including applications and limitations. These models allow for the investigation of complex digestive processes that are relevant to in vivo digestion. However, they may be delicate and need intense labor and time for preparation as well as post-cleaning. Repeatability and in vitro in vivo correlation is another concern. For example, parameters are often validated for one certain type of food which may not be suitable for others (Dupont et al., 2019).

4.2 Outlook

In vitro digestion models have boosted the area of digestion studies. They are often used to provide important information for prospective or pilot studies. Compared to in vivo studies, in vitro models are time- and cost-effective and do not require ethical approval. Outcomes of in vitro digestion models include the rates of food breakdown and the bioaccessibility of digested nutrients under controlled conditions. One limitation is that it is difficult to replicate the complexity of the digestive system outside the body. One example is the difficulty of replicating feedback mechanisms. Other challenges involve the vulnerability of some reagents, for example, mucins.

5 In silico digestion

Deployment of silicon-based chips in computers has enabled a means to model and simulate physiological and pathological processes quantitatively. The expression "in silico" directly translates to "within the silicon" and, therefore, to the mathematical/computational approaches used to describe and predict such processes. Modeling has been proven to be a useful tool to analyze complex

TABLE 10.1 Multicompartmental dynamic in vitro models, main dynamic features, current applications, and limitations.

Model	Dynamic features	Applications and limitations	Creator institutions
DIDGI	• pH monitors and control (pH electrode and perforated loop) • Flow of digestive secretions (redox sensors) • Rhythmic peristaltic pumps • Gastric emptying, sieving, and mass transfer (teflon membrane and StoRM software) • Anaerobic condition (nitrogen purging)	Classic examples are digestion of dairy products, human milk, and infant formula *Limitations*: • Lack of oral phase simulation • Only basic stirring (propeller) used in the stomach phase • Fundus exposed to air (open top) • No satiety signals controlling the rate of food ingestion and digestion	INRA
TIM-1 Simplified model: Tiny-TIM Large-intestinal model: TIM-2	• Pre-set peristaltic valve pumps • Rhythmic contractions for mixing (flexible walls in water pressure) • pH monitors and control • Flow of digestive secretions and gastric emptying rates are programmable in variable conditions • Digested products transit to two different systems. Semi-permeable membranes dialysis (water-soluble) and 50 nm filter (lipophilic products)	Studies on macronutrients, fat- and water-soluble vitamins, minerals, and functional compounds (such as probiotics, prebiotics, and antioxidants). Thanks to in silico modeling, TIM can further: - Predict the glycemic response curve and insulin response - Stability and bioaccessibility amino acids, water- and fat-soluble vitamins (such as folate) *Limitations*: 1. Due to no intestinal mucosa, absorption needs to be studied in cell culture or tissues 2. The bioavailability is difficult to measure, but can be predicted using in silico models	Institute of Food Research (Norwich, UK)

SHIME	• Peristaltic pumps in a defined amount of nutritional medium • Flow of digestive secretions and pH are controllable using tailor-made software • Stirring reactors are applied in duodenum, jejunum, and ileum compartments • Different reactors with fecal microbiota inoculation	The SHIME model is more applied in colon-region specific research. Thanks to modules and in silico modeling, it can further: - Conduct dialysis - Mimic conditions in host cells (HMI module with epithelial or immune cells) *Limitations*: 1. Mixing is mainly conducted by stirrers in last three compartments	Ghent University (Belgium)
ESIN	• Flow of digestive secretions, pH, transit time, and chyme mixing rates are controllable • Different gastric emptying of solids and liquids (piston movement and biphasic pumps)	The ESIN model is more applied to liquid foods and can be further used in the digestion of real-size food matrices. *Limitations*: 1. Limited applications and additional validation tests necessary	University of Auvergne (Clermont-Ferrand, France)

biological systems (Tomlin & Axelrod, 2007). Simulating the human digestive system can be achieved through empirical and mechanistic models based on previous observations/measurements and the laws of physics, respectively. The aim of developing such models is to predict the fate of any foodstuff rapidly and accurately within the GIT and the GIT responses to perturbations.

Modeling food processing requires basic rules to be followed, namely the conservation of mass, momentum, and energy, as well as the appropriate physicochemical and biological reactions. Variables influencing digestion modeling, which can be considered individually or in combination depending on the goal of the model, are "mass m*, volume V*, specific volume V*, pressure p*, temperature T*, specific internal energy u*, specific entropy s*, specific enthalpy h*, specific free energy f*, and specific free enthalpy g*" (Rauh & Delgado, 2015). Besides these variables, the scale at which processes occur and how they interact increases modeling complexity.

A complete in silico digestion model is usually composed of a computational model representing a geometrical mesh of the section to be modeled. Numerical modeling is added to that template to represent the movements of the mesh. The related physiochemical properties of the digesta and governing equations determining its behavior in the GIT are then included. Boundary conditions can be incorporated as some kinetic equations are dependent on the geometry of the model (Ferrua et al., 2011). Meshless models are suitable for biomechanical systems where movement and deformation of the boundary may occur (Cleary et al., 2015). Insights into the performance of a model can eventually be obtained using either existing data (literature validation) or dedicated studies (in vivo or in vitro). Empirical models hold little predictive power as the primary focus is on the classification of in vivo data (Muttakin et al., 2019). A non-compartmental approach usually fits a mathematical expression to obtained in vivo data, indicating that the fitted parameters are only applicable to the analyzed system, whereas a compartmental approach can describe the movement of, for example, foodstuff from one compartment (digestion phase) to another and maybe a more suitable approach for modeling the entire digestive system (Moxon et al., 2016).

5.1 Mouth

Modeling mastication, the mechanical breakdown of food particles in the mouth, has thus far been the primary focus of the oral phase (Marze, 2017; Cleary et al., 2015). Anatomical and structural content of the mouth has been observed using processed data obtained from medical imaging techniques such as MRI or CT scans. These data can be relied upon to construct 3D computational models (Cleary et al., 2015). Harrison and colleagues developed a meshless model to simulate mastication and release of taste and aroma compounds into saliva based on the principles of coupled biomechanics-smoothed particle hydrodynamics (SPH). The 3D computational representation of the mouth focused on jaw and tongue motions, chewing movements, and deformation/fracture behavior. Two major challenges they faced when simulating mastication were the characterization of the inter-individual tongue and jaw motion variation and the implementation of a feedback system in a computational model (Harrison & Cleary, 2014; Harrison et al., 2014a,b).

Measurements of chemical and physical reactions inside the oral cavity are challenging; as such, there is little information, particularly regarding the flow of saliva and its effect on softening and lubrification. Chewing profiles obtained from in vivo and in vitro experiments can be used to evaluate computational simulations of mastication. In vivo chewing experiments also offer a means to validate predicted aroma release (Harrison & Cleary, 2014).

5.2 Stomach

Mixing, food disintegration, and gastric emptying are major challenges when it comes to modeling digestion in the stomach. The stomach has a complex geometry presenting significant challenges in simulating gastric function, for example, characterization of mechanical and contractile properties; intra- and inter-individual variation; effects of chemical reactions on flow behavior and material responses (Ferrua & Singh, 2015; Cleary et al., 2015).

Mechanical breakdown and mixing of gastric acid, meal, and saliva in the stomach are achieved by antral contraction waves (ACWs) (Bornhorst et al., 2016). One of the first models to investigate the role of ACWs on gastric flow and mixing used MRI data in a 2D presentation of stomach geometry (Pal et al., 2004). The gastric flow dynamics have been investigated further by using a 3D numerical model assuming laminar flow. The development of an experimental system to evaluate model performance was essential to the study since imaging techniques cannot quantify local flow dynamics, and the use of an in vitro system for this purpose was unattainable (Ferrua et al., 2011). The authors had previously identified that highly viscous meals reduced the gastric flow patterns (retropulsive jet-like motions and eddy structures) before reaching a threshold where flow dynamics were independent of meal viscosity. In the same study, the authors highlighted that the absolute pressure field in the stomach could not be computed due to limited experimental data available (Ferrua & Singh, 2010). The numerical model established by Ferrua and coworkers was refined to include the characterization of advective flow performance. As a first step approach, the stomach was modeled as a closed system, excluding the impact of, for example, gastric emptying. Thus, future efforts need to address the assessment of pressure history of specific regions within the stomach as well as advective flow performance following in vivo and in vitro gastric processes (Ferrua et al., 2014; Ferrua & Singh, 2010).

Gastric secretion was modeled by Moxon et al. (2017) based on data from an echo-planar MRI study assessing gastric volume and gastric chyme viscosity. The model assumed complete mixing of gastric content and that secretions were a function of the meal viscosity based on the concentrations of locust bean gum. Gastric emptying of meals with varying viscosity was also accurately predicted based on the influence of the gastric secretion rate and a feedback system. However, the effect of the gastric viscosity was not considered (Moxon et al., 2017). The effect of viscosity on gastric mixing and emptying of liquid foods was studied by the coupled biomechanical-SPH model where a nonlinear relationship between gastric emptying rate and gastric chyme viscosity was established. Acquiring validation data for this model is still one of its major challenges due to the existing gap in developing techniques that can evaluate in vivo contractile patterns (Harrison et al., 2018).

Complete models require the addition of enzymatic digestion and other chemical reactions occurring in the stomach to be more representative of the human stomach, such as kinetics of protein digestion (Sicard et al., 2018). The high variability of the motility velocity inside the stomach and the lack of methods to analyze the local 3D behavior of foods are major limitations for validated stomach computational models using in vivo studies (Ferrua et al., 2011). In vitro methods are thus preferred for validation, highlighting that these models offer only a fundamental understanding of processes involved in food digestion.

5.3 Small intestine

Modeling the small intestine requires the consideration of enzymatic hydrolysis, intense mixing, and nutrient absorption as major steps. Mathematical models focusing on digesta structural parameters without including intestinal anatomy have been developed and offer a partial quantitative explanation of the fate of some nutrients in the intestine. Starch hydrolysis mathematical modeling by first-order kinetics seems to be the best fit when simulating complete amylolysis (Nguyen & Sopade, 2018; Cleary et al., 2015; Marze, 2017). Kinetic equations of lipid digestion have also been developed from mass transfer between emulsion droplets and the aqueous phase (Li & McClements, 2010; Gaucel et al., 2015; Giang et al., 2015). Experimental data obtained from in vitro kinetic studies of lipid digestion and carotenoid bioaccessibility in emulsions may also further improve the predictive power of in silico models (Verkempinck et al., 2018).

Moxon et al., 2016 developed a mechanistic model to investigate mass transfer and absorption in the human gut, where gastric emptying generally is considered one of the rate-limiting steps. Decreased gastric chyme viscosity, increased gastric emptying, and starch hydrolysis rates resulted in increased absorption of glucose in the small intestine when assuming the intestine could be described as a plug flow reactor (Moxon et al., 2016). Further validation by in vivo experiments is, however, required to enhance the performance of the model. Modeling the absorption of minerals has also been a matter of interest. One example is modeling zinc absorption based on saturation response kinetics, which may accurately interpret the experimental bioavailability of zinc when linked to Michaelis-Menten kinetics (Tran et al., 2015). An updated model could, in that case, consider dose-response studies.

The two main motor patterns that can be ascribed to the small intestine include peristalsis (propulsive wave-like contractions of longitudinal and circular muscles) and segmentation (non-propulsive slow-wave contractions of circular muscles) (Guyton & Hall, 2010). A particle-based computational fluid dynamic (CFD) model was used to describe the peristaltic flow patterns in the duodenum. The model aimed to develop a 3D framework for calculations of fluid-structure interactions between the digesta and the intestinal wall and to predict hydrodynamic interactions between solid and liquid phases computationally. The coupling of the solid and liquid phases was achieved by combing the discrete element (DEM) and SPH methods. The DEM method represented the solid phase as a continuous field in the SPH multiphase calculation. The authors highlighted that the coupled model is a new proposition in modeling peristalsis, and thus that the model requires validation, for example, by in vitro studies (Sinnott et al., 2017).

Based on MRI data and lattice Boltzmann CFD methods, Brasseur et al. (2009) introduced a 2D model to study the contribution of intestinal motor patterns on transport, mixing, and absorption. It includes both lumen and villi motility, highlighting the multiscale interactions. This model has been upgraded by studying the effect of villi motion on nutrient absorption in-depth (Wang et al., 2010). A 3D lattice-Boltzmann model was used to study the flow patterns caused by the contractional formation of mucosal micro folds and the effects on absorption and to mix in the small intestine (Lim et al., 2015). The model focused on the use of numerical methods validated in the literature. The impact of intestinal wall protrusions, such as villi and microvilli, was further studied in the rat duodenum using a 2D CFD simulation, indicating non-uniform absorption flux (Zhang et al., 2020). Presenting the duodenum in three dimensions, including the intestinal wall protrusions, may be of future interest.

Peristalsis is, however, not only controlled by the mass of the content within the digestive system, but also by feedback signals from the brain at different stages (Norton et al., 2014). This illustrates the complexity of simulating the intestinal phase as processes occur at different length scales, dependent on movements of lumen and villi as well as feedback control.

5.4 Colon

Water absorption, microbial breakdown of undigested nutrients, and movement along the colon are the main processes for modeling digestion in the large intestine. Microbiota models started with the development of differential equation models that rely mostly on in vitro fermentation data of monocultures or simple co-cultures (Kumar et al., 2019). A knowledge-based mathematical model was developed that includes carbohydrate fermentation and mass transport between the mucus and lumen. This model uses knowledge about colon physiology and in vitro insight from anaerobic digestion modeling to predict, for example, the effect of dietary fibers on short-chain fatty acids production (Muñoz-Tamayo et al., 2010). Microbiome models can also rely on RNA or DNA sequencing data, which can be used to study bacteria association within the colon. Genome-scale metabolic models (GEMs), which represent the metabolic network of an organism, are particularly relevant to model the variation of the gut microbiome through life, and how food nutrients are metabolized. For example, the CASI-NO framework used GEMs to study diet-microorganism and host-microorganism interactions and to gain insight into the role of each major bacterial species (Shoaie et al., 2015).

A framework established by (Cremer et al., 2017) sought to provide a mechanistic understanding of flow-dependent bacterial growth, indicating that water uptake affects microbial composition. The modeling parameters describing bacterial physiology were obtained from experimental data or analysis of published data without fitting the parameters (Cremer et al., 2017). Regional differences in the microbial composition of the gut may be a result of microbial variation and digestive function (Cleary et al., 2015).

Elastic deformations of the intestinal wall can be predicted as a response of peristaltic motor patterns and pressure distributions within the colon by using an SPH model. These deformations can be defined about the elasticity of the surrounding tissue and the rest length at equilibrium when the surrounding tissue is in the rest state. This framework, proposed by Sinnott et al. (2012), is a first approach in understanding colonic motility, requiring further extension and validation by assessing variations such as colonic diameter, fecal content, and muscular tone. In a 2D model by Labarthe et al. (2019), fluid dynamics and spatial distributions of the content entering the colon were coupled to the mucus layer and the colon microbiota. This was achieved by integrating the fluid mechanics of colon content with previous models investigating the mucus layer (El Bouti et al., 2016) and bacterial metabolism (Muñoz-Tamayo et al., 2010). Peristalsis had the most prominent effect in the proximal colon, and the study suggested that low bacterial active swimming is sufficient in promoting microbial growth. The spatial distributions obtained by the model were in good agreement with experimental data. However, the authors highlighted the importance of addressing the gap between the modeled predictions and biological observations in future work. To improve the fit with specific experimental settings/bacterial populations, the model structure needs to be adjusted accordingly (Labarthe et al., 2019).

The enormous number of bacteria species within the colon is a great modeling challenge as it creates a huge complexity when interaction parameters are being considered. Future modeling perspectives need to address realistic presentations of the colon, such as colonic geometry, fermentation, and diffusion processes, gas generation, and further elucidate the motor patterns responsible for mass transfer, mixing, and absorption.

5.5 Outlook

The representation of the entire GIT in a computational or mathematical model is rare. The parameters to take into consideration are numerous, including variables within and in between compartments, and require a great amount of data. Strathe et al., 2008 refined a comprehensive model for the pig GIT that models the fate of major macronutrients in the GIT and includes gastric emptying, nutrient degradation, and absorption. Its evaluation using published in vivo studies show accurate predictability of nutrient absorption patterns, but not of lipid digestibility, emphasizing the great potential of in silico models and their continuous need of upgrading. Some compartmentalized models of multiple sections of the human digestive system have also been developed, but usually only include the stomach and the small intestine and focus on one specific aspect of the digestion, such as starch digestion (Trusov et al., 2016; Moxon et al., 2016).

Complementary to experimental studies, in silico models, expand our knowledge of multiple processes and reduce the need for long and expensive in vivo studies. To improve the strength of computational predictions, correlations between in vitro, in vivo, and in silico approaches need to be established. Harmonization between research groups regarding the use of these models is also necessary to ensure their reproducibility. Despite these limitations, computational models are promising tools in prediction of responses of the GIT under various conditions in health and disease alongside contribute noticeably to the development of new sustainable food products.

6 Conclusions

Digestion is a highly complex process involving physiochemical and physiological events that cannot be fully simulated using in vitro methods. Ideally, in vivo methods should be used to study the fate of food during digestion, but this is often not practical mostly due to the increasing ethical constraints associated with in vivo studies, and the high cost of conducting these studies. In vivo methods that utilize medical brain imaging techniques offer some new opportunities that can now allow researchers to gain new insights into brain-gut interactions through the use of brain and gut imaging in a single MRI Scan. These techniques are non-invasive and, therefore, do not interfere with the digestion process. Although in vitro methods have some limitations, there has been considerable progress to improve their accuracy and reliability in predicting the fate of food during digestion. One important attempt to standardize in vitro models is the static digestion model by the INFOGEST network (Minekus et al., 2014; Brodkorb et al., 2019). Personalized and/or tailored nutrition is increasingly gaining interest as the role of food in preventing disease and maintaining healthy well-being is being widely recognized. There is a need to establish in vitro methods that account for specific groups of the population, such as infants, the elderly, and patients with various conditions. Some progress has already been made in the development of in vitro models for infants, the elderly, and patients with GI disorders (Shani-Levi et al., 2017). The use of in silico models has great promise in the study of food digestion under different conditions of the GIT and as a tool to predict the effect of digestion on different aspects of food. However, the accuracy of in silico models is underpinned by the extent and quality of the information in publicly available databases as it requires the input of accurate and reliable data from in vivo and in vitro models. The standardization of in vitro methods and their validation with in vivo methods are needed for the establishment of robust in vitro methods that can accurately predict the correct direction and magnitude of response. The use of in silico models is likely to expand in the future.

References

Abuajah, C. I., Ogbonna, A. C., & Osuji, C. M. (2015). Functional components and medicinal properties of food: a review. *Journal of Food Science and Technology*, *52*(5), 2522–2529.

Alyami, J., Spiller, R. C., & Marciani, L. (2015). Magnetic resonance imaging to evaluate gastrointestinal function. *Neurogastroenterology & Motility*, *27*(12), 1687–1692.

Baker, D. H. (2008). Animal models in nutrition research. *The Journal of Nutrition*, *138*(2), 391–396.

Barbé, F., Ménard, O., Le Gouar, Y., Buffière, C., et al. (2013). The heat treatment and the gelation are strong determinants of the kinetics of milk proteins digestion and of the peripheral availability of amino acids. *Food Chemistry*, *136*(3–4), 1203–1212.

Batterham, R. L., Cohen, M. A., Ellis, S. M., Le Roux, C. W., et al. (2003). Inhibition of food intake in obese subjects by peptide YY3-36. *New England Journal of Medicine*, *349*(10), 941–948.

Benjamin, O., Silcock, P., Beauchamp, J., Buettner, A., et al. (2012). Tongue pressure and oral conditions affect volatile release from liquid systems in a model mouth. *Journal of Agricultural and Food Chemistry.*, *60*(39), 9918–9927.

Bohn, T., Carriere, F., Day, L., Deglaire, A., et al. (2018). Correlation between in vitro and in vivo data on food digestion. What can we predict with static in vitro digestion models? *Critical Reviews in Food Science and Nutrition*, *58*(13), 2239–2261.

Boland, M. (2016). Human digestion—a processing perspective. *Journal of the Science of Food and Agriculture*, *96*(7), 2275–2283.

Bornhorst, G. M., Gouseti, O., Wickham, M. S., & Bakalis, S. (2016). Engineering digestion: Multiscale processes of food digestion. *Journal of Food Science*, *81*(3), R534–R543.

Bornhorst, G. M., & Singh, R. P. (2014). Gastric digestion in vivo and in vitro: How the structural aspects of food influence the digestion process. *Annual Review of Food Science and Technology*, *5*(1), 111–132.

Boulby, P., Gowland, P., Adams, V., & Spiller, R. C. (1997). Use of echo planar imaging to demonstrate the effect of posture on the intragastric distribution and emptying of an oil/water meal. *Neurogastroenterology & Motility*, *9*(1), 41–47.

Boutrou, R., Gaudichon, C., Dupont, D., Jardin, J., et al. (2013). Sequential release of milk protein–derived bioactive peptides in the jejunum in healthy humans. *The American Journal of Clinical Nutrition*, *97*(6), 1314–1323.

Braden, B., Lembcke, B., Kuker, W., & Caspary, W. F. (2007). 13C-breath tests: current state of the art and future directions. *Digestive and Liver Disease*, *39*(9), 795–805.

Brandon, E. F. A., Oomen, A. G., Rompelberg, C. J. M., Versantvoort, C. H. M., et al. (2006). Consumer product in vitro digestion model: Bioaccessibility of contaminants and its application in risk assessment. *Regulatory Toxicology and Pharmacology*, *44*(2), 161–171.

Brasseur, J. G., Banco, G. G., Ailiani, A. C., Wang, Y., et al. (2009). Motility and absorption in the small intestines: Integrating mri with lattice boltzmann models. In *Proceedings—2009 IEEE International Symposium on Biomedical Imaging: From Nano to Macro* (pp. 374–377). (2009). Boston, MA: IEEE.

Brodkorb, A., Egger, L., Alminger, M., Alvito, P., et al. (2019). INFOGEST static in vitro simulation of gastrointestinal food digestion. *Nature Protocols*, *14*, 991–1014.

Bruen, C. M., O'Halloran, F., Cashman, K. D., & Giblin, L. (2012). The effects of food components on hormonal signalling in gastrointestinal enteroendocrine cells. *Food & Function*, *3*(11), 1131–1143.

Chen, J. (2009). Food oral processing-A review. *Food Hydrocolloids*, *23*(1), 1–25.

Cleary, P. W., Sinnott, M. D., Hari, B., Bakalis, S., et al. (2015). Modelling food digestion. *Modeling Food Processing Operations*, 255–305.

Cremer, J., Arnoldini, M., & Hwa, T. (2017). Effect of water flow and chemical environment on microbiota growth and composition in the human colon. *Proceedings of the National Academy of Sciences of the United States of America*, *114*(25), 6438–6443.

Deferme, S., Annaert, P., & Augustijns, P. (2007). In vitro screening models to assess intestinal drug absorption and metabolism. In *Drug Absorption Studies* (pp. 182–215). Springer, Boston, MA.

Dupont, D., Alric, M., Blanquet-Diot, S., Bornhorst, G., et al. (2019). Can dynamic in vitro digestion systems mimic the physiological reality? *Critical Reviews in Food Science and Nutrition*, *59*(10), 1546–1562.

Edwards, C. H., Grundy, M. M., Grassby, T., Vasilopoulou, D., et al. (2015). Manipulation of starch bioaccessibility in wheat endosperm to regulate starch digestion, postprandial glycemia, insulinemia, and gut hormone responses: a randomized controlled trial in healthy ileostomy participants. *The American Journal of Clinical Nutrition*, *102*(4), 791–800.

Egger, L., Schlegel, P., Baumann, C., Stoffers, H., et al. (2017). Physiological comparability of the harmonized INFOGEST in vitro digestion method to in vivo pig digestion. *Food Research International*, *102*, 567–574.

El Bouti, T., Goudon, T., Labarthe, S., Laroche, B., et al. (2016). A mixture model for the dynamic of the gut mucus layer. *E.S.A.I.M. Proceedings and Surveys*, *55*, 111–130.

Eldeghaidy, S., Marciani, L., McGlone, F., Hollowood, T., et al. (2011). The cortical response to the oral perception of fat emulsions and the effect of taster status. *Journal of Neurophysiology*, *105*(5), 2572–2581.

English, L. K., Fearnbach, S. N., Wilson, S. J., Fisher, J. O., et al. (2017). Food portion size and energy density evoke different patterns of brain activation in children. *The American Journal of Clinical Nutrition*, *105*(2), 295–305.

Evrensel, A., & Ceylan, M. E. (2015). The gut-brain axis: The missing link in depression. *Clinical Psychopharmacology and Neuroscience*, *13*(3), 239–244.

Ferrua, M. J., Kong, F., & Singh, R. P. (2011). Computational modeling of gastric digestion and the role of food material properties. *Trends in Food Science & Technology*, *22*(9), 480–491.

Ferrua, M. J., & Singh, R. P. (2015). Computational modelling of gastric digestion: current challenges and future directions. *Current Opinion in Food Science*, *4*, 116–123.

Ferrua, M. J., & Singh, R. P. (2010). Modeling the fluid dynamics in a human stomach to gain insight of food digestion. *Journal of Food Science*, *75*(7), R151–R162.

Ferrua, M. J., Xue, Z., & Singh, R. P. (2014). On the kinematics and efficiency of advective mixing during gastric digestion—A numerical analysis. *Journal of Biomechanics*, *47*(15), 3664–3673.

Flint, A., Raben, A., Blundell, J. E., & Astrup, A. (2000). Reproducibility, power and validity of visual analogue scales in assessment of appetite sensations in single test meal studies. *International Journal of Obesity*, *24*(1), 38–48.

Freitas, D., & Le Feunteun, S. (2019). Oro-gastro-intestinal digestion of starch in white bread, wheat-based and gluten-free pasta: Unveiling the contribution of human salivary α-amylase. *Food Chemistry*, *274*, 566–573.

Gaucel, S., Trelea, I. C., & Le Feunteun, S. (2015). Comment on new mathematical model for interpreting pH-stat digestion profiles: Impact of lipid droplet characteristics on in vitro digestibility. *Journal of Agricultural and Food Chemistry*, *63*(47), 10352–10353.

Giang, T. M., Le Feunteun, S., Gaucel, S., Brestaz, P., et al. (2015). Dynamic modeling highlights the major impact of droplet coalescence on the invitro digestion kinetics of a whey protein stabilized submicron emulsion. *Food Hydrocolloids*, *43*, 66–72.

Gill, S., Chater, P. I., Wilcox, M. D., Pearson, J. P., et al. (2018). The impact of dietary fibres on the physiological processes of the large intestine. *Bioactive Carbohydrates and Dietary Fibre*, *16*, 62–74.

Goñi, I., Garcia-Alonso, A., & Saura-Calixto, F. (1997). A starch hydrolysis procedure to estimate glycemic index. *Nutrition Research*, *17*(3), 427–437.

Gouseti, O., Bornhorst, G. M., Bakalis, S., & Mackie, A. (2019). In O. Gouseti, G. M. Bornhorst, S. Bakalis, & A. R. Mackie (Eds.), *Interdisciplinary approaches to food digestion*Springer International Publishing.

Grimes, D. A., & Schulz, K. F. (2002). An overview of clinical research: the lay of the land. *The Lancet*, *359*(9300), 57–61.

Guyton, A. C., & Hall, J. E. (2010). Chapter 63: Propulsion and Mixing of Food in the Alimentary Tract. In *Textbook of Medical Physiology* (12th ed., pp. 1120). Sauders.

Harrison, S. M., & Cleary, P. W. (2014). Towards modelling of fluid flow and food breakage by the teeth in the oral cavity using smoothed particle hydrodynamics (S.P. H.). *European Food Research and Technology*, *238*(2), 185–215.

Harrison, S. M., Cleary, P. W., Eyres, G., Sinnott, D., et al. (2014). Challenges in computational modelling of food breakdown and flavour release. *Food Function*, *5*(11), 2792–2805.

Harrison, S. M., Cleary, P. W., & Sinnott, M. D. (2018). Investigating mixing and emptying for aqueous liquid content from the stomach using a coupled biomechanical-SPH model. *Food Function*, *9*(6), 3202–3219.

Harrison, S. M., Eyres, G., Cleary, P. W., Sinnott, M. D., et al. (2014). Computational modeling of food oral breakdown using smoothed particle hydrodynamics. *Journal of Texture Studies*, *45*(2), 97–109.

Hartung, D. M., & Touchette, D. (2009). Overview of clinical research design. *American Journal of Health-System Pharmacy*, *66*(4), 398–408.

Helander, H. F., & Fändriks, L. (2014). Surface area of the digestive tract-revisited. *Scandinavian Journal of Gastroenterology*, *49*(6), 681–689.

Huda, M. S. B., Wilding, J. P. H., & Pinkney, J. H. (2006). Gut peptides and the regulation of appetite. *Obesity Reviews*, *7*(2), 163–182.

Hussain, M., Bakalis, S., Gouseti, O., Zahoor, T., et al. (2015). Dynamic and shear stress rheological properties of guar galactomannans and its hydrolyzed derivatives. *International Journal of Biological Macromolecules*, *72*, 687–691.

Jacka, F. N., O'Neil, A., Opie, R., Itsiopoulos, C., et al. (2017). A randomised controlled trial of dietary improvement for adults with major depression (the 'SMILES' trial). *B.M.C. Medicine*, *15*(23), 1–13.

Koziolek, M., Schneider, F., Grimm, M., Modeβ, C., et al. (2015). Intragastric pH and pressure profiles after intake of the high-caloric, high-fat meal as used for food effect studies. *Journal of Controlled Release*, *220*, 71–78.

Kumar, M., Ji, B., Zengler, K., & Nielsen, J. (2019). Modelling approaches for studying the microbiome. *Nature Microbiology*, *4*(8), 1253–1267.

Labarthe, S., Polizzi, B., Phan, T., Goudon, T., et al. (2019). A mathematical model to investigate the key drivers of the biogeography of the colon microbiota. *Journal of Theoretical Biology*, *462*, 552–581.

Le Huërou-Luron, I., Bouzerzour, K., Ferret-Bernard, S., Ménard, O., et al. (2018). A mixture of milk and vegetable lipids in infant formula changes gut digestion, mucosal immunity and microbiota composition in neonatal piglets. *European Journal of Nutrition*, *57*(2), 463–476.

Li, Y., & McClements, D. J. (2010). New mathematical model for interpreting ph-stat digestion profiles: Impact of lipid droplet characteristics on in vitro digestibility. *Journal of Agricultural and Food Chemistry*, *58*(13), 8085–8092.

Lim, Y. F., de Loubens, C., Love, R. J., Lentle, R. G., et al. (2015). Flow and mixing by small intestine villi. *Food Function*, *6*(6), 1787–1795.

Little, T. J., Pilichiewicz, A. N., Russo, A., Phillips, L., et al. (2006). Effects of intravenous glucagon-like peptide-1 on gastric emptying and intragastric distribution in healthy subjects: relationships with postprandial glycemic and insulinemic responses. *The Journal of Clinical Endocrinology & Metabolism*, *91*(5), 1916–1923.

Lucas-González, R., Viuda-Martos, M., Pérez-Alvarez, J. A., & Fernández-López, J. (2018). In vitro digestion models suitable for foods: Opportunities for new fields of application and challenges. *Food Research International*, *107*, 423–436.

Ludvigsson, J. F., & Green, P. H. (2011). Clinical management of coeliac disease. *Journal of Internal Medicine*, *269*, 560–571. doi:10.1111/j.1365-2796.2011.02379.x.

Major, G., Pritchard, S., Murray, K., Alappadan, J. P., et al. (2017). Colon hypersensitivity to distension, rather than excessive gas production, produces carbohydrate-related symptoms in individuals with irritable bowel syndrome. *Gastroenterology*, *152*(1), 124–133.

Malik, V. S., Popkin, B. M., Bray, G. A., Després, J. P., et al. (2010). Sugar-sweetened beverages, obesity, type 2 diabetes mellitus, and cardiovascular disease risk. *Circulation*, *121*(11), 1356–1364.

Maljaars, P. W. J., Peters, H. P. F., Mela, D. J., & Masclee, A. A. M. (2008). Ileal brake: A sensible food target for appetite control. A review. *Physiology and Behavior*, *95*(3), 271–281.

Maljaars, P. W. J., van der Wal, R. J. P., Wiersma, T., Peters, H. P. F., et al. (2012). The effect of lipid droplet size on satiety and peptide secretion is intestinal site-specific. *Clinical Nutrition*, *31*(4), 535–542.

Marciani, L., Gowland, P. A., Spiller, R. C., Manoj, P., et al. (2001). Effect of meal viscosity and nutrients on satiety, intragastric dilution, and emptying assessed by M.R.I. *American Journal of Physiology-Gastrointestinal and Liver Physiology*, *280*(6), G1227–G1233.

Marciani, L., Pritchard, S. E., Hellier-Woods, C., Costigan, C., et al. (2013). Delayed gastric emptying and reduced postprandial small bowel water content of equicaloric whole meal bread versus rice meals in healthy subjects: novel M.R.I. insights. *European Journal of Clinical Nutrition*, *67*(7), 754–758.

Marze, S. (2017). Bioavailability of nutrients and micronutrients: Advances in modeling and in vitro approaches. *Annual Review of Food Science and Technology*, *8*(1), 35–55.

Miller, E. R., & Ullrey, D. E. (1987). The pig as a model for human nutrition. *Annual Review of Nutrition*, *7*(1), 361–382.

Minekus, M., Alminger, M., Alvito, P., Ballance, S., et al. (2014). A standardised static in vitro digestion method suitable for food—an international consensus. *Food Function*, *5*(6), 1113–1124.

Mioche, L., & Peyron, M. A. (1995). Bite force displayed furing assessment of hardness in various texture contexts. *Archives of Oral Biology*, *40*(5), 415–423.

Mosele, J. I., Macià, A., Romero, M. -P., Motilva, M. -J., et al. (2015). Application of in vitro gastrointestinal digestion and colonic fermentation models to pomegranate products (juice, pulp and peel extract) to study the stability and catabolism of phenolic compounds. *Journal of Functional Foods*, *14*, 529–540.

Most, M. M., Ershow, A. G., & Clevidence, B. A. (2003). An overview of methodologies, proficiencies, and training resources for controlled feeding studies. *Journal of the American Dietetic Association*, *103*(6), 729–735.

Moxon, T. E., Gouseti, O., & Bakalis, S. (2016). In silico modelling of mass transfer & absorption in the human gut. *Journal of Food Engineering*, *176*, 110–120.

Moxon, T. E., Nimmegeers, P., Telen, D., Fryer, P. J., et al. (2017). Effect of chyme viscosity and nutrient feedback mechanism on gastric emptying. *Chemical Engineering Science*, *171*, 318–330.

Muñoz-Tamayo, R., Laroche, B., Walter, É., Doré, J., et al. (2010). Mathematical modelling of carbohydrate degradation by human colonic microbiota. *Journal of Theoretical Biology*, *266*(1), 189–201.

Muttakin, S., Moxon, T. E., & Gouseti, O. (2019). In vivo, in vitro, and in silico studies of the G.I. tract. In O. Gouseti, G. M. Bornhorst, S. Bakalis, & A. R. Mackie (Eds.), *Interdisciplinary Approaches to Food Digestion* (pp. 29–67). Springer International Publishing.

Nguyen, G. T., & Sopade, P. A. (2018). Modeling starch digestograms: Computational characteristics of kinetic models for in vitro starch digestion in food research. *Comprehensive Reviews in Food Science and Food Safety*, *17*(5), 1422–1445.

Norton, J. E., Wallis, G. A., Spyropoulos, F., Lillford, P. J., et al. (2014). Designing food structures for nutrition and health benefits. *Annual Review of Food Science and Technology*, *5*(1), 177–195.

Padayachee, A., Day, L., Howell, K., & Gidley, M. J. (2017). Complexity and health functionality of plant cell wall fibers from fruits and vegetables. *Critical Reviews in Food Science and Nutrition*, *57*(1), 59–81.

Paddon-Jones, D., & Rasmussen, B. B. (2009). Dietary protein recommendations and the prevention of sarcopenia. *Current Opinion in Clinical Nutrition and Metabolic Care*, *12*(1), 86–90.

Pal, A., Indireshkumar, K., Schwizer, W., Abrahamsson, B., et al. (2004). Gastric flow and mixing studied using computer simulation. *Proceedings of the Royal Society of London. Series B: Biological Sciences*, *271*(1557), 2587–2594.

Picó, C., Serra, F., Rodríguez, A. M., Keijer, J., et al. (2019). Biomarkers of nutrition and health: New tools for new approaches. *Nutrients*, *11*(5), 1092.

Rauh, C., & Delgado, A. (2015). Different modelling and simulation approaches for food processing operations. In S. Bakalis, K. Knoerzer, & P. J. Fryer (Eds.), *Modeling Food Processing Operations* (pp. 3–33). Woodhead Publishing Series in Food Science, Technology and Nutrition.

Riedijk, M. A., Stoll, B., Chacko, S., Schierbeek, H., et al. (2007). Methionine transmethylation and transsulfuration in the piglet gastrointestinal tract. *Proceedings of the National Academy of Sciences*, *104*(9), 3408–3413.

Rodríguez, M. G., Saldaña, M. R., Leyva, J. M. A., Rojas, R. M., et al. (2019). Design and validation of a food frequency questionnaire (F.F.Q.) for the nutritional evaluation of food intake in the Peruvian Amazon. *Journal of Health, Population and Nutrition*, *38*(1), 47.

Roura, E., Koopmans, S. -J., Lallès, J. -P., Le Huerou-Luron, I., et al. (2016). Critical review evaluating the pig as a model for human nutritional physiology. *Nutrition Research Reviews, 29*(1), 60–90.

Russo, F., Clemente, C., Linsalata, M., Chiloiro, M., et al. (2011). Effects of a diet with inulin-enriched pasta on gut peptides and gastric emptying rates in healthy young volunteers. *European Journal of Nutrition, 50*(4), 271–277.

Salles, C., Tárrega, A., Mielle, P., Maratray, J., et al. (2007). Development of a chewing simulator for food breakdown and the analysis of in vitro flavor compound release in a mouth environment. *Journal of Food Engineering, 82*, 189–198.

Santos, D. I., Saraiva, J. M. A., Vicente, A. A., & Moldão-Martins, M. (2019). Methods for determining bioavailability and bioaccessibility of bioactive compounds and nutrients. In F. J. Barba, J. M. A. Saraiva, G. Cravotto, & J. M. Lorenzo (Eds.), *Innovative thermal and non-thermal processing, bioaccessibility and bioavailability of nutrients and bioactive compounds* (pp. 23–54). Woodhead Publishing Series in Food Science, Technology and Nutrition.

Shani-Levi, C., Alvito, P., Andrés, A., Assunção, R., et al. (2017). Extending in vitro digestion models to specific human populations: Perspectives, practical tools and bio-relevant information. *Trends in Food Science & Technology, 60*, 52–63.

Shoaie, S., Ghaffari, P., Kovatcheva-Datchary, P., Mardinoglu, A., et al. (2015). Quantifying diet-induced metabolic changes of the human gut microbiome. *Cell Metabolism, 22*(2), 320–331.

Sicard, J., Mirade, P. -S., Portanguen, S., Clerjon, S., et al. (2018). Simulation of the gastric digestion of proteins of meat bolus using a reaction–diffusion model. *Food Function, 9*(12), 6455–6469.

Sinnott, M. D., Cleary, P. W., Arkwright, J. W., & Dinning, P. G. (2012). Investigating the relationships between peristaltic contraction and fluid transport in the human colon using smoothed particle hydrodynamics. *Computers in Biology and Medicine, 42*(4), 492–503.

Sinnott, M. D., Cleary, P. W., & Harrison, S. M. (2017). Peristaltic transport of a particulate suspension in the small intestine. *Applied Mathematical Modelling, 44*, 143–159.

Skinner, M., Eldeghaidy, S., Ford, R., Giesbrecht, T., et al. (2018). Variation in thermally induced taste response across thermal tasters. *Physiology & Behavior, 188*, 67–78.

Smout, A. J. P. M., & Mundt, M. W. (2009). Gastrointestinal motility testing. *Best Practice & Research Clinical Gastroenterology, 23*(3), 287–298.

Solem, R. C. (2015). Limitation of a cross-sectional study. *American Journal of Orthodontics and Dentofacial Orthopedics, 148*(2), 205.

Strathe, A. B., Danfær, A., & Chwalibog, A. (2008). A dynamic model of digestion and absorption in pigs. *Animal Feed Science and Technology, 143*(1–4), 328–371.

Sullivan, L. M., Kehoe, J. J., Barry, L., Buckley, M. J. M., et al. (2014). Gastric digestion of α-lactalbumin in adult human subjects using capsule endoscopy and nasogastric tube sampling. *British Journal of Nutrition, 112*(4), 638–646.

Tharakan, A., Norton, I. T., Fryer, P. J., & Bakalis, S. (2010). Mass transfer and nutrient absorption in a simulated model of small intestine. *Journal of Food Science, 75*(6), E339–E346.

Thiese, M. S. (2014). Observational and interventional study design types; an overview. *Biochemia Medica, 24*(2), 199–210.

Thursby, E., & Juge, N. (2017). Introduction to the human gut microbiota. *Biochemical Journal, 474*(11), 1823–1836.

Tomlin, C. J., & Axelrod, J. D. (2007). Biology by numbers: Mathematical modelling in developmental biology. *Nature Reviews Genetics, 8*(5), 331–340.

Tran, C. D., Gopalsamy, G. L., Mortimer, E. K., & Young, G. P. (2015). The potential for zinc stable isotope techniques and modelling to determine optimal zinc supplementation. *Nutrients, 7*(6), 4271–4295.

Trusov, P. V., Zaitseva, N. V., & Kamaltdinov, M. R. (2016). A multiphase flow in the antroduodenal portion of the gastrointestinal tract: A mathematical model. *Computational and Mathematical Methods in Medicine*, *2016*, 1–18.

Valdes, A. M., Walter, J., Segal, E., & Spector, T. D. (2018). Role of the gut microbiota in nutrition and health. *B.M.J.*, *361*, k2179.

Verkempinck, S. H. E., Salvia-Trujillo, L., Moens, L. G., Carrillo, C., et al. (2018). Kinetic approach to study the relation between in vitro lipid digestion and carotenoid bioaccessibility in emulsions with different oil unsaturation degree. *Journal of Functional Foods*, *41*, 135–147.

Wang, Y., Brasseur, J. G., Banco, G. G., Webb, A. G., et al. (2010). A multiscale lattice Boltzmann model of macro-to micro-scale transport, with applications to gut function. *Philosophical Transactions of the Royal Society A: Mathematical, Physical and Engineering Sciences*, *368*(1921), 2863–2880.

Willett, W., Rockström, J., Loken, B., Springmann, M., et al. (2019). Food in the Anthropocene: the EAT–Lancet Commission on healthy diets from sustainable food systems. *The Lancet*, *393*(10170), 447–492.

Woolnough, J. W., Monro, J. A., Brennan, C. S., & Bird, A. R. (2008). Simulating human carbohydrate digestion in vitro: A review of methods and the need for standardisation. *International Journal of Food Science and Technology*, *43*(12), 2245–2256.

Wu, P., Deng, R., Wu, X., Wang, Y., et al. (2017). In vitro gastric digestion of cooked white and brown rice using a dynamic rat stomach model. *Food Chemistry*, *237*, 1065–1072.

Zhang, Y., Wu, P., Jeantet, R., Dupont, D., et al. (2020). How motility can enhance mass transfer and absorption in the duodenum: Taking the structure of the villi into account. *Chemical Engineering Science*, *213*, 115406.

Ziegler, A., Gonzalez, L., & Blikslager, A. (2016). Large animal models: The key to translational discovery in digestive disease research. *Cellular and Molecular Gastroenterology and Hepatology*, *2*(6), 716–724.

Index

Note: Page numbers followed by "f" indicate figures, "t" indicate tables.

D

E

O

P

U

V

W

Y

Z

FSC
www.fsc.org
MIX
Papier | Fördert gute Waldnutzung
FSC® C083411

Kolibri 360 GmbH
Ferdinand-Jühlke-Straße 7
99095 Erfurt, Deutschland
produktsicherheit@kolibri360.de